Advanced Physics with Vernier – Mechanics

Larry Dukerich

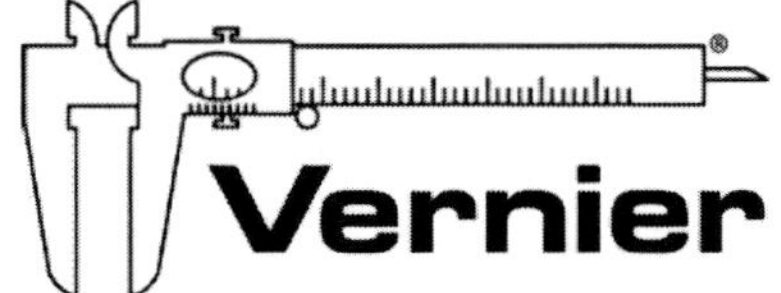

Measure. Analyze. Learn.™
Vernier Software & Technology
13979 S.W. Millikan Way • Beaverton, OR 97005-2886
Toll Free (888) 837-6437 • (503) 277-2299 • FAX (503) 277-2440
info@vernier.com • www.vernier.com

Advanced Physics with Vernier – Mechanics

Published by
Vernier Software & Technology
13979 SW Millikan Way
Beaverton, OR 97005-2886
(888) 837-6437
(503) 277-2299
FAX (503) 277-2440
info@vernier.com
www.vernier.com

ISBN 978-1-929075-64-5
First Edition
First Printing
Printed in the United States of America

About the Author

Larry Dukerich received his B.S. in Chemistry from Michigan State University and his Masters of Natural Science from Arizona State University. He taught high school chemistry and physics, including regular, honors and AP courses, in Michigan and Arizona for 34 years. He is currently a Faculty Associate at Arizona State. Since 1995, he has conducted numerous summer workshops for physics and chemistry teachers as part of the Modeling Instruction Program at ASU, and later at LaSalle University, North Carolina State University and Mansfield (PA) University. He has also conducted chemistry workshops for the Ministry of Education in Singapore. He was a Woodrow Wilson Dreyfus Fellow in Chemistry in 1986 and a Presidential Awardee for Excellence in Science Teaching in 2000.

Proper safety precautions must be taken to protect teachers and students during experiments described herein. Neither the authors nor the publisher assumes responsibility or liability for the use of material described in this publication. It cannot be assumed that all safety warnings and precautions are included.

Contents

Activities

Experiments

Appendices

Sensors and Accessories Used in Experiments

		Sensor				Vernier Accessory							
		Motion Detector	Photogate	Force Sensor	Rotary Motion	Dynamics System	Bumper and Launcher Kit	Cart Friction Pad	Picket Fence	Cart Picket Fence	Ultra Pulley and Bracket	Centripetal Force Apparatus	Rotary Motion Accessory Kit
A1	An Exploration of Graphical Methods	No sensor or accessory needed											
A2	Investigating Motion	1				1							
A3	Working with Analytical Tools	No sensor or accessory needed											
A4	Introduction to the Vernier Photogate		1						1	1			
1	Motion on an Incline	1				1							
2	Error Analysis		1						1				
3	Newton's First Law	1				1	1	1					
4	Newton's Second Law		1	1		1				1	1		
5	Newton's Third Law			2		1	1						
6	Projectile Motion	Video analysis: No sensor or accessory needed											
7	Energy Storage and Transfer: Elastic Energy			1		1	1						
8	Energy Storage and Transfer: Kinetic Energy		1			1	1			1			
9	Energy Storage and Transfer: Gravitational Energy		1*			1	1			1*			
10A	Impulse and Momentum	1		1		1	1						
10B	Impulse and Momentum		1	1		1	1			1			
11A	Momentum and Collisions	2				1							
11B	Momentum and Collisions		2			1				2			
12A	Centripetal Acceleration		1	1								1	
12B	Centripetal Acceleration		1	1		No accessory needed							
13	Rotational Dynamics				1								1
14	Conservation of Angular Momentum				1								1
15	Simple Harmonic Motion: Mathematical Model	1				No accessory needed							
16	Simple Harmonic Motion: Kinematics and Dynamics	1		1		No accessory needed							
17	Pendulum Periods				1								1
18	Physical Pendulum				1								1
19	Center of Mass	Video analysis: No sensor or accessory needed											

* For use in the Extension

Preface

This book contains four Introductory Activities and 19 student experiments and using Vernier LabQuest, Vernier LabQuest Mini, or Vernier LabPro for collecting, displaying, analyzing, graphing, and printing data. These experiments represent most of the experiments included in the mechanics portion of an advanced physics course, including what one would teach in an AP or IB Physics course.

Vernier is convinced of the importance of *hands-on* experiments. As a general rule, experiments serve two quite different functions in a science course: they can allow students to verify concepts they have learned in the lecture/recitation portion of a course, or they can provide students the opportunity to carefully examine phenomena and try to make sense of their findings. This book adopts the latter approach to the role of the laboratory. We feel that the easy-to-use, yet powerful analytical tools in Logger *Pro* and LabQuest App software enable students to discern quantitative relationships between the variables they investigate. You should find that the experiments in this book are aligned with the recommendations for the role of laboratory made by the College Board and the International Baccalaureate.

This book is unlike other collections of physics laboratory experiments. Most activity books include self-guided instructions that can be handed to students who then go away, do the lab with varying degrees of success, and hand in some kind of summary for evaluation. The experiments in this book are different in that they assume that there will be regular interactions with an instructor and between student groups. It is important that you read the instructor's notes before using an experiment with your students, so that you will be prepared to intervene at the appropriate times with the critical information.

While the experiments are designed around these interactions, they are not open-ended experiments that lack closure or direction. (There are some extensions that are intended to be open-ended.) The activities nevertheless address the key concepts of a rigorous introductory physics course in a way that ensures all students will be prepared to move forward in the curriculum.

This book is also unlike the more basic *Physics with Vernier* in that the instructions assume that students (and instructors) know how to use the software as a tool for doing science. Many students will already know how to interpret graphs or to create a new graph of calculated quantities. Students and instructors unsure of these skills will want to perform the introductory activities. Doing so provides students with the tools needed to do their experiments, focusing on the physics, and not the software. However, when unusual or difficult calculations are performed, appropriate assistance is provided.

Vernier sensors make possible accurate measurements of distance, velocity, acceleration, force, and rotary motion in the physics lab. Your students can perform many new experiments with measurements not previously practical in the lab. These sensors, supported by the intuitive, high-quality software, make it relatively simple for instructors of advanced high school or university physics to integrate probeware into their physics classes.

Experiments in this book can be used unchanged or they can be modified using the word-processing files provided on the CD. In writing these experiments we made the instructions to the student less detailed than those found in the *Physics with Vernier* book. As a general rule, the instructions are intended to help students to effectively use the sensors and software to collect and analyze the data without telling them relationships that they can, with guidance, uncover themselves. Each of the experiments begins with a pre-lab investigation designed to set the stage for the main portion of the experiment. Opportunities are provided for students to pause and

discuss their thinking either in a whole-class or small group setting. Students are encouraged to use inductive reasoning to develop general equations from their specific findings. The instructor can then help students build on what they have learned to develop a solid understanding of the physics concepts underlying the phenomena they have examined.

Here are some ways to use the experiments in this book:

- Unchanged, but with guidance. You can photocopy the student sheets, and after the pre-lab discussion, distribute them to the students to use. There are places, designated by the symbol at right in the teacher's version of the experiment, where you may need to provide some guidance in the collection and evaluation of data. For some of the experiments, Logger *Pro* files are provided to help students evaluate their data.

- Slightly modified. The CD accompanying this book is for this purpose. Before producing student copies, you can change the directions to adjust them to your teaching circumstances. Experienced Logger *Pro* or LabQuest App users may need less help in the design of their experiment files.

- Extensively modified. Using the CD accompanying this book, some teachers may choose to decrease the detail provided in the student instructions. We expect experienced physics teachers will significantly modify the instructions given in this manual.

Following each student experiment is an extensive Teacher Information section with recommendations for setting up equipment and helping students collect useful data, sample results, and suggestions to guide the post-lab discussion as well as the extensions.

The computer-centered instructions in this book assume that Logger *Pro* 3.8.4 or newer is used. The LabQuest instructions assume that version 1.5 or newer is used. Updates to both software titles are available at www.vernier.com/downloads.

Most activities may be performed with either Logger *Pro* or LabQuest App. Due to the complexity of some analysis, most users will prefer to use a computer; some activities require a computer.

I am grateful to David Braunschweig and Rex Rice for their contributions of experiment ideas to this book, to Rick Sorensen and John Gastineau for their advice on physics and technical matters, and to Gretchen Stahmer DeMoss for making sure that it made sense.

Larry Dukerich

An Exploration of Graphical Methods

Graphs are very useful representations of the relationship between variables of interest. The data collection and analysis software Logger *Pro* is a powerful tool that assists you in your analysis of graphs of experimental data. This exploration affords you the opportunity to practice using Logger *Pro* to analyze relationships with which you are already familiar. If you are new to Logger *Pro*, consider exploring the tutorials (especially 5 and 10) found in the Experiments folder in the Logger *Pro* folder before you attempt this exploration activity.

OBJECTIVES

In this activity, you will

- Practice manual entry of data in Logger *Pro*.
- Perform linear fits to data and analyze the resulting equations.
- Linearize data to find the relationship between the variables.
- Perform a curve fit to data and analyze the resulting equation.

MATERIALS

computer
Logger *Pro*
centimeter-ruled graph paper
variety of circular objects: cans, jars, glasses, bowls, plates
flexible metric tape measure

PROCEDURE

Part 1 Circumference *vs.* diameter

You should have available to you a number of objects that have a circular cross section. Ideally, the largest of these should have a diameter at least ten times as great as the smallest. Using your metric tape, measure and record the diameter and circumference of at least 6 of these objects.

Part 2 Area *vs.* radius

1. Place one of these objects as close to the center of the centimeter-ruled graph paper as you can. Trace the circular cross section of the object on the paper. Measure and record the diameter of this circle.

2. Count the number of squares enclosed by this circle. Since your circle cuts through some squares, count only those squares that are completely enclosed or have at least half of the square enclosed. If a portion of the circle consistently encloses less than half a square, estimate how many squares should be added to your total.

3. Repeat Steps 1 and 2 for five other circular objects.

EVALUATION OF DATA

Part 1 Circumference *vs.* diameter

1. Start Logger *Pro*. Double-click on the header of the *x*-axis in the data table. This brings up a Manual Column Options box. Enter **diameter** as the name, **d** as the short name and **cm** as the units. Select Done.

2. Double-click on the header of the y-axis in the data table. Enter **circumference** as the name, **C** as the short name and **cm** as the units. Select Done.

3. Manually enter the data you have recorded. Press Return or Enter after typing the value to move the cursor to the next cell in the data table.

4. Choose Autoscale From 0 from the Analyze menu.

5. Choose Linear Fit from the Analyze menu to have Logger *Pro* draw a line of best fit through your data.

6. Write the equation for your best-fit line. After examining the value and units of the slope, write a general expression for the relationship between circumference and diameter. Compare your findings with those of other groups in class.

Part 2 Area *vs.* radius

1. Choose Add Page from the Page menu. Select New Data Set and Graph and give the page an appropriate name.

2. Note that the data table now shows Data Set 2. As you did in Part 1, re-name the column headers for the *x* and *y* axes. To choose appropriate units for area, you can use the pull down menu to the right of the Units field to choose '2' as the superscript for cm (see Figure 1).

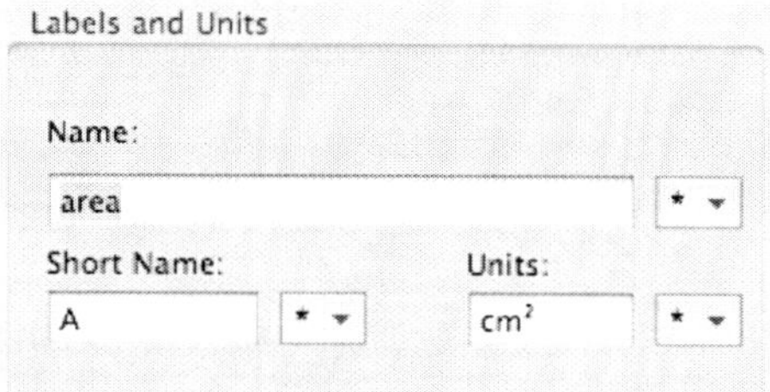

Figure 1

3. As you did in Part 1, manually enter your diameter and area data. Be careful after each entry for the area to make sure you return to the column for diameter for Data Set 2.

4. You can use Logger *Pro* to calculate and display the value of the radius of each of the circles in Part 2. Choose New Calculated Column from the Data menu. In the window that is displayed, name the column and choose in which Data Set it should appear. The window also provides a place where you can specify the equation used to calculate the values. Insert the cursor in the Equation field, then, rather than enter the variable name yourself, select Choose Specific Column from the Variables (Columns) menu. Specify Data Set 2|diameter and enter **/2** to divide the diameter by 2 (see Figure 2). Click Done.

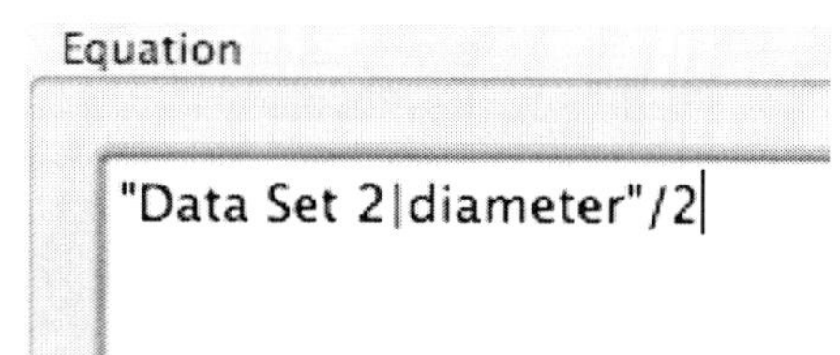

Figure 2

5. At this point, you should have a graph of area *vs.* diameter. Click the horizontal axis label, then select **radius** as the variable for this axis. Autoscale the graph as you did in Part 1. You can also use the icon shown at left in the Toolbar to do this task.

6. What relationship appears to exist between area and the radius of your circles? While your first impulse might be to fit a curve to the data, you will first explore "linearizing" the graph[1]. As you did in Step 4, create a new calculated column. Enter **radius^2** as the title and choose appropriate units. Then position the cursor in the Equation field, select radius from Variables (Columns) and enter **^2** to square the radius.

7. As you did in Step 5, click the horizontal label, select More, then, from the pull down menu, choose radius^2 for the horizontal axis of your graph. Examine your graph. If the plot appears to be linear, choose a linear fit for your graph.

8. Write the equation for your best-fit line. After examining the value and units of the slope, write a general expression for the relationship between area and the square of the radius of your circles.

Now that you have analyzed the relationship between area and radius through linearization, you will now try a different approach using the curve-fitting tool in Logger *Pro*.

9. Choose Graph from the Insert menu. A small graph of circumference *vs.* diameter should appear on top of your first graph. Choose Auto Arrange from the Page menu; this re-sizes both graphs and arranges them nicely on the page.

10. Click the vertical axis label and choose More. Uncheck Circumference and check Area from Data Set 2 for this axis. In a like manner, choose Radius for the horizontal axis, then autoscale the graph.

11. Choose Curve Fit from the Analyze menu. The Curve Fit dialog box will be displayed. Under the test plot of your data are a number of general equations from which you could choose to fit your data. Scroll down until you find Power (*Ar^B*), select that equation, then increase the value of the *B* coefficient by clicking the up arrow next to the field until the value 2 appears. Note how the test plot changes with the value of *B*.

12. Now, gradually increase the *A* coefficient until the curve on the test plot best matches your data, then click OK. You have now performed a manual curve fit to your plot of area *vs.* radius. In what ways is the information provided by the two methods the same; how does it differ?

EXTENSION

Account for the fact that the constant of proportionality you obtained in your two linear relationships may have differed somewhat from the expected value.

[1] If you have never done this before, now might be a good time to explore Tutorial 10-2 Linearization. Save your current file and open the tutorial. After completing it, return to the file for your activity.

An Exploration of Graphical Methods

Graphs are very useful representations of the relationship between variables of interest. The data collection and analysis software LabQuest App is a powerful tool that assists you in your analysis of graphs of experimental data. This exploration affords you the opportunity to practice using LabQuest App to analyze relationships with which you are already familiar. If you are new to using the LabQuest App, consider reviewing the LabQuest Reference Guide, available at the Vernier website, before you attempt this exploration activity.

OBJECTIVES

In this activity, you will

- Practice manual entry of data in LabQuest App.
- Perform linear fits to data and analyze the resulting equations.
- Linearize data to find the relationship between the variables.
- Perform a curve fit to data and analyze the resulting equation.

MATERIALS

LabQuest
variety of circular objects: cans, jars, glasses, bowls, plates
centimeter-ruled graph paper
flexible metric tape measure

PROCEDURE

Part 1 Circumference *vs.* diameter

You should have available to you a number of objects that have a circular cross section. Ideally, the largest of these should have a diameter at least ten times as great as the smallest. Using your flexible metric tape, measure and record the diameter and circumference of at least six of these objects.

Part 2 Area *vs.* radius

1. Place one of these objects as close to the center of the centimeter-ruled graph paper as you can. Trace the circular cross section of the object on the paper. Measure and record the diameter of this circle.

2. Count the number of squares enclosed by this circle. Since your circle cuts through some squares, count only those squares that are completely enclosed or have at least half of the square enclosed. If a portion of the circle consistently encloses less than half a square, estimate how many squares should be added to your total.

3. Repeat Steps 1 and 2 for five other circular objects.

EVALUATION OF DATA

Part 1 Circumference *vs.* diameter

1. Turn on your LabQuest; this starts LabQuest App. Tap the Table tab to display the data table. Tap the header of the x-axis in the data table to display the Column Options dialog.

2. Enter **diameter** as the name and **cm** as the units. Choose to display 3 significant figures, then tap OK.

3. Tap the header of the y-axis in the data table. Enter **circumference** as the name and **cm** as the units.

4. Tap in the first cell below diameter in the table. Manually enter the data you have recorded. Tapping the Return key on the keyboard after each entry moves the cursor to the next cell in the data table. When you are finished, tap the graph tab to view your graph. Choose Autoscale Once from the Graph menu.

5. Choose Graph Options from the Graph menu. Uncheck Connect Points and check Point Protectors. Tap OK. Choose Curve Fit from the Analyze menu and check Circumference. From the Fit Equation menu, choose Linear and tap OK. LabQuest App will draw a line of best fit through your data. Note that the slope and intercept are given in the box to the right of the graph.

6. Record the equation for your best-fit line. After examining the value and units of the slope, write a general expression for the relationship between circumference and diameter. Compare your findings with those of other groups in class.

Part 2 Area *vs.* radius

1. Choose New from the File menu. LabQuest App asks you if you wish save the existing file. Enter an appropriate name and select Save.

2. Tap the Table tab as you did in Part 1, then re-name the column headers for the x and y axes. To enter appropriate units for area, tap on the 2nd shift key (lower-left part of the keyboard) to be able to choose '2' as the superscript for cm (see Figure 1).

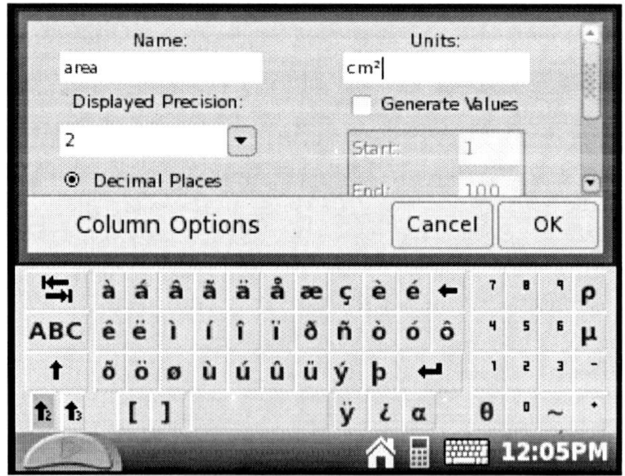

Figure 1

3. As you did in Part 1, manually enter your diameter and area data. Uncheck Connect Points in the Graph Options window.

4. You can use LabQuest App to calculate and display the value of the radius of each of the circles in Part 2. Return to your data table; choose New Calculated Column from the Table menu. This brings up a window in which you can name the column and units and choose the precision you wish to display. This window also provides the place where you can specify the equation used to calculate the values. Select X/A as the Equation Type from the options in the drop down menu, diameter as the column for X, and then enter **2** as the value of A (see Figure 2).

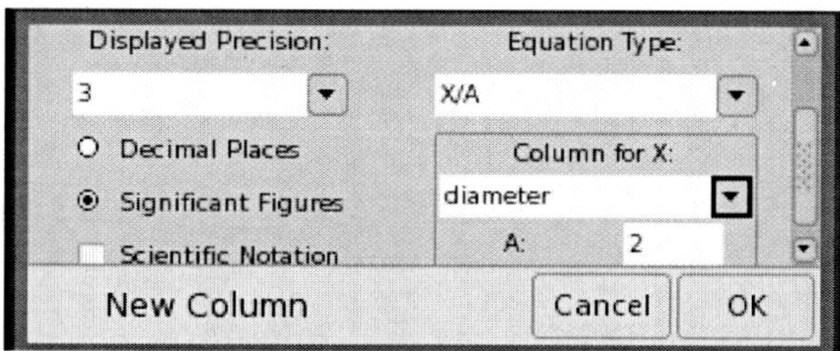

Figure 2

5. When you tap the Graph tab, you should see a graph of area *vs*. radius. If not, click the horizontal axis label. Select radius as the variable for this axis from the menu. You may not see all the data points marked with symbols; however, if you tap on the screen roughly where a point should appear, you will find that the point appears.

6. What relationship appears to exist between area and the radius of your circles? While your first impulse might be to fit a curve to the data, you will first explore "linearizing" the graph. As you did in Step 4, create a new calculated column. Enter **radius^2** as the title and choose appropriate units. Then choose AX^B as the Equation Type and select radius as the Column for X. Enter **1** for A and **2** for B to square the radius.

7. When you return to the Graph window, you should see that radius^2 is now the horizontal axis label of your graph. Examine your graph. If the plot appears to be linear, choose a linear fit for your graph as you did in Part 1.

8. Write the equation for your best-fit line. After examining the value and units of the slope, write a general expression for the relationship between area and the square of the radius of your circles.

Now that you have analyzed the relationship between area and radius through linearization, you will now try a different approach using LabQuest App's modeling function.

9. Choose radius as the horizontal axis label for your graph. Choose Model from the Analyze menu and check Area. Choose Ax^B as the Equation for the model and enter **2** as the value for B. This changes the model function to a parabola (see Figure 3).

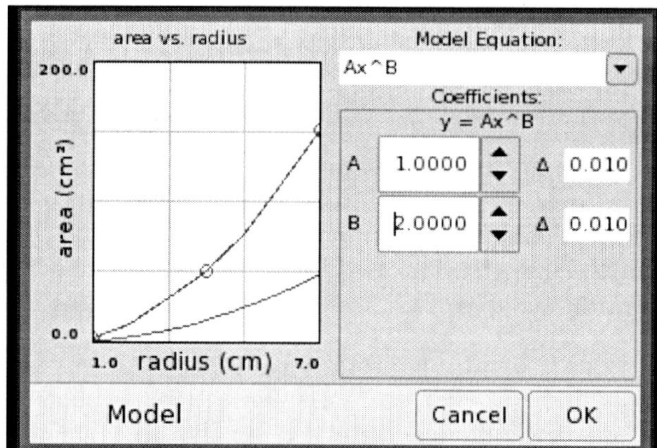

Figure 3

10. Now, adjust the value of A using the up and down arrows until the model function best matches the plot of your data. Then tap OK.

11. You have now performed a manual curve fit to your plot of area *vs*. radius. In what ways is the information provided by the two methods the same; how does it differ?

EXTENSION

Account for the fact that the constant of proportionality you obtained in your two linear relationships may have differed somewhat from the expected value.

Activity 1

INSTRUCTOR INFORMATION

An Exploration of Graphical Methods

This activity is designed to provide students with the opportunity to use Logger *Pro* or LabQuest App to perform graphical analysis techniques on data they have collected. While students are certain to know the equations used to find the circumference of a circle from its diameter and the area of a circle from its radius, many students have not determined these relationships empirically. Inform students that the purpose of this activity is to familiarize them with features of Logger *Pro* or LabQuest App that they will use throughout this manual.

The Microsoft Word files for the student pages can be found on the CD that accompanies this book. See *Appendix A* for more information.

OBJECTIVES

In this activity, the student objectives include

- Practice manual entry of data into Logger *Pro* and LabQuest App.
- Perform linear fits on data and analyze the resulting equations.
- Linearize data to find the relationship between the variables.
- Perform a curve fit to data and analyze the resulting equation.

During this activity, you will help the students

- Approximate the area of a circle by counting the number of unit squares enclosed.
- Recognize that the slope of the linear fits to the graphs of circumference *vs.* diameter and area *vs.* radius^2 is nearly the accepted value of pi.
- Recognize that limits of precision in measurement and approximation of the area account for the fact that the constant of proportionality is not exactly pi.

EQUIPMENT TIPS

For this activity, students must have access to a variety of objects that have a circular cross section. For Part 1, the size of an object used is limited only by the length of the flexible metric tape measures available to the students. Be careful with spherical objects because it is not easy to determine the diameter precisely. For Part 2, the objects can range from a dry erase marker up to a plate small enough to fit on the centimeter-ruled graph paper provided (See the file 01 cm grid.pdf available on the CD that accompanies this book. See *Appendix A* for more information.).

PRE-LAB DISCUSSION

Inform your students that in this activity they will examine the relationships between the circumference and diameter and between area and radius of circles. They are likely to say that they already know these. Respond by saying that they will examine these relationships empirically, as a way of familiarizing themselves with Logger *Pro* software or LabQuest App.

LAB PERFORMANCE NOTES

Part 1 should pose little difficulty for the students. If they choose to try to measure the diameter of a sphere, it would be best to place the sphere on the tape measure and sight from above where each end appears to fall on the scale.

In Part 2, students may be tempted to *calculate* the area of the circle, since that is the way they are familiar with determining its area. Instead, they need to find a way to *measure* the area. Suggest that we can define area as the number of unit squares that are enclosed by the boundary of the object. For anything rectilinear, counting the number of squares would be very simple. For the circle, they will have to determine an approximate value by counting the number completely enclosed and then estimating the number of squares that are only partially enclosed. They could obtain this estimate by counting those squares where the majority falls within the circle and ignoring those where the majority is outside. This is likely to produce an underestimate. They can refine their estimate by adding in extra squares when there appears to be more squares ignored than included. For larger circles, the task of counting is made less tedious by sketching rectangles within the circle and multiplying to determine the number of squares (see Figure 1).

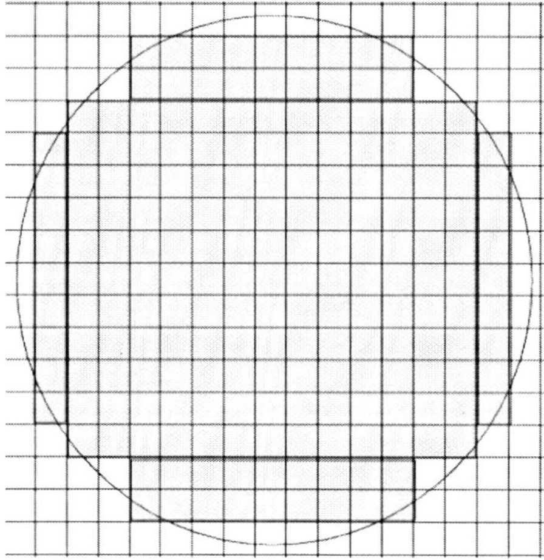

Figure 1

SAMPLE RESULTS AND POST-LAB DISCUSSION – PART 1

If this is the first opportunity students have had to use Logger *Pro* or LabQuest App to help them analyze data, then you might consider suggesting that students work through the examples in the tutorials that accompany the software. If they feel comfortable trying to figure things out on their own, the instructions provided in the student version should get them through Part 1. Figure 2 shows a graph of sample data.

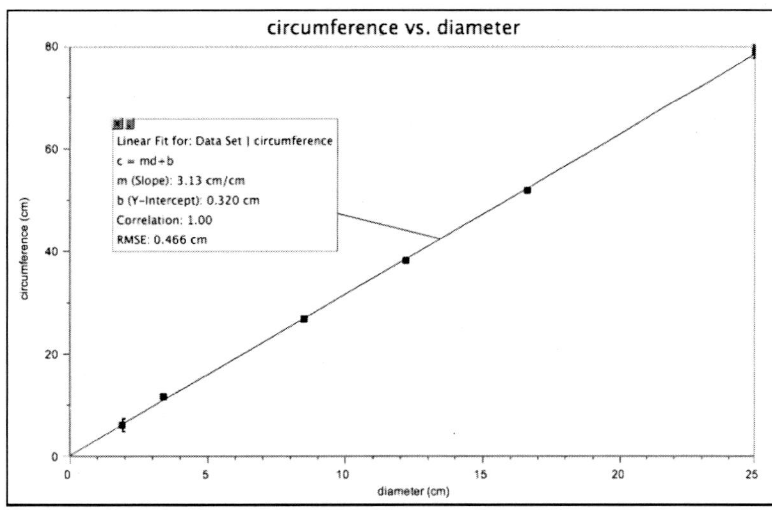

Figure 2

Step 6

Despite the fact that they know that $C = \pi d$, many students are surprised that the slope of the line is very close to the value they know for pi. Logger *Pro* reports the units of the slope as cm/cm. From this, students should conclude that pi has no units.

SAMPLE RESULTS AND POST-LAB DISCUSSION – PART 2

Steps 1–2

While students could do all of this analysis on a single page, adding a page in Logger *Pro* allows them to toggle back and forth between their two graphs.

Step 3

When there is more than one data set, students need to exercise caution when entering data so as not to type over data from the existing data set. They can continue to hit Enter to return to the next row, or simply click or tap in the desired column to enter the next data pair.

Step 4

If students feel they need more help with how to create a new calculated column, they can review Tutorial 08-Stats,Tangents,Integral in the Tutorials folder. Once they have values for the radii of their circles, they can place that variable name on the horizontal axis of the graph on Page 2 in Logger *Pro*.

Step 5

This step shows students how to choose the desired variable for the horizontal axis of their new graph in Logger *Pro*. Clicking icons in the Toolbar is a shortcut to making selections from the menu. When students complete Step 5 they should obtain a graph like that shown in Figure 3.

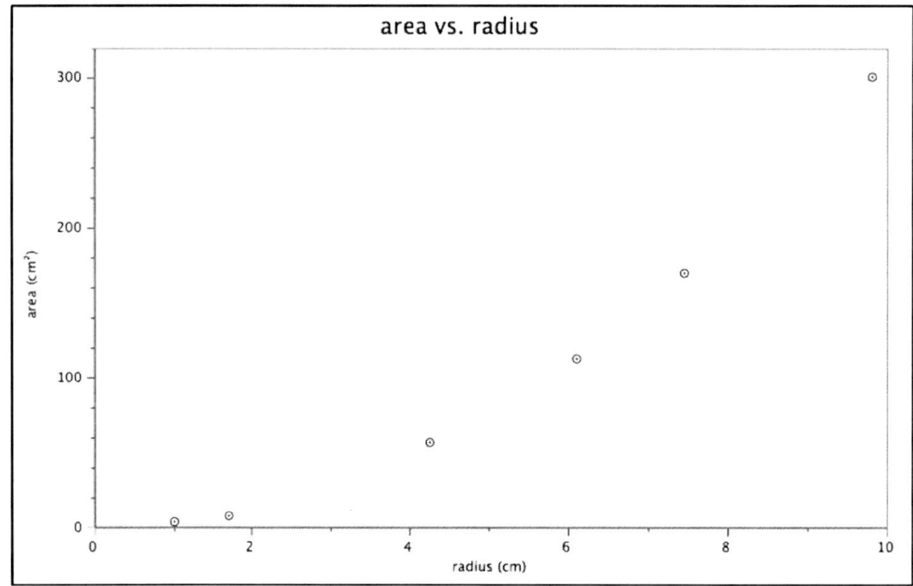

Figure 3

Steps 6–7

If students are uncertain how to interpret the relationship between the variables when they see a top-opening parabola, they might consider first working through Tutorial 10-2–Linearization. After they have created the new calculated column **radius²** and selected this for the horizontal axis label, they should perform a linear fit to their data. Performing this step produces a graph like the one shown in Figure 4.

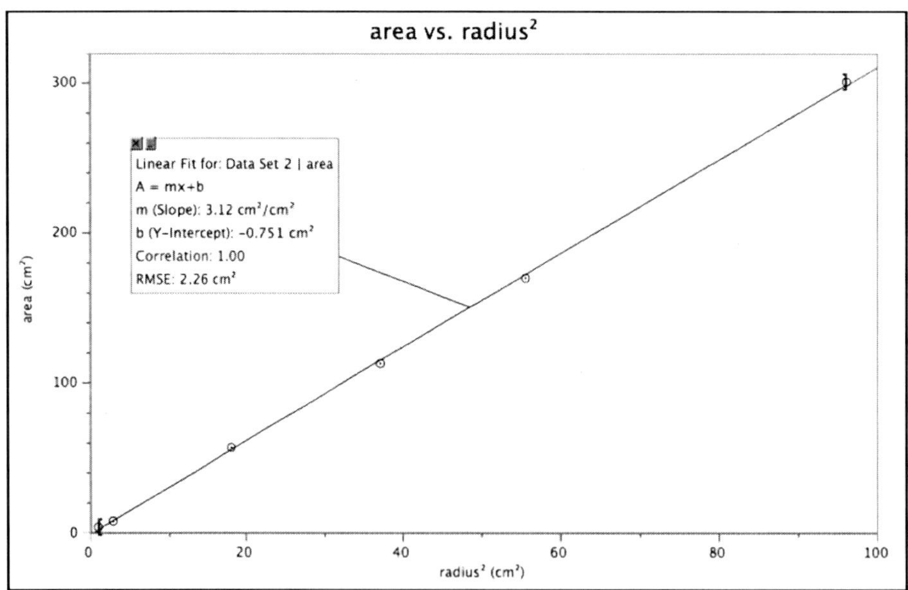

Figure 4

Step 8

The equation of the line of best fit in this graph is $A = \left(3.12\frac{\text{cm}^2}{\text{cm}^2}\right)r^2 - 0.751\ \text{cm}^2$. The y-intercept is negligible and the units of the slope cancel, leaving $A = \pi r^2$. This should come as no surprise, but students usually find it satisfying to determine a known relationship empirically.

Steps 9–10

In these steps, students obtain the graphs in Figures 3 and 4 on the same page.

Steps 11–12

When they see the parabolic shape of the plot of area *vs.* radius, students might want to choose Quadratic *(Ar^2 + Br + C)* as their equation. Doing so produces a very good fit to the data, but leaves students with the dilemma of trying to provide physical meaning to the *Br* + *C* terms in the equation. Students might be surprised to learn that the higher order polynomials (cubic, quartic, etc.) also provide excellent fits to the data, but suffer from the same problem as the quadratic.

Students should note that while they obtain essentially the same relationship via linearization and curve-fitting, the former has the advantage of displaying the units of the variables that are plotted.

EXTENSION

The more careful students are when they make their measurements in Part 1, the closer their values should come to the accepted value for pi. Variations should be evenly distributed on either side of 3.14.

In Part 2, the technique suggested for estimating the area is more likely to produce a value that is somewhat lower than it should be. As a result, the slope in the graph of area *vs.* radius^2 is likely to be smaller than 3.14. If students wanted to improve their estimate, they could add to their count any square intercepted by the perimeter of the circle. They could average this too-high value with the too-low value to obtain a median value of the area that could improve the precision of their value of pi.

Activity
2

Investigating Motion

Investigating Constant Velocity with a Motion Detector

The study of how things move and why they move is an important part of an introductory physics course. The use of technology greatly aids the collection and analysis of data. In this activity you will have the opportunity to learn how to use a Motion Detector to examine aspects of the motion of an object.

The motion detector works by emitting and detecting ultrasound. It generates a series of ultrasound pulses, and then detects the echoes returning from an object. The time between the emitting of a pulse and the reception of its echo is used to calculate the position of the object using the speed of sound. Position and time data are then used in software to determine the velocity of the object.

OBJECTIVES

In this activity, you will

- Practice using the motion detector in a variety of experiments.
- Use a variety of analysis tools on graphs of position-time and velocity-time data.
- Interpret the equations resulting from the analysis of these graphs.

MATERIALS

Vernier data-collection interface	Motion Detector and bracket
Logger *Pro* or LabQuest App	Vernier Dynamics Track

PROCEDURE

1. Connect the Motion Detector to the interface and start the data-collection program. Two graphs: position *vs.* time and velocity *vs.* time will appear in the graph window. For now, you need only consider the position *vs.* time graph.

   - In Logger *Pro*, delete the velocity graph, then choose Auto Arrange from the Page menu.
   - In LabQuest App, tap the Graph tab, then choose Show Graph and select Graph 1.

   Later, during the analysis of data, you will add the velocity *vs.* time graph back to your view.

2. Attach the motion detector to the bracket that will allow you to position it near one end of the track.

3. If your motion detector has a switch, set it to the Track setting.

4. Place the cart approximately 20 cm[1] in front of the motion detector. The live readout on the display in Logger *Pro* or LabQuest App will tell you the position of the cart. Note the position of the back end of the cart on the scale on the track.

5. Position the end stop on the track so that when the cart runs into it, the cart will have moved a known distance (~70–85 cm) from its initial position[2] (see Figure 1).

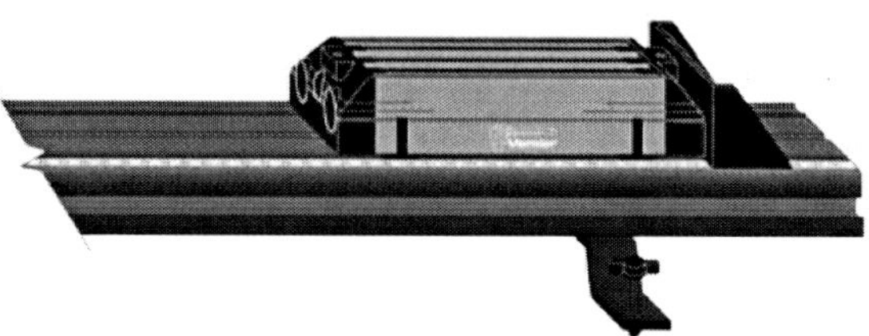

Figure 1

6. Return the cart to its original position near the motion detector. Start data collection, then, once you hear the motion detector clicking, give the cart a gentle push. Data collection stops automatically after 5 seconds.

7. The motion detector sends out its signal in a cone, and it detects the first echo from the nearest object in the cone. This object may or may not be the object of interest. Examine the position *vs.* time graph. If there are jagged dips in the graph once the cart began moving, it could be that your hand or some other object was picked up by the motion detector. If this is the case, repeat Step 6, but be sure that your hand or other stray objects do not interfere with the signal returned by the cart. If you have a smooth graph, store this run.
   - In Logger *Pro*, choose Store Latest Run from the Experiment menu.
   - In LabQuest App, tap the filing cabinet icon.

8. Repeat Step 6, but this time, launch the cart somewhat faster than you did the first time. If you have a smooth graph, store this run. Note differences in the appearance of the position-time graph for the two runs. You will examine this in greater detail in the Evaluation of Data section.

Zeroing the motion detector

9. In the analysis of position-time (x-t) data it is convenient to consider the initial position as zero. This can be done with the motion detector. Place the cart at the starting position you used in the previous run, then zero the detector.
   - In Logger *Pro*, choose Zero from the Experiment menu.
   - In LabQuest App, tap the position reading in the Meter tab and choose Zero.

10. Repeat Step 6, then store this run as before. Compare the x-t graph for this run to that obtained in your previous runs.

[1] If you are using an older motion detector without a switch, the cart needs to be at least 45 cm from the detector.
[2] This distance will be less if you are using an older motion detector.

Reversing the detector

11. Now, position the cart at the far end of the track. Start data collection, then give the cart a gentle push toward the motion detector. Be sure to catch the cart before it runs into the detector. If your x-t graph is smooth, store this run.

12. The default setting for the motion detector is to designate the direction of motion *away* from the detector as positive. In this run, the object was moving in the opposite direction; note that the x-t graph has a positive vertical intercept and a negative slope. It is sometimes useful to consider the direction of motion as positive. You can set the motion detector to treat motion *toward* the detector as positive.

    - In Logger *Pro*, choose Set Up Sensors from the Experiment menu, then select your interface (LabQuest, Lab Pro, etc.). Click the icon representing the motion detector and choose Reverse Direction in the pop-up dialogue box, then close the box.
    - In LabQuest App, tap the position reading in the meter tab and choose Reverse Direction.

13. Now, position the cart at the far end of the track. Zero the motion detector as you did in Step 9, then give the cart a gentle push toward the detector. Be sure to catch the cart before it runs into the detector. If your x-t graph is smooth, store this run. Compare the appearance of your x-t graph to that obtained in the previous run. You need not store this run; however, you should save this experiment file.

EVALUATION OF DATA

1. Examine the position *vs*. time graph for your first run.

    - In Logger *Pro*, the color of the entries in the data table matches the trace of the line in the graph. You can choose to view only one run by clicking the vertical axis label, choosing More, then selecting Position for the run of interest and de-selecting the other runs.
    - In LabQuest App, you can choose to view any of your four runs by tapping on the number for that run.

2. Determine the rate of change of position of the cart while it was moving at nearly constant speed. To do this, select the portion of the graph in which the plot appears linear by dragging your cursor (or stylus) across this region. Then perform a linear fit on that portion of the graph.

    - In Logger *Pro*, choose Linear Fit from the Analyze menu. You can adjust the segment over which the linear fit was performed by dragging the bounds, marked by [and] symbols.

    - In LabQuest App, choose Curve Fit from the Analyze menu, check the box marked position, then, in the Choose Fit drop down menu, choose Linear. If you decide you wish to adjust the region over which you have performed the linear fit, repeat the process.

3. What information about the motion of the cart is provided by the slope of the graph? How do the units confirm your answer?

4. Now choose to view the position *vs.* time graph for your second run. Explain how you can tell, by looking at the graph alone, how the speed of the cart compares to that in the first run. Now, repeat Step 2 and compare the value of the slope of the linear region to that obtained for your first run.

5. Now choose to view the position *vs.* time graph for the run in which you first launched the cart *toward* the motion detector (Run 4). What is the significance of the sign of the slope of the linear portion of this graph?

6. Now return to your first run. Choose Velocity as the vertical axis label. How does the plot of velocity *vs.* time correspond to the motion that you observed? Can you account for the fact that the plot may not be entirely horizontal?

7. Select an interval during which the velocity was nearly constant by dragging your cursor (or stylus) across this region. Choose Statistics from the Analyze menu. **Note:** Either application will display a number of statistical measures relating to velocity over this interval. From the information provided determine the percent decrease in the velocity over this interval.

8. In addition to slope, the area under a curve often has physical meaning. To determine the area, select the portion of the graph corresponding to when the cart was moving, then choose Integral from the Analyze menu.

9. What are the units of the area under the curve? What information about the motion of the cart does the area provide?

Activity
2

INSTRUCTOR INFORMATION

Investigating Motion

Investigating Constant Velocity with a Motion Detector

This activity provides students the opportunity to use a Vernier Motion Detector to obtain position-time and velocity-time information for a cart on a track. It is designed to both familiarize students with the operation of the Motion Detector and to apply graphical analysis techniques to the study of constant velocity.

The Microsoft Word files for the student pages can be found on the CD that accompanies this book. See *Appendix A* for more information.

OBJECTIVES

In this activity, the student objectives include

- Practice using the motion detector in a variety of experiments.
- Use a variety of analysis tools on graphs of position-time and velocity-time data.
- Interpret the equations resulting from the analysis of these graphs.

During this activity, you will help the students

- Learn to zero and reverse the direction of a motion detector.
- Navigate through the various menus in Logger *Pro* or LabQuest App.
- Learn to use Toolbar shortcuts for various menu items.

EQUIPMENT TIPS

While there are a number of ways one could collect suitable data for the analysis called for in this experiment, the best results are obtained by using a Dynamics Track, standard cart, and motion detector.

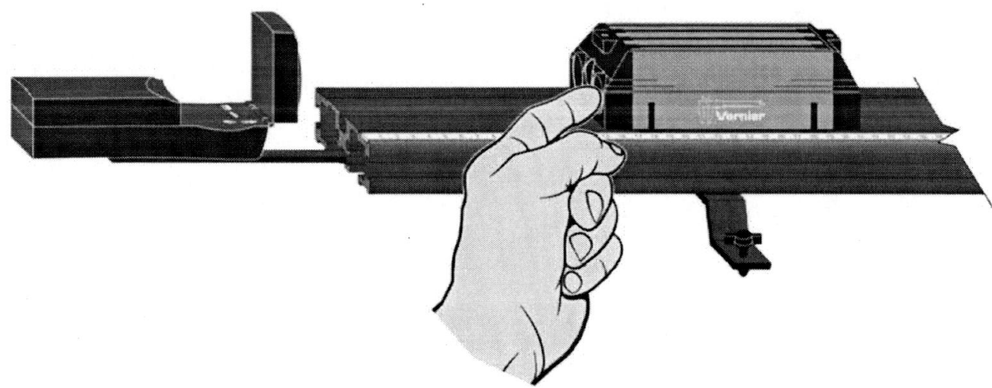

Figure 1

Attach the motion detector to the dynamics track using the bracket. If you set the tracks up in advance, it would help if you raised the end with the detector slightly to help counter frictional effects when students push the cart toward the other end. This will help make the velocity-time graph appear nearly horizontal. Make sure that the motion detector is set to Track mode. Students should remove objects (backpacks, books, etc.) from the area near the track so that they do not interfere with the signal from the motion detector.

PRE-LAB DISCUSSION

Inform your students of the two-fold purpose of this activity: they are learning to use the motion detector as well as applying graphical analysis techniques to the study of the behavior of an object moving at constant velocity.

LAB PERFORMANCE NOTES

Demonstrate applying a gentle push to the cart so that it moves smoothly down to the end of the track in approximately two seconds. Remind students that when they launch the cart toward the motion detector, they should catch it before it collides with the detector. If students are using Logger *Pro*, remind them to check Live Readouts under the Experiment menu so that the motion detector will display the current position of the cart. In LabQuest App, this is the default setting for the sensor tab.

SAMPLE RESULTS AND POST-LAB DISCUSSION

Detailed instructions for collecting and evaluating the data in either Logger *Pro* or LabQuest App are provided in the student version of this lab. If this is the first opportunity students have had to use Logger *Pro* to help them analyze data, then you might consider suggesting that students work through the examples in the tutorials that accompany this software.

Step 2

Figure 2 shows a graph of sample position-time data for the first run.

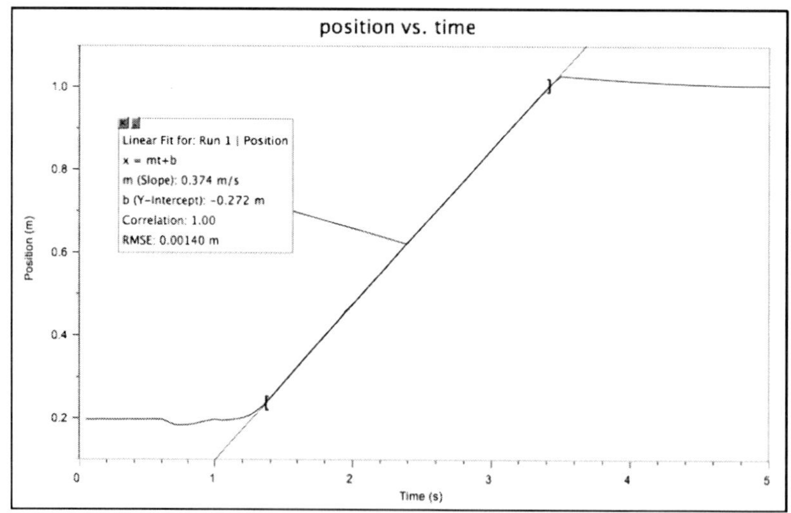

Figure 2

Step 3

The rate of change of position with respect to time is the average velocity, $\bar{v} = \Delta x/\Delta t$.

Logger *Pro* reports the units of the slope as $\frac{\text{m}}{\text{s}}$, consistent with units of velocity. LabQuest App does not give units for the slope or intercept, so students have to deduce these from the axis labels.

Step 4

If students launched the cart more rapidly than in the first run, the position *vs.* time graph will be steeper than that of the previous run. The magnitude of the slope is greater, confirming that the cart moved more rapidly.

Step 5

The negative sign of the slope indicates that the cart was moving in the opposite direction (see Figure 3).

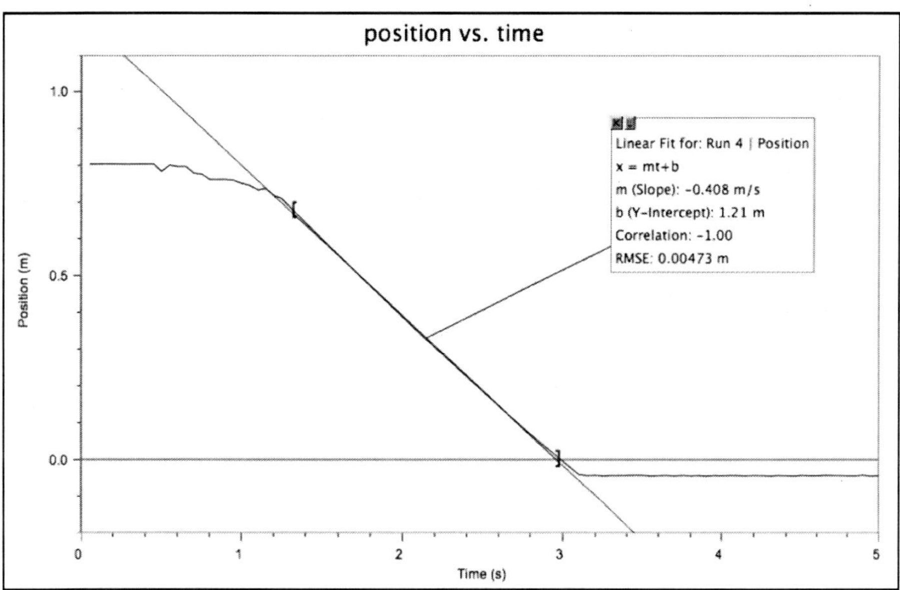

Figure 3 Moving in the opposite direction

Steps 6–7

Except for the initial part of the motion (when the cart was accelerated by the push) the velocity is nearly constant. If it were not for minor frictional losses, the plot would be horizontal (see Figure 4). When this graph was first auto-scaled, the maximum value of the vertical axis was 0.5 m/s, and the slope of the middle region was slightly negative. Increasing the vertical axis maximum to 1.0 m/s de-emphasized the effect of the frictional losses on the velocity. The statistics show that the decrease in velocity over the selected interval was $0.013/0.376 = 3.5\%$.

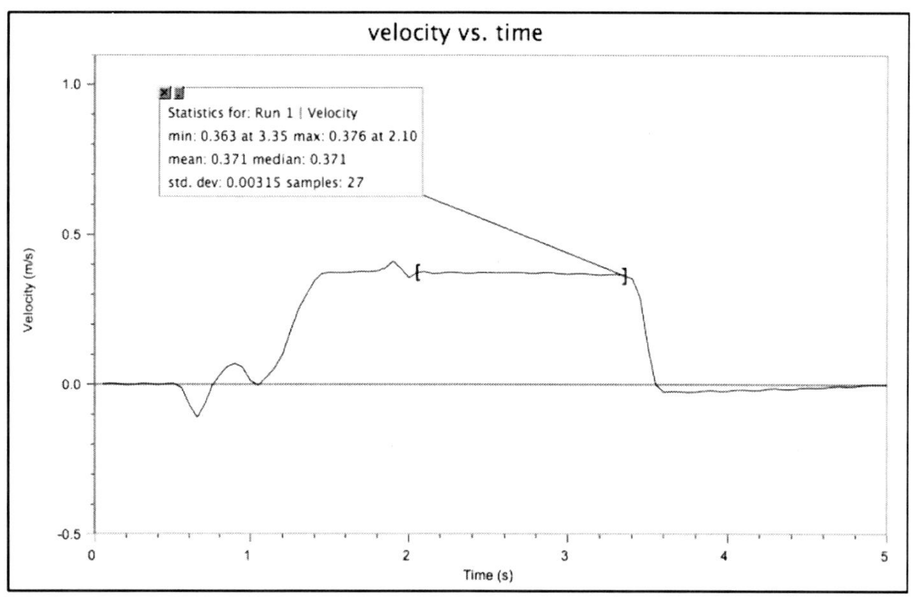

Figure 4

Steps 8–9

The region under velocity *vs.* time graph is nearly rectangular. Using the formula $A = l \times w$, one obtains the unit for length from the product of the units of time and velocity: $\mathrm{m} = \mathrm{s} \cdot \frac{\mathrm{m}}{\mathrm{s}}$. If you use the Integral tool to find the area under the curve (see Figure 5), you can obtain a value of this length that is quite close to the *displacement* of the cart from its original position. This is consistent with the equation for displacement: $\Delta x = \bar{v} \cdot \Delta t$.

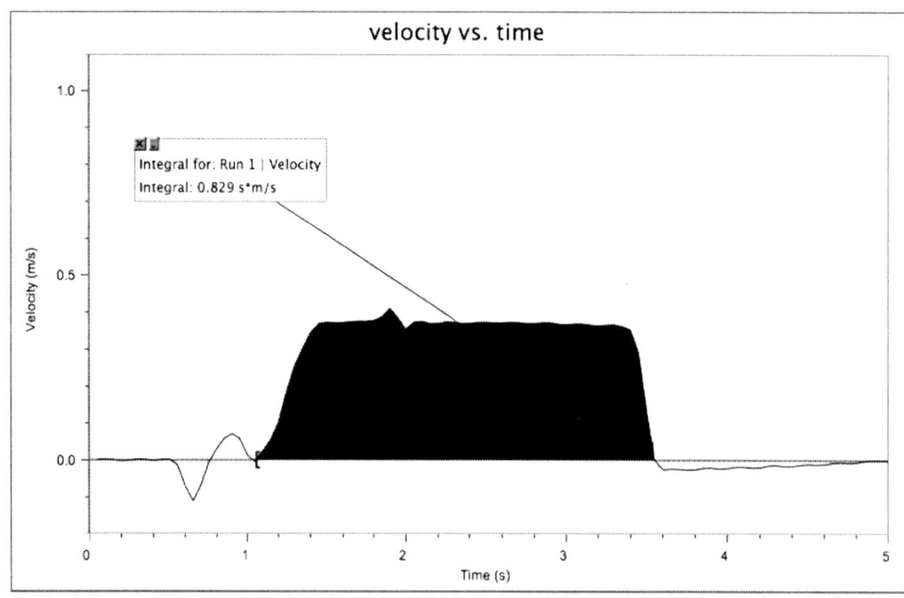

Figure 5 Area under the curve

Activity
3

Working with Analytical Tools

It is commonly said that at age 3 a child will be half of his or her adult height. You may also have heard that teenagers grow "like weeds." Both Logger *Pro* and LabQuest App have a set of tools you can use in your analysis of experimental data. This activity give you practice using these tools in Logger *Pro* and LabQuest App to answer these questions.

OBJECTIVES

In this activity, you will

- Add a calculated column to an existing data set.
- Re-size and group graphs in Logger *Pro*.
- Practice using the Examine, Tangent, Statistics, and Integral tools in Logger *Pro* or LabQuest App.
- Interpret the information obtained by the use of these tools.

MATERIALS

Logger *Pro* or LabQuest App
Sample data file: A3 Height v age.cmbl (computer)
or A3 Height v age.qmbl (LabQuest)

EVALUATION TOOLS IN LOGGER *PRO*

1. Open the Logger *Pro* file, A3 Height v age.cmbl, provided on the CD that accompanies this book. Note how Colin's height increases as he grows older. To better describe how his growth depends on his age you can use Logger *Pro* to determine the rate of change of height with age.

2. Choose New Calculated Column from the Data menu. This brings up a dialogue box in which you can name the column (e.g., growth rate) and units (m/yr) as well as specify the equation used to calculate the values. Insert the cursor in the Equation field, then, choose calculus▶ derivative from the Function list. Choose height as the Variable, then click Done (see Figure 1). The newly created column should now appear in your data table.

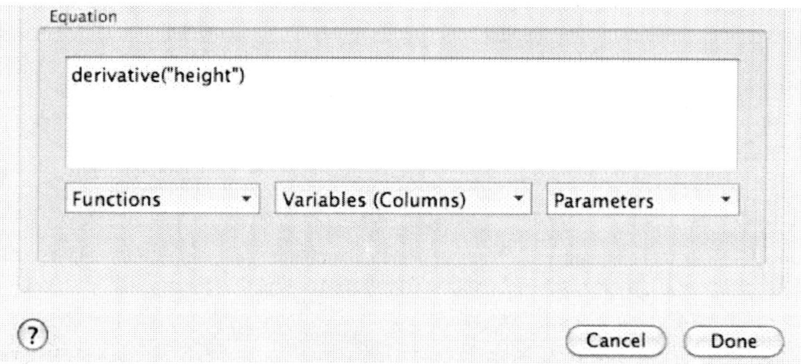

Figure 1

3. By adding a graph you can view how both Colin's height and his growth rate vary with time. Choose Graph from the Insert menu; a small graph of growth rate *vs.* age appears superimposed on your data table and graph. Choose Auto Arrange from the Page menu; this re-sizes both graphs and arranges them nicely on the page.

4. Select the height *vs.* age graph and activate the Tangent tool. You can do this either by choosing Tangent from the Analyze menu or by clicking the button in the toolbar shown on the right. Move the cursor across the graph to see how the software displays a portion of the tangent line drawn at a given point and gives the value of its slope.

5. Now select the growth rate *vs.* age graph and activate the Examine tool. You can do this either by choosing Examine from the Analyze menu or by clicking the button in the toolbar shown on the right. Move the cursor across the graph to see how the software displays a vertical line on the graph. In the information box the values of both variables for the nearest point are displayed.

6. To link both graphs so that these tools work simultaneously, select both graphs then choose Group Graphs (X-Axes) from the Page menu. As you move your mouse, note that the tangent and examine lines move in tandem. How does the value of the slope of the tangent in the top graph compare to the value of the growth rate in the bottom graph?

7. Deactivate these tools on each graph either by selecting a graph and clicking the tool you chose (it behaves like a toggle switch), or by clicking the little 'x' in the upper left corner of the information box.

8. Now select an age interval (e.g., from age 5 to 10) on the top graph by drag-selecting your mouse across a portion of the graph. Note that this action highlights a portion of the graph and the relevant entries in the data table as well. Choose Statistics from the Analyze menu or click the button in the toolbar shown to the right. From the maximum and minimum values determine Colin's change in height during this interval.

9. Now select this same age interval on the bottom graph. Choose Integral from the Analyze menu or click the button in the toolbar shown on the right. Note how Logger *Pro* shades in the area under the curve for this interval. Examine the units of the variables on the axes of this graph; what would you expect this area to represent? How does the value of the selected area compare to the value you obtained in Step 8?

10. Close the Logger *Pro* file, but do not save changes. Assuming Colin reached his maximum height at age 20, does the generalization made in the introduction apply? Was he growing most rapidly as a teenager? If not, at what age was he growing most rapidly?

EVALUATION TOOLS IN LABQUEST APP

1. To evaluate the data in LabQuest App, you will need to transfer the file, A3 Height v age.qmbl, provided on the CD that accompanies this book to a USB flash drive, connect that to the LabQuest, then open the file from the flash drive. Tap the graph tab. Note how Colin's height increases as he grows older. To better describe how his growth depends on his age you can use LabQuest App to determine the *rate* of change of height with age.

2. Tap the Table tab. Choose New Calculated Column from the Table menu. This brings up a window in which you can name the column (growth rate) and units (m/yr) and choose the precision you wish to display. This window also provides the place where you can specify

the equation used to calculate the values. Select 1st Derivative (Y,X) as the Equation Type from the options in the drop down menu. Accept the default choices of variables for X and Y by tapping the OK button (see Figure 2 below).

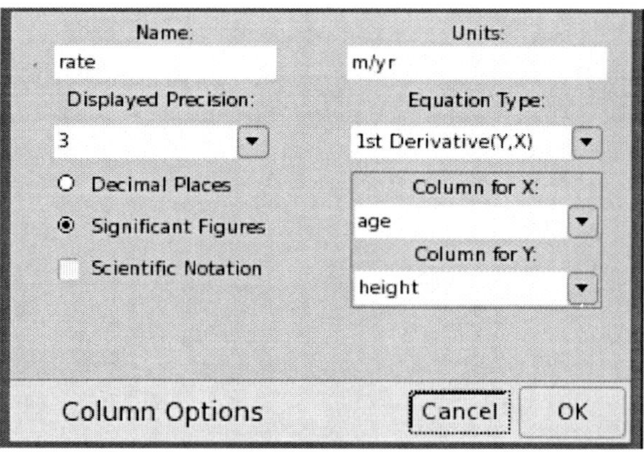

Figure 2

3. When you tap OK, LabQuest App returns you to the graph window. From the Graph menu choose Show Graph ▶ All Graphs; you should now see two graphs. Activate the Tangent tool by choosing Tangent from the Analyze menu. As you tap the stylus on different parts of the screen, see how the software displays a portion of the tangent line drawn at a given point on both graphs. To the right of the graphs are boxes displaying the x and y values of that location on each graph and the values of the slopes of the tangents. How does the value of the slope of the tangent in the top graph compare to the value of the growth rate in the bottom graph?

4. Deactivate the Tangent tool by choosing Tangent from the Analyze menu (tapping a selected tool deactivates it like a toggle switch). Now select an age interval (e.g., from age 5 to 10) on the top graph by dragging your stylus across a portion of the graph. From the Analyze menu choose Statistics and then check the **height** box. LabQuest App provides various statistics for the selected interval. Use the up and down arrows to view all the information in the window (see Figure 3).

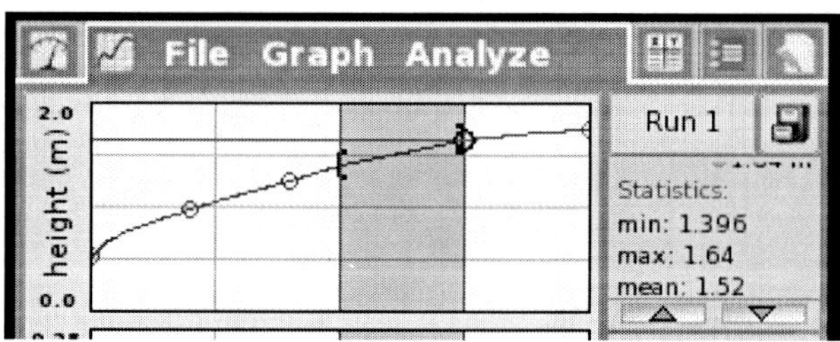

Figure 3

From the maximum and minimum values determine Colin's change in height during this interval.

5. Now select the same age interval on the bottom graph. Choose Integral from the Analyze menu and check the box next to **rate**. Note how LabQuest App shades in the area under the curve for this interval. Examine the units of the variables on the axes of this graph; what would you expect this area to represent? How does the value of the selected area compare to the value you obtained in Step 4?

6. Assuming Colin reached his maximum height at age 20, does the generalization made in the introduction apply? Was he growing most rapidly as a teenager? If not, at what age was he growing most rapidly?

Activity
3

INSTRUCTOR INFORMATION

Working with Analytical Tools

It is presumed that your students will have either performed the first activity or have used Logger *Pro* or LabQuest App before. This activity is designed to provide students further opportunity to become familiar with some of the analytical tools available in both Logger *Pro* and LabQuest App. Understanding how to use these tools will help students when they analyze their data in the experiments in this manual.

The Microsoft Word files for the student pages can be found on the CD that accompanies this book. See *Appendix A* for more information. If your students are using only one of data collection, you might wish to eliminate the directions for the software your students will not be using.

OBJECTIVES

In this experiment, the student objectives include

- Add a calculated column to an existing data set.
- Re-size and group graphs in Logger *Pro*.
- Practice using the Examine, Tangent, Statistics, and Integral tools in Logger *Pro* or LabQuest App.
- Interpret the information obtained by the use of these tools.

During this experiment, you will help the students

- Recognize that the slope of the tangent to a point on a curve gives the rate of change of the function at that point.
- Recognize that the area under a rate *vs*. time curve is the change in that quantity over the selected interval.

EQUIPMENT TIPS

Students do not collect any data in this experiment. Instead they must have access to the file A3 Height v age.cmbl if they are using Logger *Pro* on a computer, or A3 Height v age.qmbl on LabQuest as a standalone device.

SAMPLE RESULTS AND POST-LAB DISCUSSION – LOGGER *PRO*

The sample data in the file provided are not ideal or made up. They are actual measurements of a real person. This should give students confidence that they can obtain useful results with somewhat messy data.

Step 2

Encourage students to choose a short name as well as the name when they create their new calculated column. Logger *Pro* always uses the long name for the graph axes but will use the short name if the width of the column in the data table does not allow the full name to be displayed. If they leave this field blank, then Logger *Pro* will use CC (calculated column) for the data table.

Steps 4–5

Use of the buttons in the Toolbar in Logger *Pro* provides a faster way to call upon these tools than accessing them from the Analysis menu. These buttons work like toggle switches.

Step 6

With the graphs grouped, the same value for the x-axis is chosen on both graphs. Students should see that the slope of the tangent to the curve in the top graph is the same as the growth rate at that age on the bottom graph (see Figure 1).

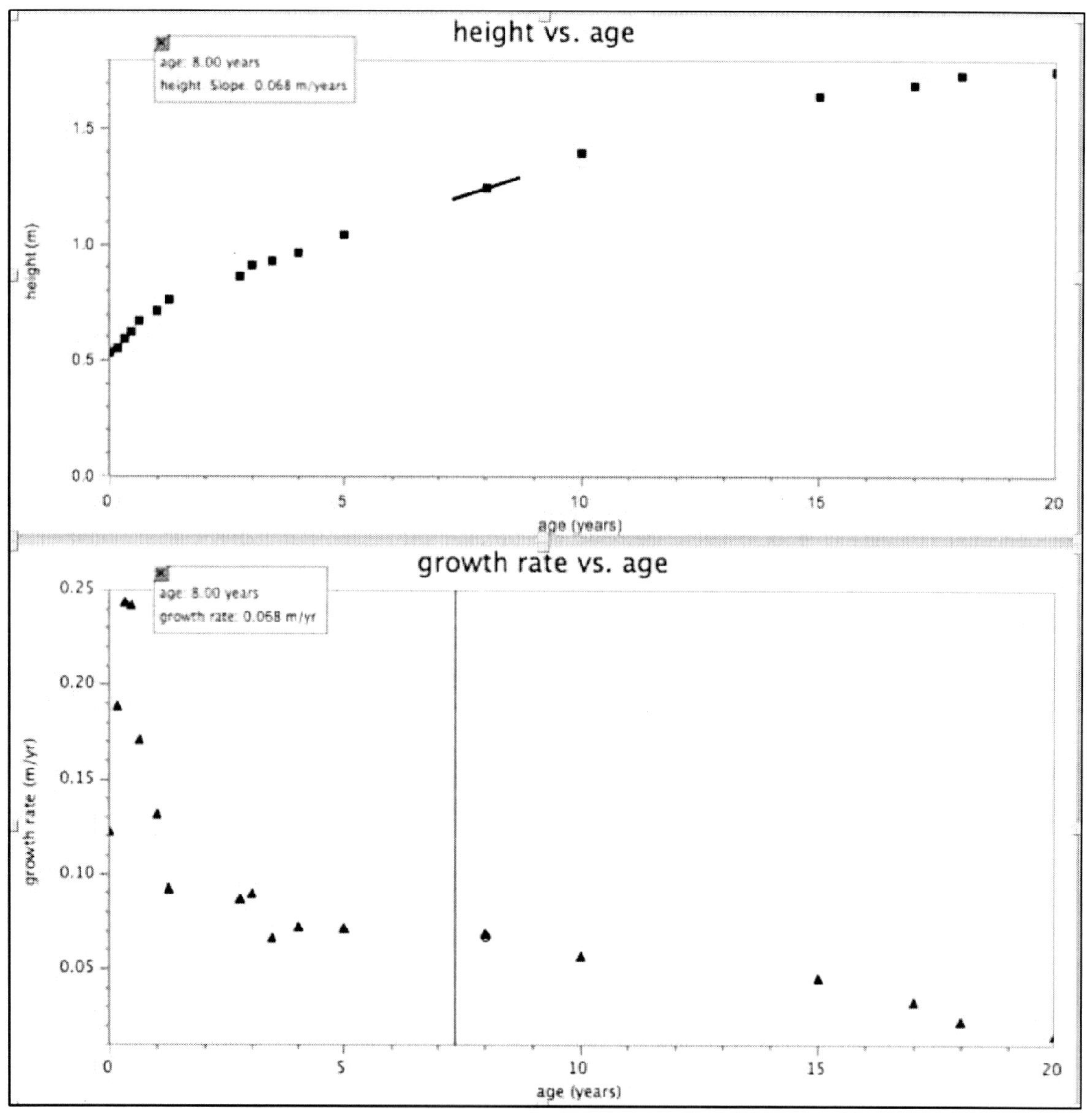

Figure 1 Grouped graphs

Steps 8–9

From the Statistics provided by Logger *Pro*, one can see that between ages 5 and 10 Colin's height increased by 1.396 – 1.042 = 0.354 m. When the same age interval is selected on the bottom graph and Integral is activated, Logger *Pro* reports the area under the curve to be 0.3346 years*m/yr. The software leaves it up to the user to simplify the units of the area (see Figure 2). You should point out that while the unit for the area under the curve is meters, this quantity very nearly equals Colin's *change* in height.

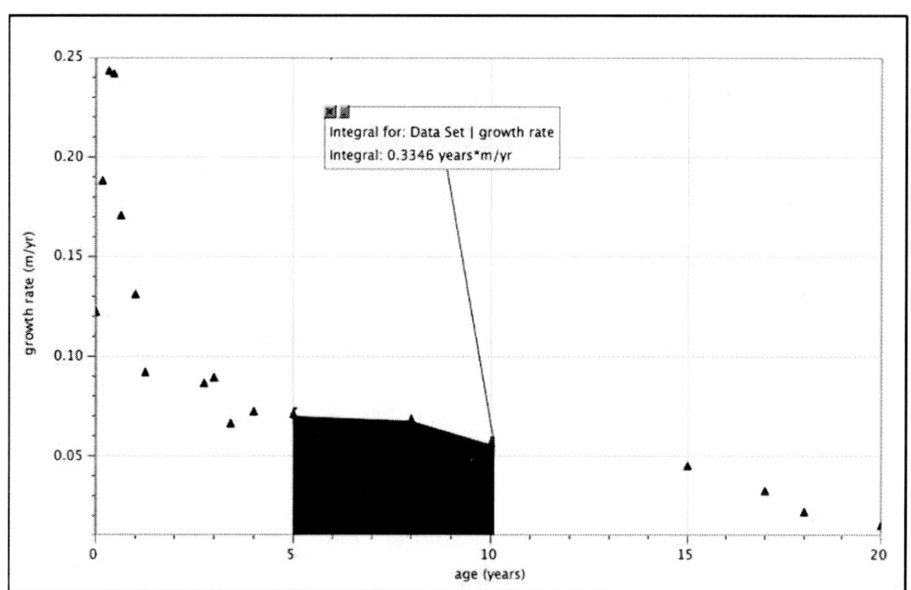

Figure 2 Integrating area under curve

Step 10

Colin's height at age 3 was 0.908 m, very close to half of his adult height of 1.750 m. Students should find that Colin was growing most rapidly when he was about 4 months old.

SAMPLE RESULTS AND POST-LAB DISCUSSION – LABQUEST APP

Step 2

When students add a calculated column and then begin to name it, the keyboard takes up about half of the window (see Figure 3). Point out that they can either use the scrollbar to reveal portions of the window covered by the keyboard or tap the keyboard icon to make it go away.

Figure 3

Step 3

When students choose Tangent from the Analyze menu, LabQuest App shows tangents to the curves on both graphs simultaneously (see Figure 4). Students are instructed to compare the slope of the tangent to the height *vs.* age graph to the value of the growth rate at that same age on the bottom graph. They should see that these values are the same.

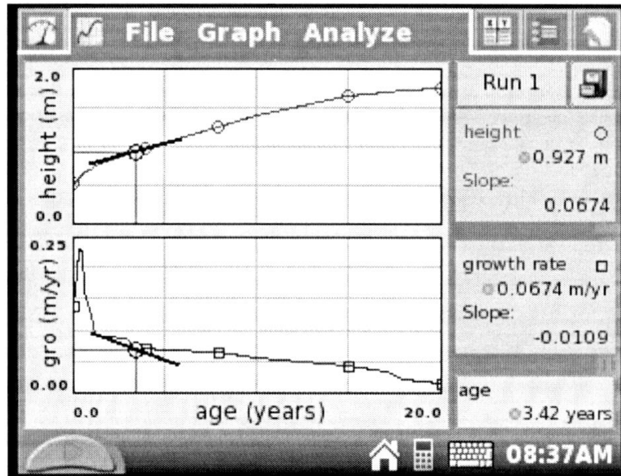

Figure 4 Tangents

Steps 4–5

From the Statistics provided by LabQuest App, one can see that between ages 5 and 10 Colin's height increased by 1.396 – 1.042 = 0.354 m. When Integral is selected for this same age interval, LabQuest App reports the area to have a value of 0.32128 (see Figure 5). Students will have to determine the unit for the area by examining the units for the x and y axes.

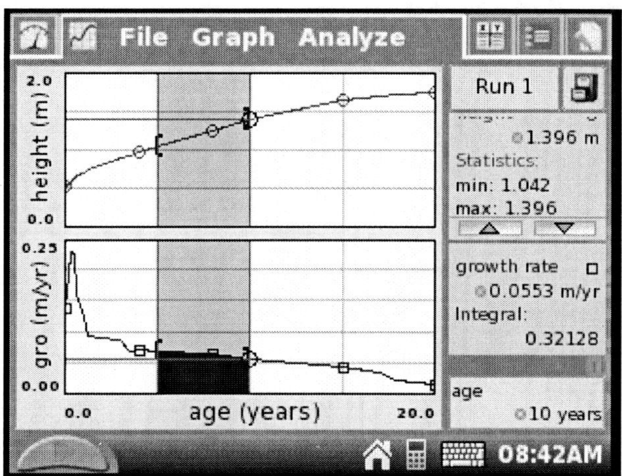

Figure 5 Integral

Students should conclude that the area under the curve is very nearly the *change* in Colin's height during this interval.

Step 6

Colin's height at age 3 was 0.908 m, very close to half of his adult height of 1.750 m. Students should find that Colin was growing most rapidly when he was about 4 months old. Having completed this activity, students should feel more comfortable using some of the analytical tools they will use frequently in the experiments in this manual.

Activity
4
Introduction to the Vernier Photogate

The Vernier Photogate is used to collect a wide variety of motion data. It can also be used to study the motion of toy cars, dynamics carts, objects in freefall, pendulums, projectiles, and much more.

The photogate works by projecting an infrared beam from one arm of the sensor to the other arm. When the beam is blocked the sensor sends a signal, which illuminates an LED on the top of the gate as well as triggering the software to display a blocked message in the data-collection area.

Photogates measure times at which the gate is blocked or unblocked. There are several ways of using this timing information, depending on the goal of your experiment.

- If you need to know timing information from a pulley or a picket fence object, you will need "Motion Timing." Motion timing uses the block-to-block timing of regularly spaced marks and can be use to generate position, velocity and acceleration graphs.
- If you have an object passing through a gate, and you want to know its speed, you need "Gate" timing. Gate timing measures the block-to-unblock time interval; the ratio of the object length and the time interval is the object's speed.
- If you have an object passing from one photogate to another you will need "Pulse Timing." Pulse timing measures the block-to-block time for a pair of gates. For example, you might measure the average speed of a cart passing through one gate and then through a second gate. The ratio of the gate spacing and the block-to-block time is the cart's speed. Pulse timing is not used in this book.
- If you need to measure the period of a pendulum, which is the time from the first to the third block of a photogate, you need "Pendulum Timing."

There are also other, less common timing modes. The default timing mode is Motion Timing.

This tutorial shows you some of the ways you can use the Vernier Photogate to collect data. Refer to the part you need for the specific application in your experiment.

OBJECTIVES

In this activity, you will collect and interpret data from a Vernier Photogate in the following modes:

- Motion Timing mode
- Gate Timing mode
- Pendulum Timing mode

MATERIALS

Vernier data-collection interface
Logger *Pro* or LabQuest App
Cart Picket Fence
Vernier Photogate and bracket
Picket Fence

PART 1 – MOTION TIMING

When you start the data-collection software, the default mode for the photogate is Motion Timing. In this mode the software records a series of blocking events from an object that has a repeating transparent and opaque pattern.

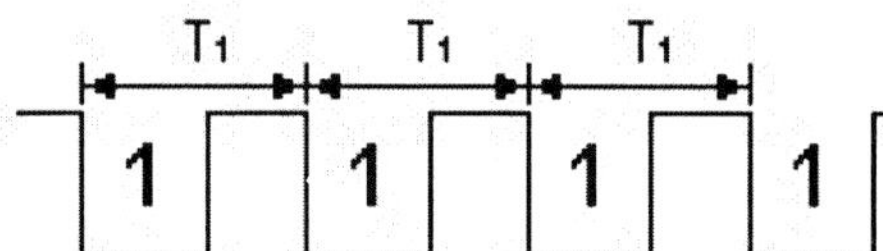

Two good examples of objects that utilize this mode are the Vernier Picket Fence (a plastic bar with a series of opaque and transparent bars) or the Vernier Ultra Pulley.

When a picket fence (see right) passes completely through a photogate, the equipment records eight blocking events, one for each black bar. The distance from the leading edge of one black bar to the next is 5 cm. Using the times from the blocking events and the 5 cm spacing, position, speed and acceleration data are determined. A smaller version of this picket fence is available for use on dynamics carts. The spacing between leading edges of the black bars on the smaller fence is 1 cm.

If you are using a pulley instead (see right), the pulley spokes block and unblock the beam. Using the blocking times and knowing the circumference of the pulley, the software can again determine position, speed and acceleration. A good example of the use of the pulley is with an Atwood's machine.

Using Logger *Pro*

1. Connect a Vernier Photogate to the interface and start Logger *Pro*.

2. Verify that the photogate and software are working by placing your hand in the photogate beam. The red LED on the top of the gate should illuminate and the software should display Blocked next to GateState above the data table in the data-collection area. If this does not happen, check all of your connections.

3. From the Experiment menu choose Set Up Sensors, specify your interface, then click on the icon of the photogate in the window. Note that Motion Timing is the default mode. When you select Set Distance or Length you can choose from a number of pre-set values or choose your own length by selecting User Defined (see right). The value and unit entered determine the steps in the distance column, which in turn scales the velocity and acceleration columns.

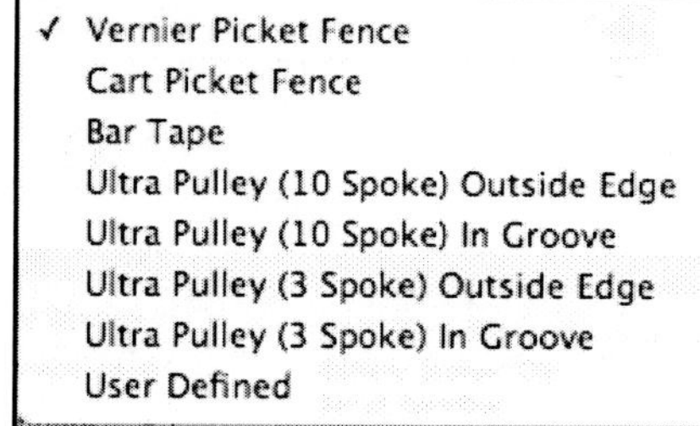

4. Begin collecting data, then, holding a picket fence by the edge, pass it through the photogate in a steady motion. Then stop collecting data. This can be accomplished either by clicking the Stop button or by pressing the spacebar.

5. An examination of the graph of distance *vs.* time shows that the picket fence was moving at nearly constant velocity (see Figure 1).

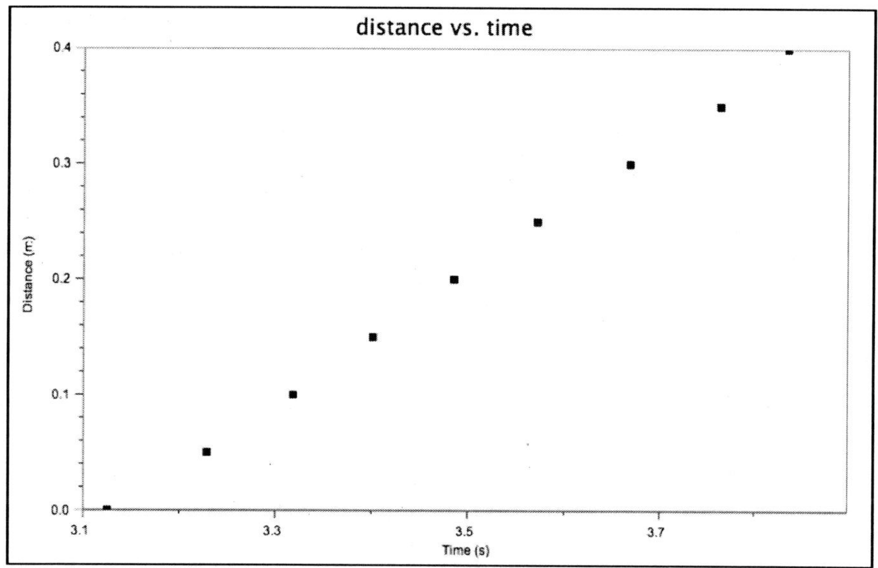

Figure 1

Using LabQuest as a standalone device

1. Connect a Vernier Photogate to LabQuest and turn on the interface or select New from the File menu.

2. Verify that the photogate and software are working by placing your hand in the photogate beam. The red LED on the top of the gate should illuminate and the software should display Blocked in the sensor window. If this does not happen, check all of your connections.

3. Tap Mode; note that Motion is the default Photogate Mode and Vernier Picket Fence is the default object to block the beam (see right). You can change the distance setting in this window. User defined allows you to set the spacing to 1 cm for a cart picket fence.

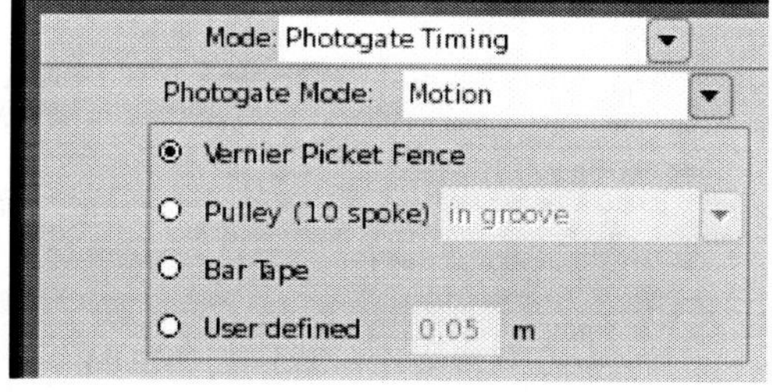

4. Begin collecting data, then, holding a picket fence by the edge, pass it through the photogate in a steady motion. The software automatically stops collecting data.

5. An examination of the graph of distance *vs.* time shows that the picket fence was moving at nearly constant velocity (see Figure 2). Unless you re-scale the graph to set 0 as the bottom of the y-axis (from the Graph Options menu), the velocity-time graph is likely to exaggerate any variation in the speed of the picket fence through the photogate.

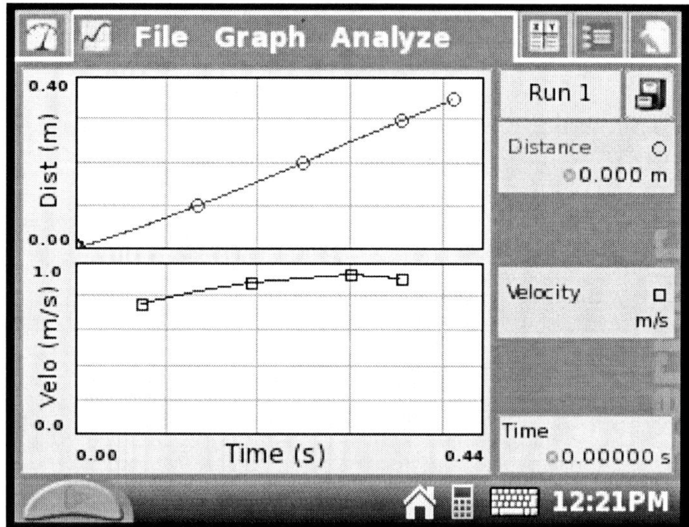

Figure 2

PART 2 – GATE TIMING

Gate timing begins when the photogate is first blocked. The timing continues until the gate is unblocked. The duration of the interruption is recorded by the software. If the length of the object is entered in the Length of Object field, the speed is calculated. The diagram at right represents the gate state during a blocking event. Initially the gate is unblocked, then blocked, and then unblocked again.

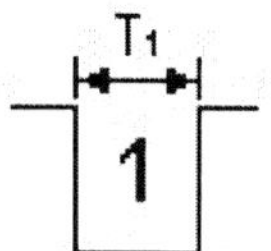

Speed calculation in Gate mode relies upon knowing the length of the object going through the gate. For some objects (the flag on the cart picket fence or note cards) this is not a problem, but other objects might be more challenging. For example, if you rolled a marble through the gate, you would need to know the diameter of the ball that went through the beam. Inaccuracies in positioning the photogate can introduce errors into the speed measurements; therefore, alignment in this type of experiment is important.

When using the Gate mode, you can also add a second photogate to measure the speed of an object through that gate. Use of two photogates in this mode would allow you to determine the change in the speed of a single object or to determine the speed of two objects.

Using Logger *Pro*

1. Connect a Vernier Photogate to the interface and start Logger *Pro*.

2. Verify that the photogate and software are working by placing your hand in the photogate beam. The red LED on the top of the gate should illuminate and the software should display

Blocked next to GateState above the data table in the data-collection area. If this does not happen, check all of your connections.

3. From the Experiment menu choose Set Up Sensors, specify your interface, then click on the icon of the photogate in the window. Select Gate Timing as the mode.

4. Click on the icon of the photogate again and select Set Distance or Length. Note that the default setting is 0.050 m. This is the width of the flag on the cart picket fence. Choose an object (your finger, a ruler, etc.) you can use to block the beam. Measure its width, then set the distance to that value.

5. Begin collecting data, pass the object slowly through the photogate, then back through the beam again, then stop the data-collection. Examine the data table. You should see something like that in Figure 3.

	Latest			
	Time (s)	State	GT (s)	Velocity (m/s)
1	1.833258	1		
2	1.854904	0	0.021646	0.924
3	3.123231	1		
4	3.165193	0	0.041962	0.477

Figure 3

The time column indicates when the beam was blocked (State = 1) and then unblocked (State = 0). The GT (Gate Time) column displays the duration. The Velocity is calculated from the distance you set and the duration of the block. These data show that the object moved more slowly in the second pass through the photogate than in the first.

Using LabQuest as a standalone device

1. Connect a Vernier Photogate to LabQuest and turn on the interface or select New from the File menu.

2. Verify that the photogate and software are working by placing your hand in the photogate beam. The red LED on the top of the gate should illuminate and the software should display Blocked in the sensor window. If this does not happen, check all of your connections.

3. Tap Mode and select Gate as the Photogate Mode. Note that the default setting in the Length of object field is 0.050 m. This is the width of the flag on the cart picket fence. Choose an object (your finger, a ruler, etc.) you can use to block the beam. Measure its width, then set the distance to that value. Tap OK.

4. Begin collecting data, pass the object slowly through the photogate, then back through the beam again, then stop the data-collection. The software displays the Graph window, but it is easier to understand the data if you tap the Table tab to examine the data table. You should see something like that in Figure 4.

File Table

Run 1

Time (s)	Gate Sta	Gate (s)	Ve (m/s)
0.00000	Blocked		
0.05675	Unblocked	0.056755	0.352
2.56039	Blocked		
2.66768	Unblocked	0.107288	0.186

Figure 4

The time column indicates when the beam was blocked and then unblocked. The Gate column displays the duration. The Velocity is calculated from the distance you set and the duration of the block. These data show that the object moved more slowly in the second pass through the photogate than in the first.

In either application, if a second photogate is connected to the interface, data for this gate will also be displayed in the table as Gate 2.

PART 3 – PENDULUM TIMING

The Pendulum Timing mode uses a single photogate attached to an interface. The timing will begin when the photogate is first interrupted. The timing will continue until the photogate is interrupted twice more, so that you get the time for a complete swing of a pendulum or other oscillating object.

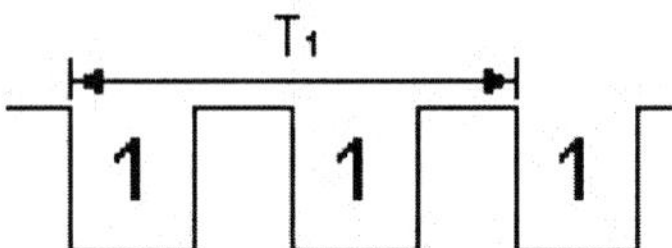

Use of this mode will give you very accurate measurements of pendula periods.

Using Logger *Pro*

1. Connect a Vernier Photogate to the interface and start Logger *Pro*.

2. Verify that the photogate and software are working by placing your hand in the photogate beam. The red LED on the top of the gate should illuminate and the software should display Blocked next to GateState above the data table in the data collection area. If this does not happen, check all of your connections.

3. From the Experiment menu choose Set Up Sensors, specify your interface, then click on the icon of the photogate in the window. Select Pendulum Timing as the mode.

4. Move your finger through the photogate, count "one thousand one", move your finger back through the photogate, count "one thousand one" again and now pass your finger back through the gate a third time. Stop collecting data. The period of this motion is displayed in

the third column of the data table. This value should be close to 2 seconds if your timing was correct.

5. Try collecting data again, but this time move your finger through the photogate several times in a consistent back and forth motion. Stop collecting data. Judge how uniform was the period of your "pendulum motion" (see Figure 5).

	Time (s)	Latest State	Period (s)
1	1.306657	1	
2	1.329886	0	
3	2.378946	1	
4	2.410011	0	
5	3.470532	1	2.163875
6	3.502938	0	
7	4.701334	1	
8	4.732922	0	
9	5.775299	1	2.304767
10	5.811170	0	
11	7.028317	1	
12	7.061474	0	
13	8.145139	1	2.369840
14	8.183325	0	

Figure 5

Using LabQuest as a standalone device

1. Connect a Vernier Photogate to LabQuest and turn on the interface or select New from the File menu.

2. Verify that the photogate and software are working by placing your hand in the photogate beam. The red LED on the top of the gate should illuminate and the software should display Blocked in the sensor window. If this does not happen, check all of your connections.

3. Tap Mode and select Pendulum as the Photogate Mode. Tap OK.

4. Move your finger through the photogate, count "one thousand one", move your finger back through the photogate, count "one thousand one" again and now pass your finger back through the gate a third time. Stop collecting data.

5. This value should be close to 2 seconds if your timing was correct. Now try collecting data again, but this time move your finger through the photogate several times in a consistent back and forth motion. Stop collecting data.

6. Tap the Table tab to see the uniformity of the period of your "pendulum motion" (see Figure 6).

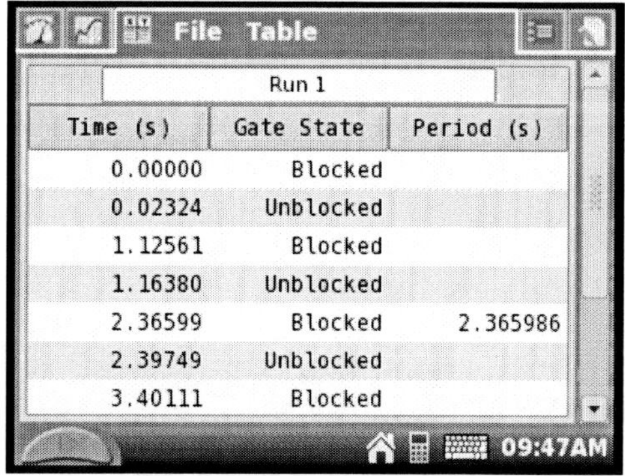

Time (s)	Gate State	Period (s)
0.00000	Blocked	
0.02324	Unblocked	
1.12561	Blocked	
1.16380	Unblocked	
2.36599	Blocked	2.365986
2.39749	Unblocked	
3.40111	Blocked	

Figure 6

Experiment

1

Motion on an Incline

INTRODUCTION

When you examined an object moving with constant velocity in introductory Activity 2, you learned two important points about the line of best fit to the graph of position *vs.* time:

1. The slope (rate of change) of the graph was constant, and gave the velocity of the object.

2. The intercept gave the initial position of the object.

In this experiment, you will examine a different kind of motion and contrast features of the position-time and velocity-time graphs with those you have studied earlier.

OBJECTIVES

In this experiment, you will

- Collect position, velocity, and time data as a cart rolls up and down an inclined track.
- Analyze the position *vs.* time and velocity *vs.* time graphs.
- Determine the best fit equations for the position *vs.* time and velocity *vs.* time graphs.
- Distinguish between average and instantaneous velocity.
- Use analysis of motion data to define instantaneous velocity and acceleration.
- Relate the parameters in the best-fit equations for position *vs.* time and velocity *vs.* time graphs to their physical counterparts in the system.

MATERIALS

Vernier data-collection interface
Logger *Pro* or LabQuest App
Vernier Motion Detector
Vernier Dynamics Track
standard cart
Motion Detector bracket
books or blocks to elevate track

PRE-LAB INVESTIGATION

Elevate one end of the track. Place the cart at the lower end and give it a gentle push so that it moves up the track (without falling off) and returns to its starting position.

On the axes to the right, predict what a graph of the position *vs.* time would look like. Use a coordinate system in which the origin is on the left and positive is to the right.

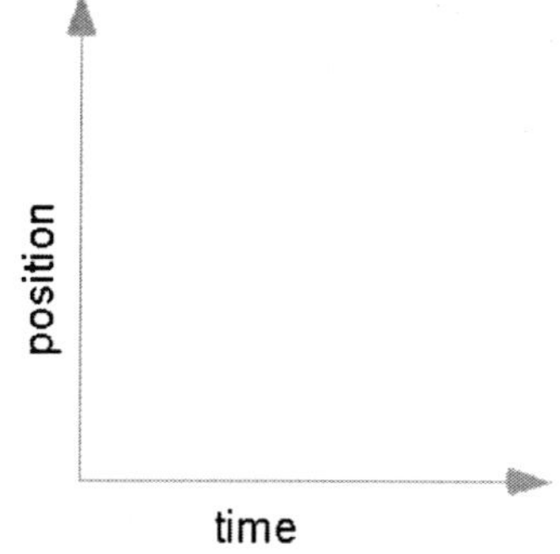

PROCEDURE

1. Connect the Motion Detector to the interface and start the data-collection program. Two graphs: position *vs.* time and velocity *vs.* time will appear in the graph window. For now, hide or remove the velocity *vs.* time graph. Later, during the analysis of data, you will add the v-t graph back to your view.

2. Attach the motion detector to the bracket that will allow you to position it near one end of the track.

3. If your motion detector has a switch, set it to Track.

4. Elevate the end of the track opposite the motion detector as directed by your instructor.

5. Practice launching the cart with your finger so that it slows to a stop at least 50 cm from its initial position before it returns to the initial position.

6. Hold the cart steady with your finger at least 20 cm from the motion detector[1], then zero the motion detector.

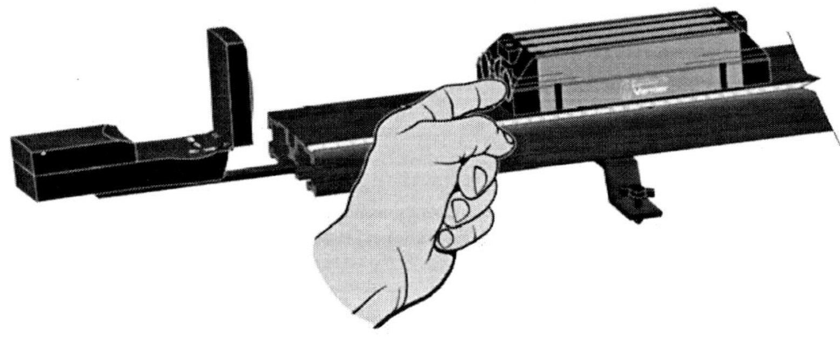

Figure 1

7. Begin collecting data, then launch the cart up the ramp. Be sure to catch it once it has returned to its starting position.

8. Repeat, if necessary, until you get a trial with a smooth position-time graph.

EVALUATION OF DATA

Part 1

1. Either print or sketch the position *vs.* time (x-t) graph for your experiment. On this graph identify:

   - Where the cart was rolling freely up the ramp
   - Where the cart was farthest from its initial position
   - Where the cart was rolling freely down the ramp

[1] If you are using an older motion detector without a switch, the cart needs to be at least 45 cm from the detector.

2. In your investigation of an object moving at constant velocity, you learned that the slope of the *x-t* graph was the average velocity of the object. In this case, however, the slope for any interval on the graph is not constant; instead, it is constantly changing. Based on your observations, sketch a graph of velocity *vs.* time corresponding to that portion of the *x-t* graph where the cart was moving freely.

3. Now, view both the position *vs.* time and velocity *vs.* time graphs. Compare the *v-t* graph to the one you sketched in Step 2.

4. Take a moment to think about and discuss how you could determine the cart's velocity at any given instant.

5. If you are using Logger *Pro*, group the two graphs (x-axis), and turn on the Tangent tool for the *x-t* graph and the Examine tool for the *v-t* graph. (In LabQuest App, simply turn on the Tangent tool). Using either program, compare the slope of the tangent to any point on the *x-t* graph to the value of the velocity on the *v-t* graph. Write a statement describing the relationship between these quantities.

Part 2

1. Perform a linear fit to that portion of the *v-t* graph where the cart was moving freely. Print or sketch this *v-t* graph. Write the equation that represents the relationship between the velocity and time; be sure to record the value and units of the slope and the vertical intercept.
   On this *v-t* graph identify:
   - Where the cart was being pushed by your hand
   - Where the cart was rolling freely up the ramp
   - The velocity of the cart when it was farthest from its initial position
   - Where the cart was rolling freely down the ramp

2. The slope of a graph represents the rate of change of the variables that were plotted. What can you say about the rate of change of the velocity as a function of time while the cart was rolling freely? In your discussion, you will give a name to this quantity. What is the significance of the algebraic sign of the slope?

3. Compare the value of your slope to those of others in the class. What relationship appears to exist between the value of the slope and the extent to which you elevate the track?

4. The vertical intercept of the equation of the line you fit to the *v-t* graph represents what the velocity of the cart would have been at time $t = 0$ had it been accelerating from the moment you began collecting data. Suggest a reasonable name for this quantity. Now write a general equation relating the velocity and time for an object moving with constant acceleration

5. The position-time graph of an object that is constantly accelerating should appear parabolic. Use the Curve Fit function of your data analysis program to fit a quadratic equation to that portion of the *x-t* graph where the cart was moving freely. Note the values of the A and B parameters in the quadratic equation. You will have to provide the units.

6. Compare these parameters (values and units) to the slope and intercept of the line used to fit the *v-t* graph. Now write a general equation relating the position and time for an object undergoing constant acceleration.

EXTENSION

Try repeating the data collection with the same apparatus, but, this time, place the Motion Detector at the top of the track. Interpret your *x-t* and *v-t* graphs as you did before.

ANIMATED DISPLAY

If you are using Logger *Pro*, inserting an animated display gives you another tool to represent both the position and velocity of the cart at a number of instants during the experiment. Your instructor will show you how to set up the point display options for such a display.

Experiment

1

INSTRUCTOR INFORMATION

Motion on an Incline

This lab is written with the assumption that the instructor will engage the students in discussions at critical junctures. These discussions can take place with the entire class or with individual lab groups. The icon at left indicates where these discussions should occur.

The Microsoft Word files for the student pages can be found on the CD that accompanies this book. See *Appendix A* for more information.

OBJECTIVES

In this experiment, the student objectives include

- Collect position, velocity, and time data as a cart rolls up and down an inclined track.
- Analyze the position *vs.* time and velocity *vs.* time graphs.
- Determine the best fit equations for the position *vs.* time and velocity *vs.* time graphs.
- Distinguish between average and instantaneous velocity.
- Use analysis of motion data to define instantaneous velocity and acceleration.
- Relate the parameters in the best-fit equations for position *vs.* time and velocity *vs.* time graphs to their physical counterparts in the system.

During this experiment, you will help the students

- Distinguish between average and instantaneous velocity.
- Recognize that instantaneous velocity is approximately the average rate of change in the position *vs.* time graph over a very small time interval.
- Define the instantaneous velocity as the slope of the tangent line at a point on the position *vs.* time graph.
- Recognize that an acceleration *vs.* time graph can be derived by examining the rate of change of a velocity *vs.* time graph.
- Relate the A and B parameters in a quadratic fit to a position *vs.* time graph to their physical counterparts in the system.

REQUIRED SKILLS

In order for student success in this experiment, it is expected that students know how to

- Zero a motion detector. This is addressed in Activity 2.
- Perform linear and curve fits to graphs. This is addressed in Activity 1.
- Manipulate graphs including deleting and adding graphs, grouping graphs in Logger *Pro*, and using the Tangent and Examine tools. This is addressed in Activity 3.

EQUIPMENT TIPS

While there are a number of ways one could collect suitable data for the analysis called for in this experiment, the best results are obtained by using a Dynamics Track, standard cart, and motion detector.

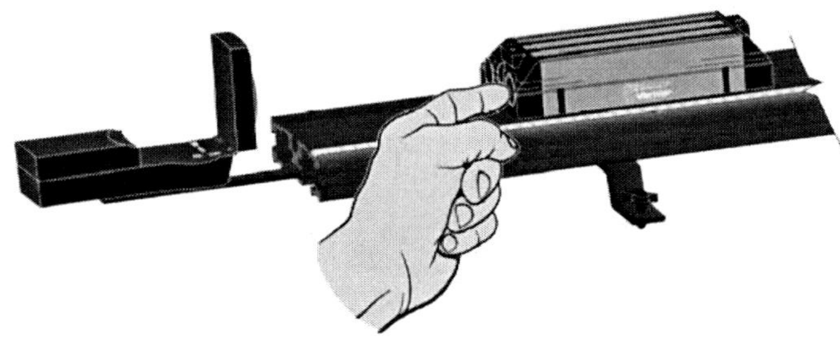

Figure 1

Attach the motion detector to the dynamics track using the bracket and position it so that it is at the lower end of the track. Make sure that the motion detector is set to Track mode. Students should remove objects (backpacks, books, etc.) from the area near the track so that they do not interfere with the signal from the motion detector.

PRE-LAB DISCUSSION

This experiment should be performed only after students have had the opportunity to explore the behavior of an object moving with constant velocity. Have students sketch position *vs.* time (*x*-*t*) and velocity *vs.* time (*v*-*t*) graphs for an object traveling with constant velocity as a review prior to exploring accelerated motion. Wait until the evaluation of data collected in this experiment to introduce such technical terms as instantaneous velocity and acceleration. As Arnold Arons recommends in *Teaching Introductory Physics*[1], "Recognize that to be understood and correctly used, such terms require careful operational definition, rooted in shared experience and in simpler words previously defined; in other words, that a scientific concept involves an idea first and a name afterwards…".

After demonstrating the motion of a cart up and down an elevated track, inform the students that they will be investigating the position *vs.* time and velocity *vs.* time behavior of this system. Allow them the opportunity to explore the behavior of the cart and ramp briefly, so they can make a prediction of the shape of the position *vs.* time graph, then distribute the lab instructions.

LAB PERFORMANCE NOTES

When the motion detector is connected to the data-collection interface and Logger *Pro* or LabQuest App is started, the default graph screen shows both position *vs.* time (*x*-*t*) and velocity *vs.* time (*v*-*t*) graphs. In order to gradually develop the concepts of instantaneous velocity and acceleration, instruct the students to hide the *v*-*t* graph. In Logger *Pro* this can be accomplished simply by dragging the corner of the *x*-*t* graph so that it covers the *v*-*t* graph. In LabQuest App, the students can choose to show Graph 1 only. Later, after they have made a prediction of the appearance of the *v*-*t* graph, they can show both graphs for the remainder of the analysis of data.

[1] A. Arons, *Teaching Introductory Physics*, John Wiley & Sons, Inc, 1997

The connection between the acceleration and the incline of the track is more easily seen if you provide uniform thickness blocks to elevate the track. The sample data were collected using 2, 3 and 4 blocks that were ~1.5 cm thick.

By positioning the motion detector at the bottom of the track, the peak of the *x-t* graph corresponds to the maximum height up the track, a result that is sensible to the students. Caution the student to catch the cart before it collides with the motion detector on its return to the starting position.

In the Extension, students place the motion detector at the top of the track and interpret the resulting *x-t* and *v-t* graphs.

SAMPLE RESULTS AND POST-LAB DISCUSSION – PART 1

The emphasis in Part 1 of the evaluation of data is qualitative, with students matching observed behavior with regions on the *x-t* graph as shown in Figure 2.

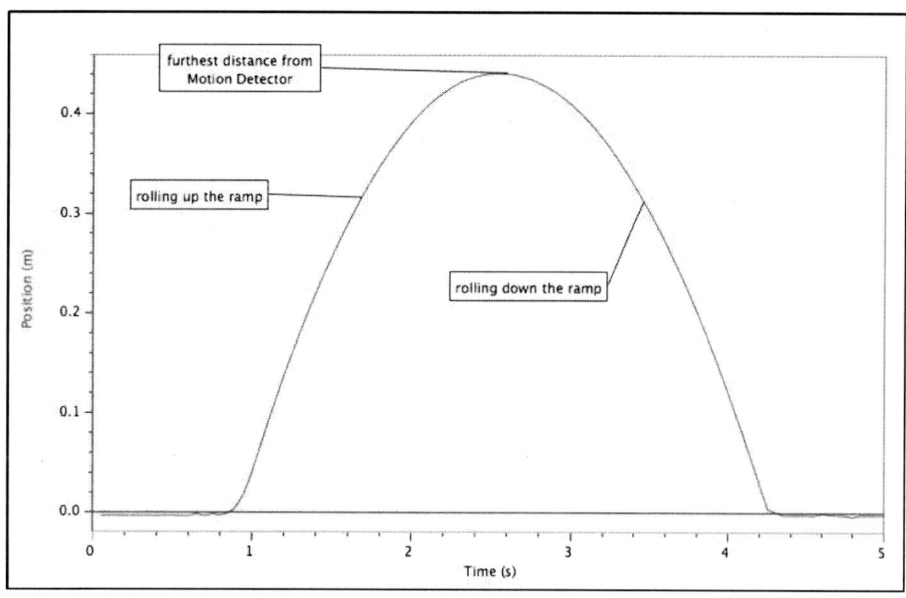

Figure 2 x-t graph using 2 blocks

Steps 2–4

These questions provide an opportunity for a discussion that helps students make the distinction between average and instantaneous velocity. While the slope of the chord connecting two points for a large Δt gives the average velocity, this value is not a particularly good descriptor of the velocity of the object at a given clock reading. Arons writes, " ...one can move to cases of speeding up and slowing down with corresponding curvature of the graphs, examine chords on the graphs and their connection to average velocities over arbitrary time intervals, and finally go to the tangents to the graphs at different clock readings." After students have sketched the *v-t* graph, have them display both graphs. (In Logger *Pro*, choose Auto Arrange from the Page menu; on the LabQuest, choose Show All Graphs.)

Step 5

The Tangent tool provides a way to compare the slope of the tangent to the *x-t* graph to the value of the velocity at that instant.[2] Such comparison should reinforce the concept that the slope of the tangent to the *x-t* graph yields the instantaneous velocity of the cart. Contrast this with average velocity as the slope of the chord connecting the two points that define the interval over which the average is taken.

SAMPLE RESULTS AND POST-LAB DISCUSSION – PART 2

The emphasis in Part 2 of the evaluation of data is quantitative.

Steps 1–2

Have the students prepare presentations of their findings (whiteboard, chart paper, etc). These should include a sketch of the graph and the equation of their line of best fit.

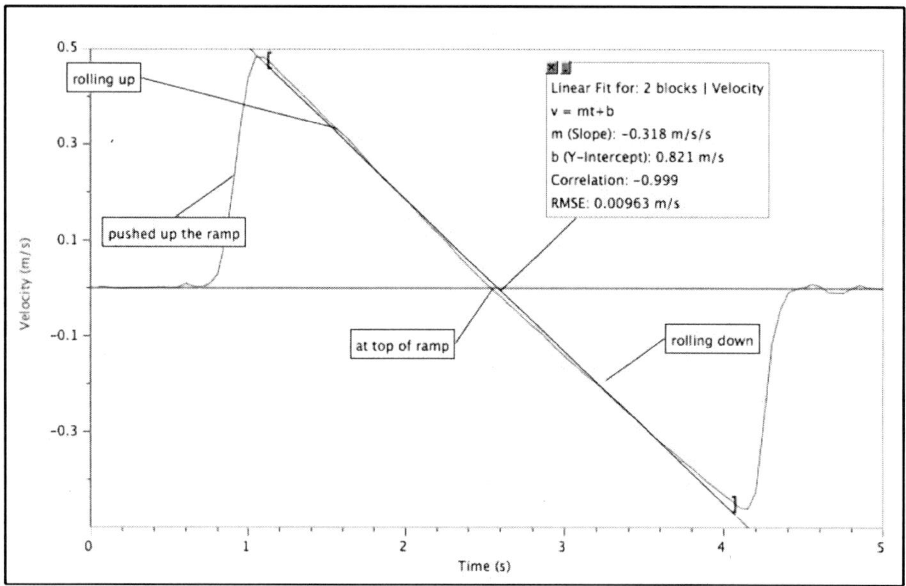

Figure 3 v-t graph using two blocks

Students may have difficulty articulating the meaning of the units of the slope of their line of best fit to the *v-t* graph. In this example, the slope indicates that the magnitude of the velocity of the cart changes by 0.318 meters per second *each second*. This is more easily seen when the units are expressed as m/s/s than as m/s^2. Once they understand that the slope gives the rate of change of the velocity, you can introduce the term *acceleration* as the name we give to this quantity.

It is important to help students understand the significance of the sign of the acceleration. In this experiment the positive direction is *away* from the motion detector. During both parts of the motion, the change in velocity is in the negative direction.

[2] You should point out that the software actually determines the slope of a chord for a very short time interval centered around the selected point on the curve. Logger *Pro* finds a weighted slope from 1, 2, 3 or more chords to the position-time graph.

Note: Discerning students might notice that the plot of the *v-t* graph is not truly linear over the range of values depicted in Figure 3. This is due to the fact that the frictional force acts in different directions when the cart moves up and down the track. The effect is more noticeable when the track angle is low. Congratulate such students for their powers of observation, but advise them that, for now, they should ignore this discrepancy; they will re-visit this effect when they learn more about the force concept.

Step 3

If you have used uniform blocks to elevate one end of the track, students can readily see the relationship between the height and the value of the acceleration.

Blocks	Acceleration (m/s/s)
2	–0.318
3	–0.482
4	–0.639

Step 4

Since the cart started at rest and then was given a push up the track, students may find it troubling to call the vertical intercept value the initial velocity, v_0. Stress that the vertical intercept of the line of best fit gives the value of the initial velocity of the cart, had it been accelerating uniformly from time $t = 0$. So, for the line of best fit, the students can write the general equation $v = at + v_0$.

Step 5

Neither Logger *Pro* nor LabQuest App supplies units to the parameters in a curve fit to a graph.

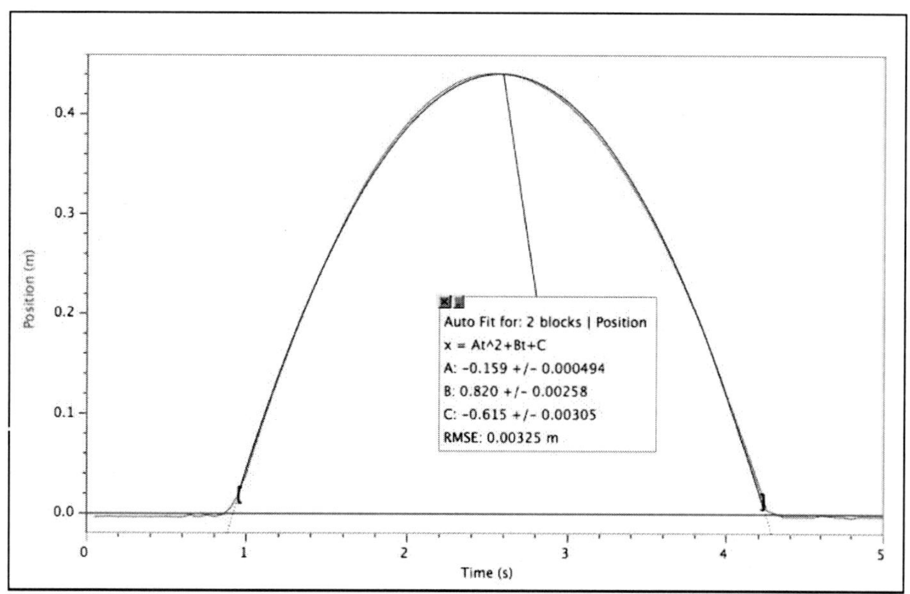

Figure 4 Curve fit to x-t graph

Encourage students to use dimensional analysis to show that the units of A must be m/s^2 and those of B must be m/s.

Step 6

By comparing the value of A to the slope of the *v-t* graph (acceleration) for several trials, students should be able to conclude that A is half of the acceleration. This is an appropriate time to point out that units of acceleration (m/s/s) can be reduced to m/s^2. Furthermore, the B parameter is the same as the intercept to the *v-t* graph (initial velocity). Once these connections have been made, students should be able to write the general equation for the position of an object moving with constant acceleration, $x = \frac{1}{2}at^2 + v_0t$.

EXTENSION

When the motion detector is placed at the top of the track, the cart reaches its minimum position when it has reached its farthest position up the track. The resulting *x-t* graph is a top-opening parabola (Figure 5).

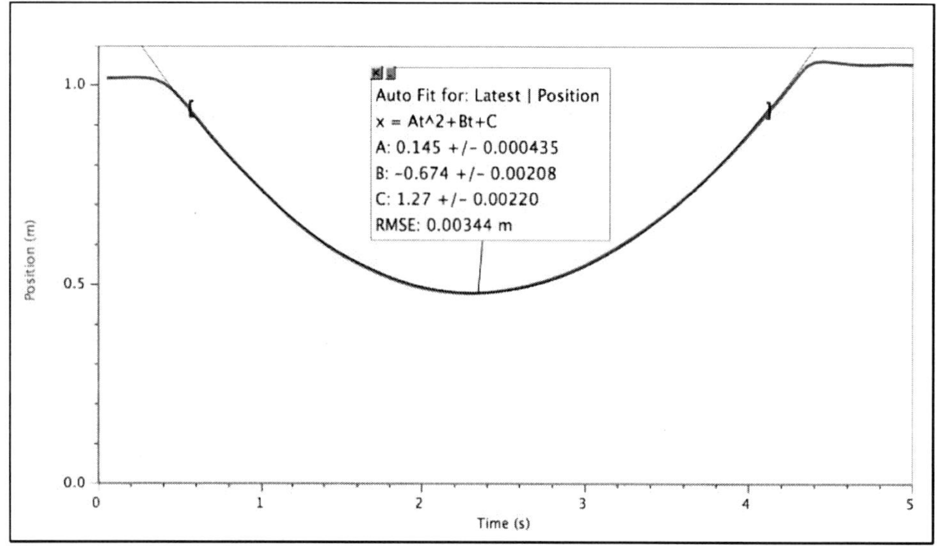

Figure 5

Ask students to interpret the sign of the velocity as it first approaches, then moves away from the motion detector. Challenge them to explain how the acceleration of the cart is positive while the cart is slowing down during the first half of its motion (Figure 6).

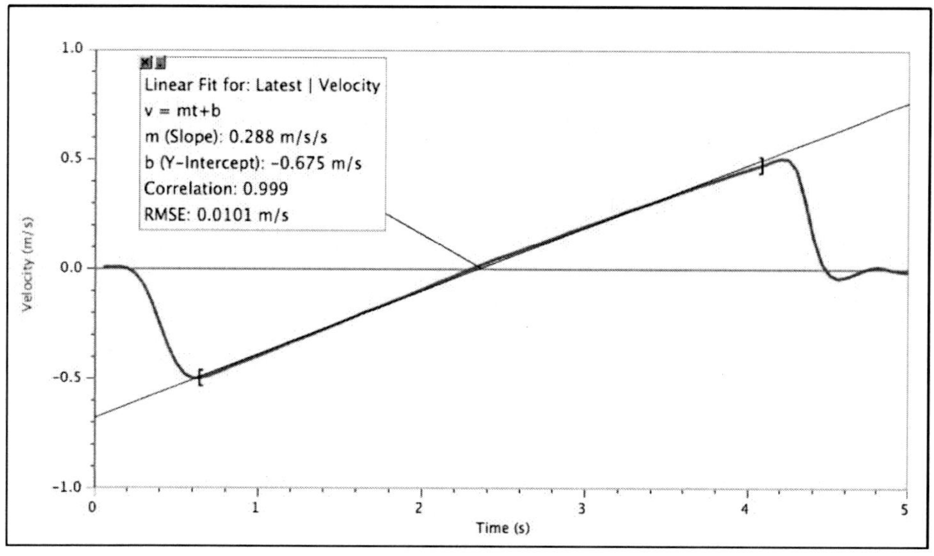

Figure 6

ANIMATED DISPLAY

Use of the Animated Display tool in Logger *Pro* allows students to produce a "motion map" to represent the motion of the cart during this experiment. Suggest that students first view the example given in the file "Exploring Animated Displays" by choosing Open from the File menu and then following the path: Experiments▶ Sample Data▶ Physics▶ Animated Display Vectors.

They should then choose Meter▶ Animated Display from the Insert menu to insert an animated display into their Logger *Pro* experiment file. Stretch the window horizontally, then double-click it to bring up the Animated Display Options window. Some sample settings are provided on the next page. Selecting Replay from the Analyze menu brings up the Replay control window. Judicious positioning of the windows allows one to view the trace of either graph at the same time as the animated display is drawn.

The settings shown in Figures 7 and 8 were used to produce the display in Figure 9.

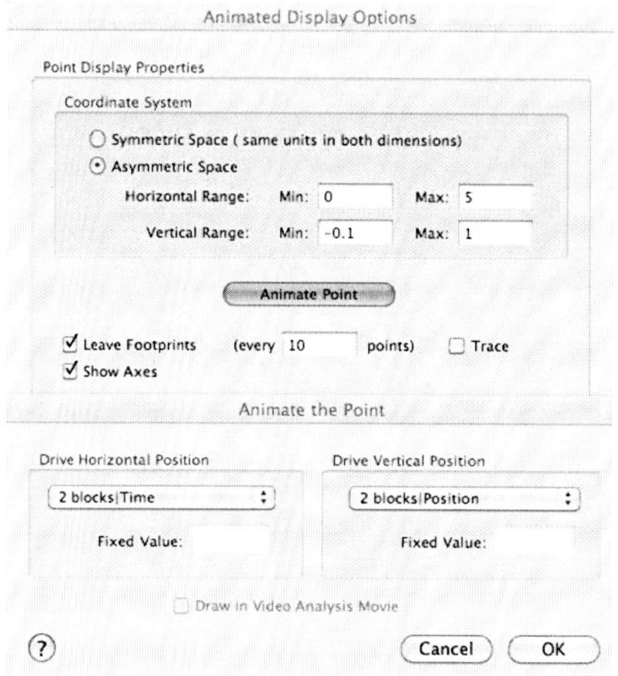

Figure 7

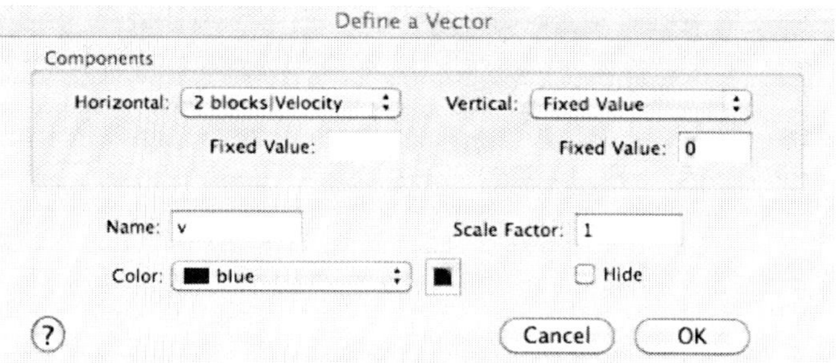

Figure 8

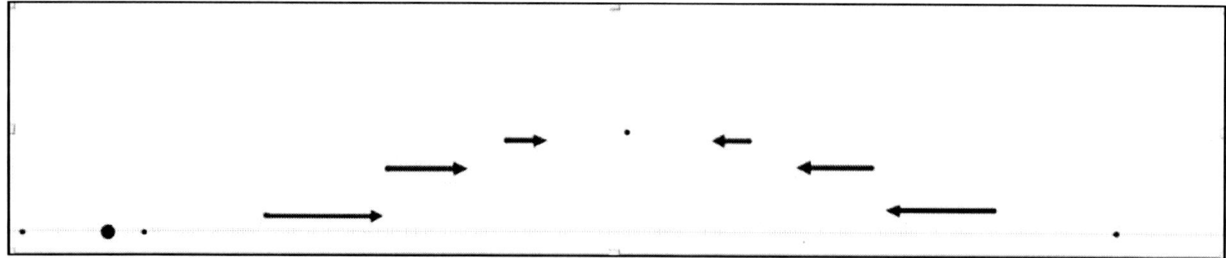

Figure 9 Animated display representing the velocity of the cart

Experiment
2

Error Analysis

INTRODUCTION

In Experiment 1, *Motion on an Incline*, you may have noticed that the slope of the v-t graph, which we call acceleration, increased as the height of the ramp increased. It seems reasonable that the maximum value of the acceleration could be obtained when the ramp was in a vertical position. In this experiment, you will use different apparatus to determine the acceleration of a freely falling object. Once you have done this, you will address the following questions:

- How do I decide if the value I obtained is "close enough" to the accepted value?
- If I were to repeat the experiment several times, within what range would I expect my values to fall?

This experiment affords you the opportunity to understand variations in experimentally determined data.

OBJECTIVES

In this experiment, you will

- Determine the value of the acceleration of a freely falling object.
- Compare your value with the accepted value for this quantity.
- Learn how to describe and account for variation in a set of measurements.
- Learn how to describe a range of experimental values.

MATERIALS

Vernier data-collection interface
Logger *Pro* or LabQuest App
Vernier Photogate
foam pad to cushion impact
Picket Fence
clamp or ring stand to secure Photogate

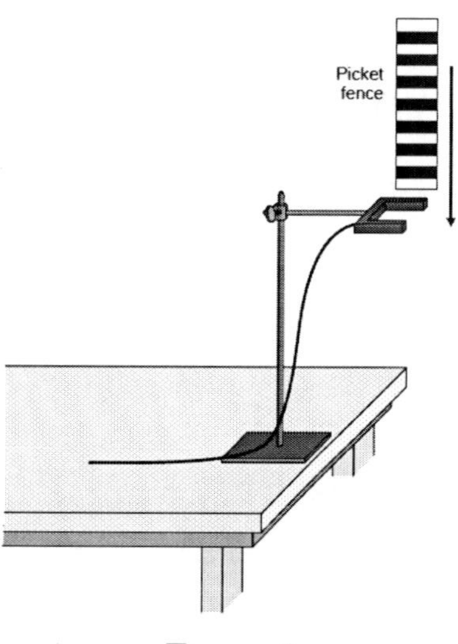

Figure 1

PRE-LAB INVESTIGATION

In the previous experiment, you used the time-based data-collection mode in which the software recorded a sensor reading (like position) at even time intervals. In this experiment, you will use a mode in which the software records the elapsed time between some regularly occurring events. When the Picket Fence (a strip of clear plastic with evenly spaced dark bands) passes through a Photogate, the device notes when the infrared beam of the photogate is blocked by a dark band and measures the time elapsed between successive "blocked" states. The software uses these times and the known distance from the leading edge of a dark band to the next to determine the velocity of the picket fence as it falls through the photogate. How does the elapsed time from Blocked state to Blocked state changes as the picket fence accelerates in free fall through the photogate?

PROCEDURE

1. Connect the photogate to one of the digital inputs on the interface and start the data-collection program. If the photogate has a sliding door, make sure it is open.

2. Check to see if the sensor is working by passing your hand between the infrared LED and the detector. The gate Status should change from "Unblocked" to "Blocked."

3. Fasten the photogate to a support rod or ring stand so that the arms of the photogate are horizontal (see Figure 1).

4. Set up data collection.

   Using Logger *Pro*

   a. Choose Data Collection from the Experiment menu.
   b. In the Mode list, select Digital Events. End timing after 16 Events.
   c. Choose Preferences from the File menu and set the Default Precision for automatic curve fits to four decimal places.

   Using LabQuest as a standalone device

   - Data collection defaults to Photogate Timing mode with four decimal places when you perform a linear fit.

5. Change the graph setup to view only the velocity *vs.* time graph.

6. Place something soft on the table or floor to cushion the picket fence as it strikes the surface.

7. Hold the picket fence vertically just above the photogate, start collecting data, and release the picket fence. Make sure that it does not strike the photogate as it passes through the arms.

8. Perform a linear fit on the graph of velocity *vs.* time. Print or sketch a copy of your graph. Take a moment to discuss what the slope and intercept of the line of best fit represent.

9. Based on your discussion, predict whether either of these quantities would change if you were to drop the picket fence through the photogate from a higher point. Test your prediction.

10. To see how repeatable the values of the slope are, repeat Steps 7 and 8 three more times. Be sure to record your values of the slope and intercept in your lab notebook.

11. You may quit the data-collection program for now but do not disassemble your apparatus. You will return to it later.

EVALUATION OF DATA

1. How do you account for the fact that the values of the slope were nearly the same, whereas the values of the intercept were much more variable?

2. It is highly unlikely that you obtained identical values of the slope of the best-fit line to the velocity *vs.* time graph for each of your trials. How might you best report a single value for the acceleration due to gravity, a_g, based on your results? Perform the necessary calculation.

3. How does your experimental value compare to the generally accepted value (from a text or other source)? One way to respond to this question is to determine the percent difference between the value you reported and the generally accepted value. Note that if you simplify your units of slope, they will match those of the reported values of a_g.

4. Your determination of the percent difference does little to answer such questions as, "Is my average value for a_g close enough to the accepted value?" or "How do I decide if a given value is too far from the accepted value?" A more thorough understanding of error in measurement is needed. Every time you make a measurement, there is some *random error* due to limitations in your equipment, variations in your technique, and uncertainty in the best-fit line to your data. Errors in technique or in the calibration of your equipment could also produce *systematic error*. We'll address this later in the experiment. In order to better understand random error in measurement, you must return to your experimental apparatus to collect more data.

5. Begin the data-collection program as you did before and drop the picket fence through the photogate another 20 times, bringing the total number of trials to 25. Since you are now investigating the variation in the values of a_g, you need only record the value of the slope of the best-fit line to the velocity-time graph for each trial.

6. To facilitate the evaluation of the data you have obtained for the 25 trials you have performed, disconnect the photogate from the interface and enter your values of slope into a new Logger *Pro* file.

7. Determine the average value of a_g for all 25 trials. How does this compare with the value you obtained for the first 5 trials? In which average do you have greater confidence? Why?

8. Sort the data in ascending order. Delete the graph of Y *vs.* slope. Now construct a histogram to display the frequency of the values of a_g you have obtained. To do this, choose Additional Graphs ► Histogram from the Insert menu. In your discussion, you will decide how best to configure the features of the histogram so as to represent the distribution of your values in the most meaningful way.

9. In what range (minimum to maximum) do the middle 2/3 of your values fall? In what range do roughly 90% of the values closest to your average fall?

10. One way to report the precision of your values is to take half the difference between the minimum and maximum values and use this result as the uncertainty in the measurement. Determine the uncertainty in this way for each range of values you determined in Step 9.

11. In what place (tenths, hundreds, thousandths) does the uncertainty begin to appear? Discuss whether it is reasonable to report values in your average beyond the place in which the uncertainty begins to appear. Round your average value of a_g to the appropriate number of digits and report that value plus the uncertainty.

EXTENSIONS

1. **Standard Deviation** When you obtained the statistics for your histogram, you may have noticed that, in addition to minimum and maximum values, and the mean and median, the standard deviation (std. dev.) was also provided. Do a web search for standard deviation. From the definition, what fraction of measurements fall within $\pm 1\ \sigma$ of the mean; what fraction would fall within $\pm 2\ \sigma$ of the mean? Write a brief explanation describing what standard deviation measures. If the standard deviation for your values of slope were four or five times as great as it was, how would that affect the shape of your histogram?

2. **Root Mean Square Error** When you performed a linear fit on your velocity-time data, you may have noticed that, in addition to the slope and intercept, the software provided the root mean square error (RMSE) for your best fit line. Visit www.vernier.com/til/1014.html to learn what this statistic tells you.

3. **Systematic Error** Suppose that the clock on the interface ran slow, say at 90% of its normal rate. How would that change the value you obtained for a_g? You may want to perform another trial and more closely examine the times in your data table to help you answer this question. Would performing more trials fix this problem? Explain.

   Consider the case of timing a race with a hand-held stopwatch. How would you expect a hand-held time to compare to one obtained by an electronic timing system? Explain.

4. The value for a_g varies according to your location. Do a web search for "Earth gravity" and explain why the value varies.

Experiment
2

INSTRUCTOR INFORMATION

Error Analysis

INTRODUCTION

This lab is written with the assumption that the instructor will engage the students in discussions at critical junctures. These discussions can take place with the entire class or with individual lab groups. The icon at left indicates where these discussions should occur.

The Microsoft Word files for the student pages can be found on the CD that accompanies this book. See *Appendix A* for more information.

This experiment is not intended to provide an exhaustive treatment of error analysis. Rather, it is designed to help students to do more than simply determine the percent difference between their experimental value and some accepted value, and use some arbitrary criterion to decide whether this difference is acceptable.

OBJECTIVES

In this experiment, the student objectives include

- Determine the value of the acceleration of a freely falling object.
- Compare your value with the accepted value for this quantity.
- Learn how to describe and account for variation in a set of measurements.
- Learn how to describe a range of experimental values.

During this experiment, you will help the students

- Learn to collect data in a mode other than Time Based.
- Learn how to express the percentage difference between their average value and the accepted value for a_g.
- Learn to construct a histogram in Logger *Pro*.
- Learn how to describe their range of values in terms of an average value ± an uncertainty.

REQUIRED SKILLS

In order for student success in this experiment, it is expected that students know how to

- Manipulate (delete and re-size) graphs in Logger *Pro*.
- Perform linear curve fits to graphs. This is addressed in Activity 1.

EQUIPMENT TIPS

Make sure that students place something soft on the table or floor to cushion the picket fence as it strikes the surface.

PRE-LAB DISCUSSION

In Experiment 1, *Motion on an Incline*, students should have noticed that the slope of the *v-t* graph, which we call acceleration, increased as the height of the ramp increased. Try to get the students to predict that the maximum value of the acceleration could be obtained when the ramp was in a vertical position. Rather than crashing a cart to determine this value, students will be using a picket fence and photogate.

Prior to distributing the lab to the students, demonstrate how the data-collection program works by slowly passing a picket fence through the arms of the photogate. Point out that the Gate State reads "Blocked" when the dark band interrupts the infrared beam and "Unblocked" when the clear portion of the picket fence passes between the LED and sensor. Instead of measuring an object's position at regular time intervals, the program determines the time elapsed between successive "blocked" states. Ask the students to predict how the blocked to blocked time would vary as the picket fence falls freely through the photogate.

After they have reached the conclusion that the times should get shorter as the picket fence speeds up, allow the students to begin the experiment.

LAB PERFORMANCE NOTES

Computer Data Collection

Students should connect the photogate to a digital port on the interface and launch Logger *Pro*. The default screen shows position-time, velocity-time, and acceleration-time graphs. Students should delete the position-time and acceleration-time graphs and expand the velocity-time graph.

To set up data collection:

a. Choose Data Collection from the Experiment menu.
b. In the Mode list, click Digital Events.
c. To end collection use the stop button or, if you explain that 16 "Events" (blocked, unblocked, blocked, etc.) occur as the picket fence passes through the photogate, they can end data collection after 16 events.
d. Explain that the default distance, 5.0 cm, is measured from the leading edge of one dark strip on the picket fence to the leading edge of the next dark strip.

LabQuest App Data Collection

The default data-collection mode for the photogate in LabQuest App (Photogate Timing) works for this experiment.

- Students can tap on the Mode window to check to see that the Photogate Mode: Motion and the Vernier Picket Fence are selected.
- Explain that the default distance, 5.0 cm, is measured from the leading edges of successive dark strips on the picket fence. Students should choose to show just Graph 2, the *v-t* graph.

With either method of data collection, students need to perform a linear fit to the *v-t* graph and record the slope and intercept (including units) for each of their first five trials. As students do this, ask them what these quantities represent. If students are using Logger *Pro*, direct them to set the preferences to four decimal place for curve fitting. LabQuest App defaults to five significant figures, which in this case, yields the same degree of precision. If students ask how many digits to record, tell them that for now, they should record all of them. This setting is intended to report greater precision than is justified; later, students will address the matter of how they can make a reasonable decision about rounding off these values.

EVALUATION OF DATA

Important: You will need to use Logger *Pro* for the Evaluation of Data beyond Step 5, whether you use computers or LabQuest to collect data. It is not possible to complete the Evaluation of Data using LabQuest because LabQuest App does not support histograms.

SAMPLE RESULTS AND POST-LAB DISCUSSION

The *v-t* graph for a typical trial is shown in Figure 1 below.

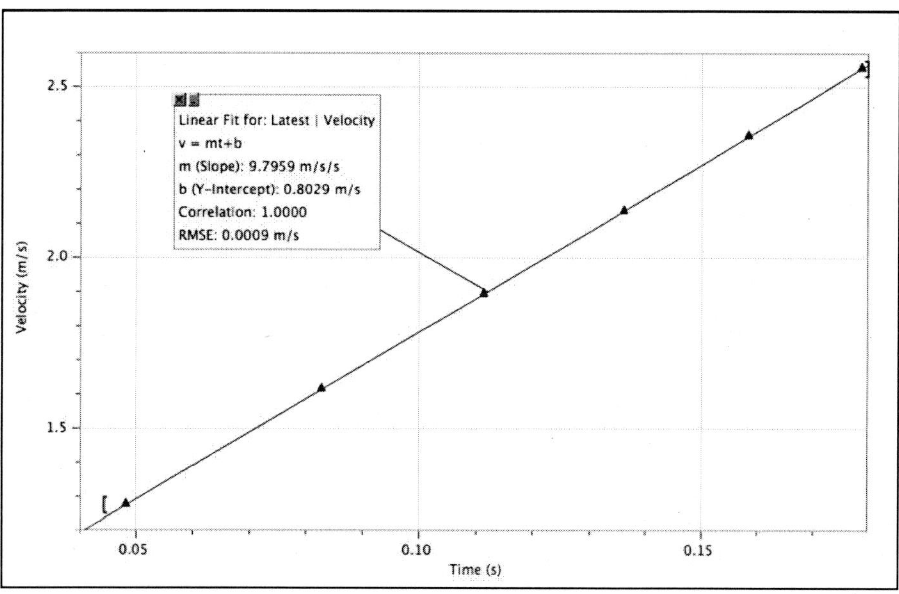

Figure 1

If the picket fence is dropped cleanly through the photogate, the value of the slope should fall within 0.5% of the accepted value of 9.80 m/s/s. After students recognize that the slope shows the rate of change of velocity, suggest that they should simplify these units to m/s^2, the standard form for acceleration. The intercept indicates the initial velocity of the picket fence, which varies, depending on the point from which it is released above the photogate.

Steps 1–3

Sample data for the first five trials are shown in Figure 2.

slope (m/s/s)	intercept (m/s)
9.7727	0.6763
9.7868	0.8626
9.8109	0.4953
9.7902	0.7314
9.8157	0.5848

Figure 2

Students should conclude that the values of the slope in their first five trials *should be* the same because the acceleration of the picket fence due to Earth's gravity does not change under these experimental conditions. On the other hand, the initial velocity depends on the point from which the picket fence is released.

Students will most likely choose to find the mean (9.7952 m/s/s for data in the table above) to report a single value for a_g.

Students' answers to Step 3 depend on the value they use as accepted. Most texts report this as 9.80 m/s^2 whereas Wikipedia gives both the "standard value" (9.8067) and the "equatorial value" (9.7803) to greater precision.

Steps 4–6

Naïve students tend to think that any value displayed on a digital device is "more accurate" than one from an analog device, especially if it is reported with more digits than one could read on an analog scale. In this portion of the experiment, students examine variation in measured values to consider both the source and the extent of this variation.

Step 7

It is highly likely that the average of the first five values of the slope will differ slightly from the average of all 25 values. Most students will answer that they have greater confidence in the average of 25 values, because a single outlying value will have a greater impact on the average when n, the number of measurements, is small.

Steps 8–9

These steps are designed to help students use a histogram to graph the distribution of their values. After they have deleted the meaningless graph of Y *vs.* slope, they should insert a histogram. From the Insert menu, choose Additional Graphs ▶, then choose Histogram. To adjust the Bin size and Bin Start values, double-click on the Histogram to open the Histogram Options dialog box. If the Bin Size is too large (say 0.1) then any sense of distribution of values is lost as they all pile into one or two bins. If the bin size is too small (say 0.001), then it is difficult to see if any range of values occurs more frequently than others. For the sample data for this experiment, a Bin Size of 0.005 and a Bin Start of 9.7 yields the histogram in Figure 3.

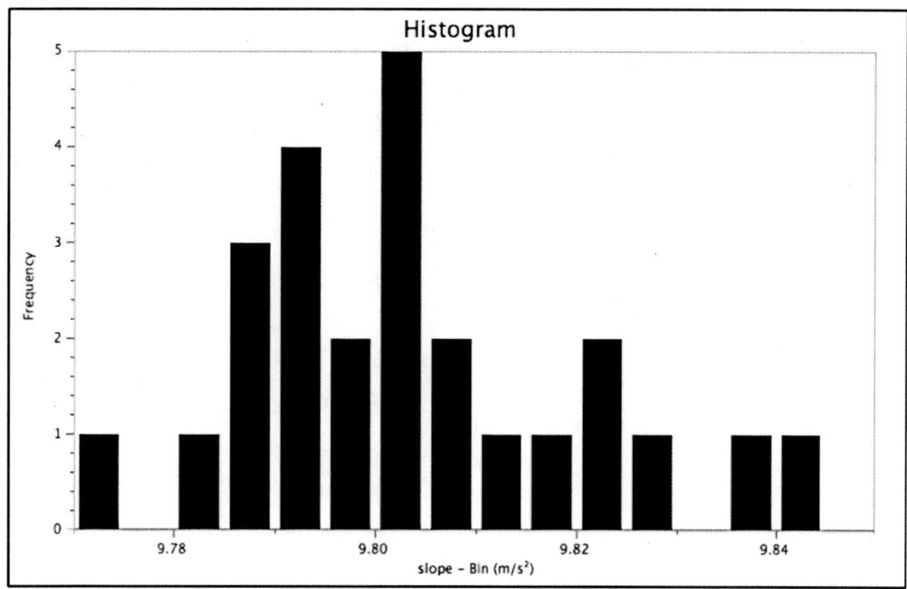

Figure 3

Sorting the data in ascending order gives students a simple way to determine the range of values called for when they answer Step 9. They should strike through the lowest four and highest four values of slope, then choose to view statistics to obtain the mean, minimum and maximum values for the remaining middle values (68%). They can restore all but the largest and smallest value to get the roughly 90% of slopes closest to the median value. These ranges were chosen to set up a possible investigation of standard deviation in the extension.

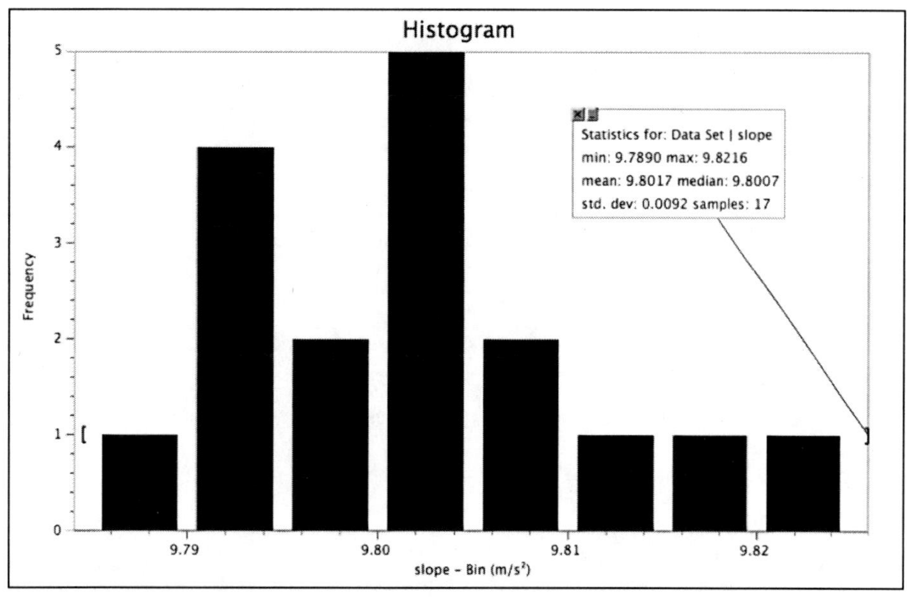

Figure 4

Step 10

For the sample set of data, the middle 68% of the values range from 9.7890 to 9.8216. The uncertainty is half of the difference between these values, or $\frac{1}{2}(9.8216-9.7890)=\pm 0.0163$. If one includes all but the highest and lowest values (the middle 92%), the range is a bit larger: 9.7814 to 9.8380. The uncertainty for this larger range is $\frac{1}{2}(9.8380-9.7814)=\pm 0.0283$. Students should conclude that the uncertainty in the mean is greater when one includes a wider range of values to calculate the mean. Nonetheless, the range of values for 92% of the data certainly includes the accepted value for a_g.

Step 11

Students may have been exposed to a variety of "rules of thumb" for deciding how to round off experimental values. One that may seem less arbitrary to them is to use the uncertainty in the data to make this determination. Whichever range of values one considers, uncertainty begins to appear in the third measured digit – in this case, in the hundredths place. Try to get students to conclude that digits beyond the one in which the uncertainty appears do not convey any useful information. In fact, reporting five digits for a value leaves the reader with the impression that the first four digits were repeatable, and the variation appeared only in the last digit. For this set of data, the value of a_g should be reported as 9.80 ± 0.03 m/s^2.

EXTENSIONS

1. **Standard Deviation** There are a number of sources that introduce standard deviation without delving into too much detail – Wikipedia is one of these. Assuming a normal distribution of values, ~68% of the values will fall within $\pm 1\ \sigma$ of the mean; ~95% fall within $\pm 2\ \sigma$ of the mean. Students should be able to state that the standard deviation is a measure of variation or spread of the data from the mean value. If the standard deviation for students' values of slope were four or five times as great as it was, their histogram would be spread over a wider range of values.

2. **Root Mean Square Error** From the Vernier Technical Information Library comes this description of RMSE.

   The Mean Squared Error (MSE) is a measure of how close a fitted line is to data points. For every data point, you take the distance vertically from the point to the corresponding y value on the curve fit (the error), and square the value. Then you add up all those values for all data points, and divide by the number of points. The squaring is done so negative values do not cancel positive values. The smaller the Mean Squared Error, the closer the fit is to the data. The MSE has the units squared of whatever is plotted on the vertical axis.

   Another quantity that we calculate is the Root Mean Squared Error (RMSE). It is just the square root of the mean square error. That is probably the most easily interpreted statistic, since it has the same units as the quantity plotted on the vertical axis.

   Key point: The RMSE is thus the distance, on average, of a data point from the fitted line, measured along a vertical line.

   The RMSE is directly interpretable in terms of measurement units, and so is a better measure of goodness of fit than a correlation coefficient. One can compare the RMSE to observed variation in measurements of a typical point. The two should be similar for a reasonable fit.

3. **Systematic Error** It is not enough for students to say that the values for a_g would be "off." A thoughtful analysis should be able to predict the direction as well as the magnitude of a systematic error. If the clock on the interface ran at 90% of its usual rate, then the times measured from one blocked state to the next would be 90% of their usual values. This would have the effect of making the velocity values appear 10% greater than what was observed in this experiment. The slope of the best-fit line to the velocity-time data would be greater. Thus, the value of a_g would be systematically greater than the accepted value.

   In the case of timing in track events, due to the delay between seeing the flash of the starter gun and pressing the start button on the stopwatch, hand-held times are shorter than electronic times. To compensate for this reaction-time delay, 0.24 s is usually added to a hand-held time to make it comparable to an electronic time. Since the error appears in the tenths place, recording digits beyond tenths in hand-held times, regardless of the precision displayed by the stopwatch, is not meaningful.

   In the first case because the error is not randomly distributed, increasing the number of trials would not make the average value any better. In the second case, greater experience might reduce the reaction-time lag, but could not entirely eliminate it.

4. In order for the value of a_g to be the same everywhere, the earth would have to be a sphere of uniform density that wasn't rotating. Clearly this is not the case as there are local variations in the density of the earth, as well as differences in altitude. In addition, the rotation of the earth produces a slight equatorial bulge. The distance between the earth's center and the surface is slightly greater at the equator than at the poles. Consequently, the force of gravity is slightly less at the equator.

Experiment

3

Newton's First Law

INTRODUCTION

Everyone knows that force and motion are related. A stationary object will not begin to move unless some agent applies a force to it. But just how does the motion of an object depend on the forces acting on it? In this experiment you will begin to examine the role that forces play in motion of an object.

OBJECTIVES

In this experiment, you will

- Collect position, velocity, and time data as a cart is launched by a spring and slowed by friction.
- Analyze the position *vs.* time and velocity *vs.* time graphs.
- Investigate the effect of varying the friction on the velocity of the cart.

MATERIALS

Vernier data-collection interface	standard cart
Logger *Pro* or LabQuest App	Cart Friction Pad
Vernier Motion Detector	(recommended) Bumper and Launcher Kit
Vernier Dynamics Track	(optional) heavy rubber band and support stands

PRE-LAB INVESTIGATION

Place the Cart Friction Pad on one end of the cart. Adjust the friction pad so that it makes contact with the track when the cart is placed on the track. Give the cart a gentle push so that it comes to a stop before reaching the end of the track.

On the axes to the right, sketch a graph of velocity *vs.* time based on what you observed. Use a coordinate system in which the origin is on the left and positive is to the right.

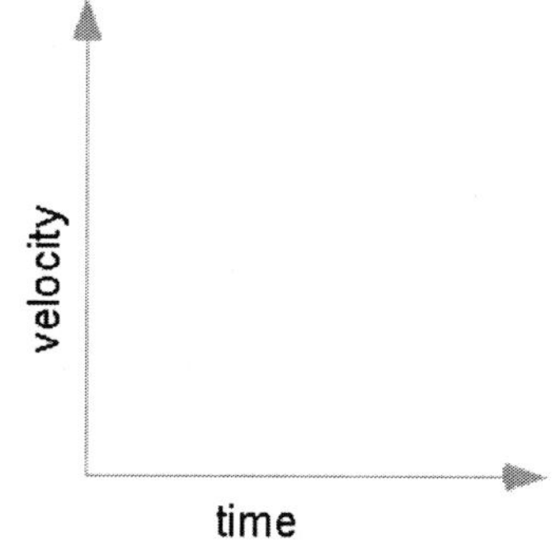

PROCEDURE

1. Attach the Launcher Assembly near the one end of the track. Use the heavier of the two hoop springs that come with the kit.[1] Adjust the screws on the feet to level the track.

2. Attach the cart friction pad to one end of a standard cart. Adjust the friction pad so that it makes slight contact with the track (see Figures 1 and 2).

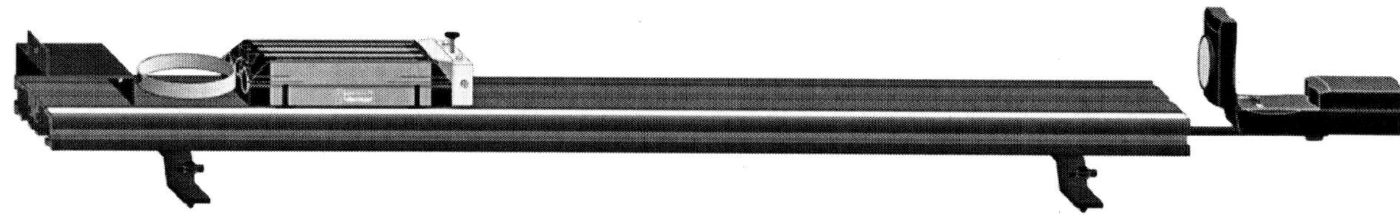

Figure 1

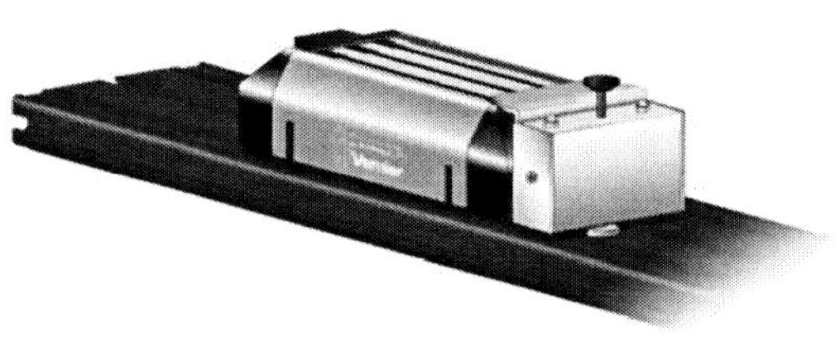

Figure 2

3. Attach the Motion Detector to the bracket that will allow you to position it near the other end of the track.

4. If your motion detector has a switch, set it to Track. 

5. Connect the motion detector to the interface and start the data-collection program. Two graphs: position *vs.* time and velocity *vs.* time will appear in the graph window.

6. Adjust both the extent to which you compress the spring and the friction pad so that the cart travels approximately 30 cm before coming to a stop. Do not compress the hoop spring more than half of its diameter.

7. Position the cart so that it is just touching the hoop spring and then zero the motion detector.

8. Push the cart against the spring, begin collecting data, wait a couple of seconds, and then release the cart.

9. Repeat Steps 7 and 8, if necessary, until you get a trial with a smooth position *vs.* time graph and a velocity *vs.* time graph that looks something like the one below.

[1] If the launcher is not available, your instructor will show you how to set up an alternative means to launch the cart.

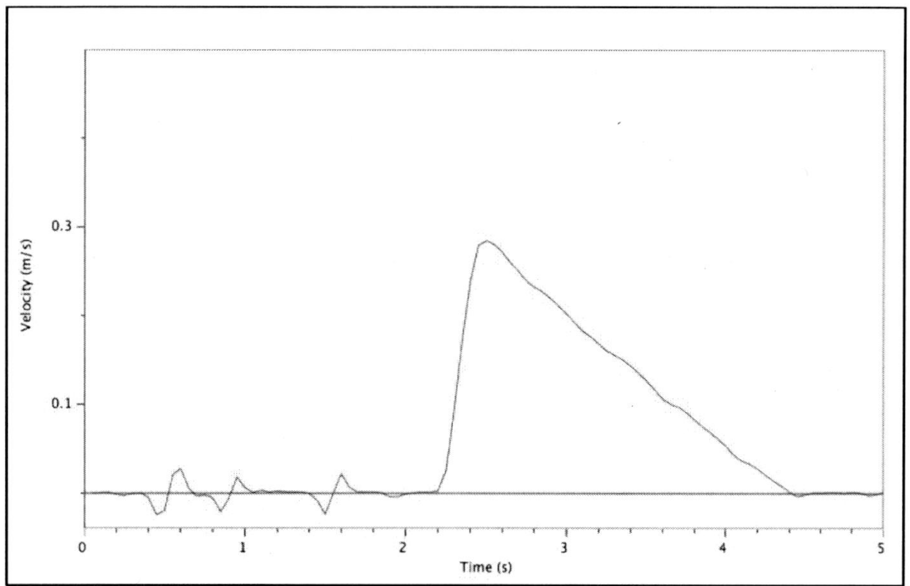

Figure 3

10. Store this run, then decrease the pressure of the friction pad on the track a bit (try a quarter turn), then repeat Step 8 until you get a smooth position graph and a velocity graph that has a less steep slope than that in the graph from the first run. Be sure to compress the spring by the same amount in each of these runs and to re-zero the motion detector from time to time.

11. Store Run 2, and continue this process, decreasing the friction slightly until you have 4–5 runs, each of which has a shallower slope in the velocity graph than that in the previous run. Be sure to store each of these runs. You may have to keep the cart from running off the end of the track in some of these runs.

12. Finally, loosen the adjusting screw until that the friction pad no longer makes contact with the track, and collect data for one last run. Be sure to save the data file.

DATA ANALYSIS

1. Hide all but your initial data set. Either print or sketch the position *vs.* time and velocity *vs.* time graphs for your first run. On the velocity graph identify:
   - Where the cart was being accelerated by the spring.
   - Where the cart was slowing to a stop.

2. Perform a linear fit on the portion of the velocity where the cart was slowing to a stop. Record the acceleration of the cart during this interval.

3. Now, show the data set for your second run. Repeat Step 2. How does the acceleration of the cart in this run compare to that in the first run?

4. Continue this process until you have analyzed the velocity graphs for all of your runs. If the window gets too cluttered, you can hide some of the earlier runs. Compare the shape of the position graph for your final run with that from the first run.

5. Recalling that acceleration is the rate of change of velocity, write a general statement describing the relationship between the rate of change of the cart's velocity and the frictional force acting on the cart.

6. Based on your statement in the previous step, how can you account for the fact that the rate of change in the cart's velocity did not reach 0 when the friction pad no longer made contact with the track?

7. Discuss what adjustment you could make to the track to reduce the change in velocity to 0. Explain why this adjustment would produce this result. Summarize your general conclusions from your discussion.

EXTENSION

Go back to your apparatus and see how close you can come to producing a run in which the change in velocity is zero.

INSTRUCTOR INFORMATION

Experiment
3

Newton's First Law

This is the first in a series of three experiments designed to help students develop an understanding of Newton's laws of motion. It is written with the assumption that the instructor will engage the students in discussions at critical junctures. These discussions can take place with the entire class or with individual lab groups. The icon at left indicates where these discussions should occur.

The Microsoft Word files for the student pages can be found on the CD that accompanies this book. See *Appendix A* for more information.

OBJECTIVES

In this experiment, the student objectives include

- Collect position, velocity, and time data as a cart is launched by a spring and slowed by friction.
- Analyze the position *vs.* time and velocity *vs.* time graphs.
- Investigate the effect of varying the friction on the velocity of the cart.

During this experiment, you will help the students

- Recognize that when the forces on an object sum to zero, its velocity remains unchanged.

REQUIRED SKILLS

In order for student success in this experiment, it is expected that students know how to

- Zero and reverse the direction of a Motion Detector. This is addressed in Activity 2.
- Perform linear fits to selected portions of graphs. This is addressed in Activity 1.
- Show only a specific data set run. This is addressed in Activity 2.

EQUIPMENT TIPS

To minimize the effect of friction on the cart, we recommend that you clean the grooves in the tracks before this experiment. This can be done by rubbing a damp paper towel through the grooves.

The Motion Detector can be attached to the Dynamics Track using the bracket, or students could simply place it on the table 10–15 cm from the end of the track. Make sure that the Motion Detector is set to Track mode.

If the Bumper and Launcher Kit is not available, you can launch the cart using a heavy rubber band stretched between ring stands clamped to the table, as shown in Figure 1. Make sure that the stretch of the rubber band is consistent for successive trials.

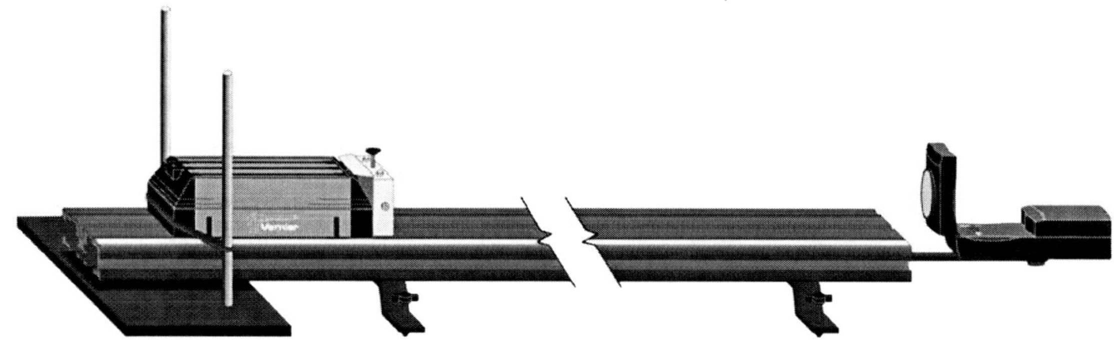

Figure 1

PRE-LAB DISCUSSION

This experiment is best performed *before* students are introduced to a formal statement of Newton's first law. The most common naïve conception students have about the role of forces on the motion of an object is that "constant force produces constant velocity." This experiment is designed to provide students with the evidence to help them confront this naïve conception. At the end of this experiment, students should be better prepared to accept the non-intuitive notion that a constant force is *not* required to maintain constant velocity.

LAB PERFORMANCE NOTES

Students can collect and evaluate the data using Logger *Pro* or LabQuest App in this experiment. At the start of the experiment, students should make sure to level the track. A quick way to check whether this condition is met is to give the cart a gentle push in both directions to see if the speed is roughly the same both ways. Students are likely to be tempted to over-tighten the screw to the friction pad at the beginning of the experiment. Doing so will bring the cart to an abrupt halt. Explain to the students that the goal is not to "slam on the brakes", but to bring the cart to a stop after 25–30 cm. It is not always easy to release the cart cleanly, so inform the students that it might take several trials at each friction setting to obtain a *v-t* graph that is not erratic. Once they get an acceptable run, they should make sure to store the run before adjusting the friction pad for the next run.

If there isn't an appreciable difference in the slope of the *v-t* graph from the previous run, advise students to turn the adjusting screw on the friction pad to lessen the pressure.

SAMPLE RESULTS AND POST-LAB DISCUSSION

Step 1

Students should obtain a velocity *vs.* time graph like the one below. When the cart is released at ~2 s, there is a brief interval during which the spring accelerates the cart, followed by nearly constant decrease in velocity produced by the friction force.

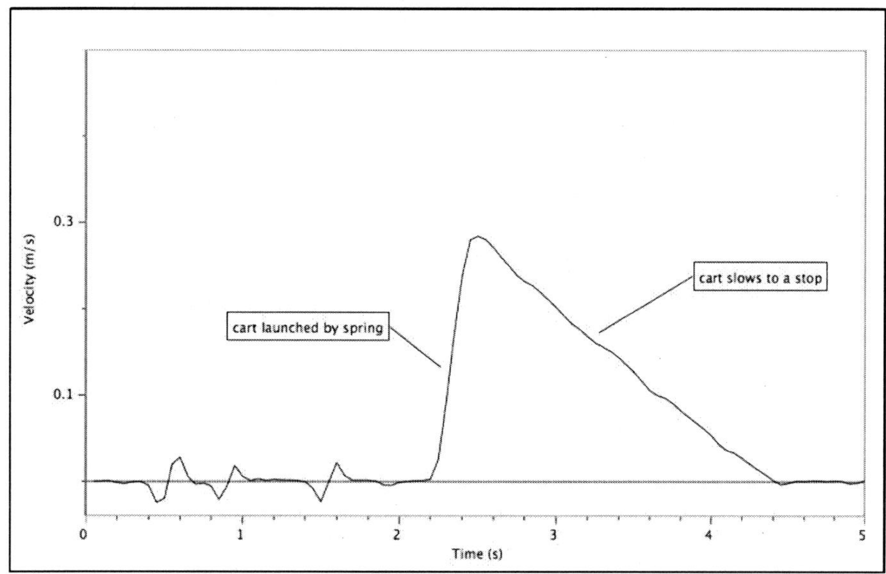

Figure 2 Velocity vs. *time graph for first run*

Steps 2–3

The key to success in these steps is for students to exercise some judgment about the portion of the velocity graph on which they try to perform a linear fit. Remind them that they are examining the effect of the friction force on the rate of change of the cart's velocity. Sample fits for Runs 1 and 2 are shown in Figure 3.

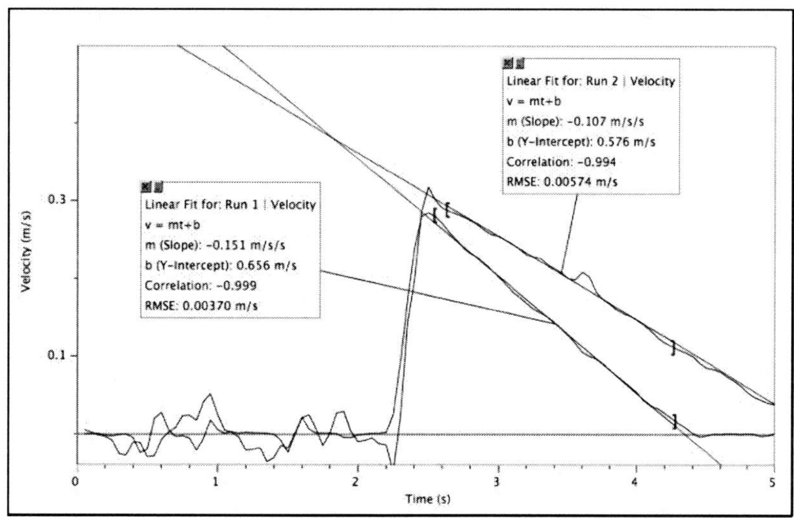

Figure 3

Students should see that the acceleration is lower in the second run than it was in Run 1.

Steps 4–5

Students should conclude that the rate of change of the velocity decreases as the friction force decreases. They should also note that the slope of the position-time graph gradually decreased to zero in the first run, but was nearly constant in the final run.

Step 6

The rate of change in velocity for the last run should be quite low, yet still greater than zero. Students ought to be able to reason that some frictional force still acts on the cart, even when the friction pad no longer applies this force.

Step 7

If students are uncertain what action to take, remind them of how they checked to make sure that the track was level at the start of the experiment. Then, advise them to raise the launcher end of the track using the leveling feet, place the cart on the track and give it a *gentle* push. When the cart can reach the end of the track without appearing to slow down, they have effectively reduced the frictional force to zero. Of course, the frictional force still acts on the cart. But by elevating the track, they have introduced another force (some component of the force of gravity) that acts to cancel out the force of friction. The net force (or sum of the forces) acting on the cart is zero when the change in the cart's velocity is zero.

This would be the ideal time to introduce this statement of Newton's First Law of Motion: When the sum of the forces acting on an object is zero, its change in velocity is zero. The converse of this statement is also true.

$$\text{If } \sum F = 0, \text{ then } \Delta v = 0$$

$$\text{If } \Delta v = 0, \text{ then } \sum F = 0$$

EXTENSION

When students understand why this adjustment works, invite them to perform another trial to see how nearly constant they have made the velocity of the cart. The velocity graph for such a run is shown in the figure below.

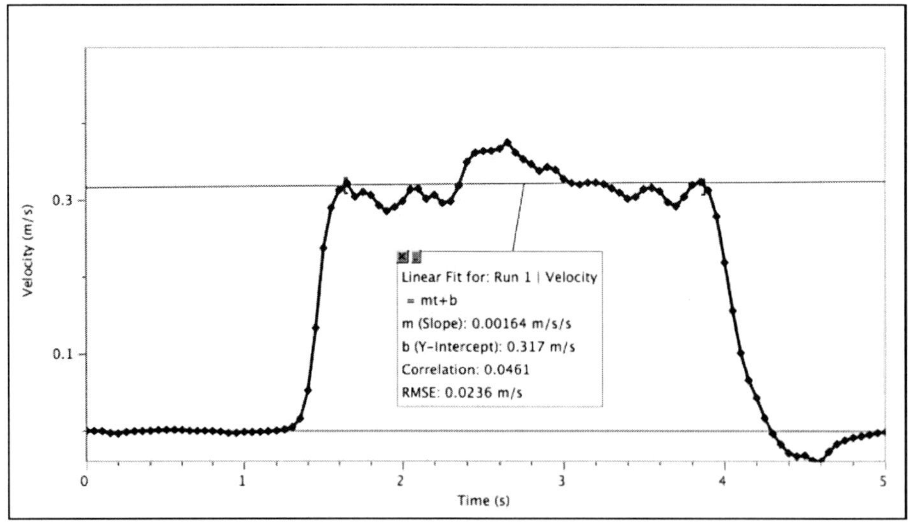

Figure 4 Nearly level

Experiment

4

Newton's Second Law

INTRODUCTION

In your discussion of Newton's first law, you learned that when the sum of the forces acting on an object is zero, its velocity does not change. However, when a net force acts on the object, it accelerates. In this experiment, you will determine the relationship between the net force acting on an object and its acceleration.

OBJECTIVES

In this experiment, you will

- Identify the forces acting on an object both when its change in velocity, Δv, is zero and when it is accelerating.
- Collect force, velocity, and time data as a cart is accelerated on a track.
- Use graphical methods to determine the acceleration of the cart.
- Determine the relationship between the cart's acceleration and the net force applied to it.
- Determine the effect of the mass on the relationship between acceleration and force.

MATERIALS

Vernier data-collection interface
Logger *Pro* or LabQuest App
Dual-Range Force Sensor **or**
  Wireless Dynamic Sensor System
Vernier Photogate and Bracket and
  Cart Picket Fence, **or**
  Motion Detector
standard hooked or slotted lab masses
Vernier Dynamics Track
standard cart
Ultra Pulley and Pulley Bracket
lightweight mass hanger

PRE-LAB INVESTIGATION

Your instructor will show you the apparatus for this experiment. It is called a "modified Atwood's machine." A weight is connected to a cart by a string over a pulley. When the weight is allowed to fall, observe the motion of the cart. A force sensor mounted on the cart enables you to measure the force acting on the cart when it is moving. Discuss how you could determine the acceleration of the cart. Then, consider how the acceleration varies as the force applied to the cart increases. On the axes to the right, draw a sketch of the acceleration *vs.* net force graph based on your observations.

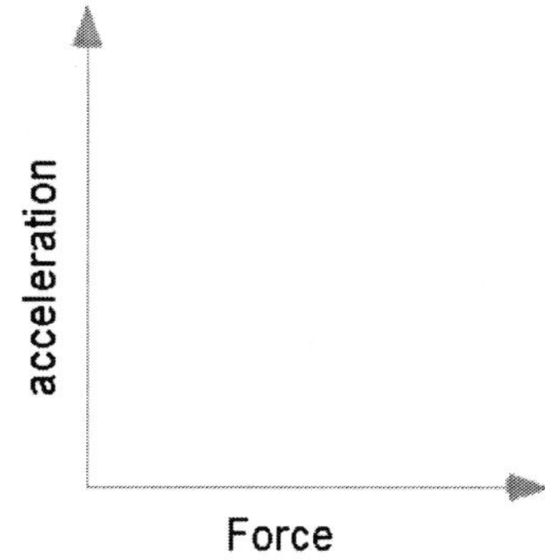

PROCEDURE

There are three methods one can use to collect velocity *vs.* time data and determine the acceleration of the cart. Choose the method that matches the apparatus you are using.

Method 1 Timing with a Photogate and Picket Fence

1. Attach an Ultra Pulley to the Pulley Bracket; attach this assembly to one end of the Dynamics Track. Make sure that you have adjusted the track to make frictional forces negligible.

2. Connect the Dual-Range Force Sensor (DFS) and a Photogate to the interface; then start the data-collection program. You will need only graphs of force *vs.* time and velocity *vs.* time, so you can delete others and then re-size graphs to make them easier to see. LabQuest App users can do this in the Graph Options menu.

3. Place the Cart Picket Fence in the slot on the cart so that the alternating dark and clear bands are on the top. Now place the force sensor on the cart.

4. Position the photogate (using a support rod and clamp) so that the alternating dark and clear bands on the picket fence will interrupt the beam, as shown in Figure 1.

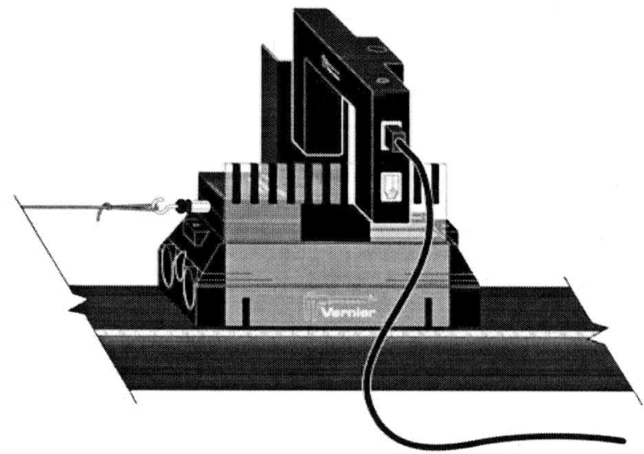

Figure 1

5. Discuss with your instructor what range of masses to use to apply the force that will accelerate the cart. Determine the total mass of your cart, force sensor, and any additional mass you may be instructed to use.

6. Set up data collection.

   Using Logger *Pro*

   a. Choose Set Up Sensors ▶ Show All Interfaces from the Experiment menu.
   b. Click the image of the Photogate, select Set Distance or Length, then select Cart Picket Fence from the list of devices.
   c. Choose Data Collection from the Experiment menu. In the Mode list, click Digital Events. End collection after 20 Events.

   Using LabQuest as a standalone device

   - Data collection defaults to Photogate Timing mode. Tap Mode, select User defined, and set the length to 0.01 m.

7. Disconnect the hanging mass from the force sensor, then zero the sensor.
8. Re-connect the hanging mass to the Force Sensor. Position the Photogate so that the cart and picket fence pass through the photogate while the cart is accelerating (before the hanging mass stops moving). Use something to cushion the hanging weight when it strikes the floor.
9. Start data collection, then release the cart.
10. To determine the force acting on the cart, select the portion of the force *vs.* time graph corresponding to the interval during which the cart's velocity was changing smoothly. Find the statistics for this interval. Manual scaling of your graph is more helpful for doing this than Autoscaling.
11. To determine the acceleration of the cart, perform a linear fit on the velocity *vs.* time graph. Be sure to record the values of force and acceleration for this hanging mass in your lab notebook.
12. Repeat Steps 9–11 until you have three pairs of force *vs.* acceleration data that are reasonably consistent for that mass.
13. Increase the hanging mass and continue until you have acceleration *vs.* force data for at least five different hanging masses.

Method 2 Timing with a Photogate and Ultra Pulley

1. Attach an Ultra Pulley to the Pulley Bracket; attach this assembly to one end of the Dynamics Track. Make sure that you have adjusted the track to make frictional forces negligible.
2. Connect the Dual-Range Force Sensor (DFS) and a Photogate to the interface; then start the data-collection program. You will need only graphs of force *vs.* time and velocity *vs.* time, so you can delete others and then re-size graphs to make them easier to see. LabQuest App users can do this in the Graph Options menu.
3. Attach the force sensor to the cart.
4. Attach the photogate to the track using a bracket so that the spokes on the pulley interrupt the beam, as shown in Figure 2.

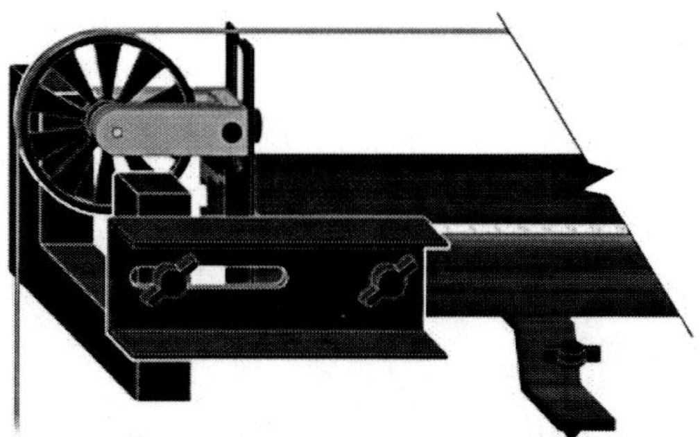

Figure 2

5. Discuss with your instructor what range of masses to use to apply the force that will accelerate the cart. Determine the total mass of your cart, force sensor, and any additional mass you may be instructed to use.

6. Set up data collection.

   Using Logger *Pro*

   a. Choose Set Up Sensors ► Show All Interfaces from the Experiment menu.
   b. Click the image of the Photogate, select Set Distance or Length, then select Ultra Pulley (10 Spoke) Inside from the list of devices.

   Using LabQuest as a standalone device

   - Data collection defaults to Photogate Timing mode. Tap Mode and select Pulley (10 spoke) ► inside edge.

7. Disconnect the hanging mass from the force sensor, then zero the sensor.

8. Re-connect the hanging mass to the force sensor. Position the cart so that the hanger has about 30 cm to fall. Use something to cushion the hanging weight when it strikes the floor.

9. Begin collecting data, then release the cart. Stop data collection after the weight hits the cushion on the floor.

10. To determine the force acting on the cart, select the portion of the force *vs.* time graph corresponding to the interval during which the cart's velocity was changing smoothly. Find the statistics for this interval. Manual scaling of your graph is more helpful for doing this than Autoscaling.

11. To determine the acceleration of the cart, perform a linear fit on the portion of the velocity-time graph during which the velocity is changing smoothly. Be sure to record the values of force and acceleration for this hanging mass in your lab notebook.

12. Repeat Steps 9–11 until you have three pairs of force *vs.* acceleration data that are reasonably consistent for that mass.

13. Increase the hanging mass and continue until you have acceleration-force data for at least five different hanging masses.

Method 3 Timing with a Motion Detector

1. Attach an Ultra Pulley to the Pulley Bracket; attach this assembly to one end of the Dynamics Track. Make sure that you have adjusted the track to make frictional forces negligible.

2. Connect the Dual-Range Force Sensor (DFS) and a Motion Detector to the interface; then start the data-collection program. You will need only graphs of force *vs.* time and velocity *vs.* time, so you can delete others and then re-size graphs to make them easier to see. LabQuest App users can do this in the Graph Options menu.

3. Attach the force sensor to the cart. Mount the motion detector at the end of the track opposite the pulley. If your motion detector has a switch, set it to Track.

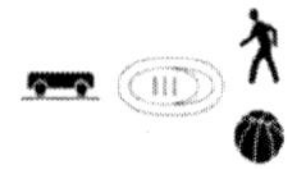

4. Discuss with your instructor what range of masses to use to apply the force that will accelerate the cart. Determine the total mass of your cart, force sensor, and any additional mass you may be instructed to use.

5. Set up data collection.

   Using Logger *Pro*

   a. Choose Data Collection from the Experiment menu.
   b. Reduce the length of time to 3 seconds.

   Using LabQuest as a standalone device

   a. Tap Length and change the length of data collection to 3 seconds.
   b. Change the time to 3 seconds on the x-axis column using Graph Options in the graph tab.

6. Position the cart so that the hanging mass is approximately 30 cm from the floor, disconnect the hanging mass from the force sensor, then zero both sensors. Make sure the starting position of the cart is at least 20 cm from the motion detector.[1]

7. Re-connect the hanging mass, begin collecting data, then release the cart. Catch the hanging mass before it strikes the floor.

8. To determine the force acting on the cart, select the portion of the force *vs.* time graph corresponding to the interval during which the cart's velocity was changing smoothly. Find the statistics for this interval. Manual scaling of your graph is more helpful for doing this than Autoscaling.

9. To determine the acceleration of the cart, perform a linear fit on the portion of the velocity *vs.* time graph during which the velocity is changing smoothly.

   Be sure to record the values of force and acceleration for this hanging mass in your lab notebook.

10. Repeat Steps 7–9 until you have three pairs of force *vs.* acceleration data that are reasonably consistent for that mass.

11. Increase the hanging mass and continue until you have acceleration *vs.* force data for at least five different hanging masses.

[1] If you are using an older Motion Detector without a position switch, the cart needs to be at least 45 cm from the detector.

EVALUATION OF DATA

1. To evaluate the relationship between acceleration and force, disconnect the sensors and choose New from the File menu.

2. Even though you investigated how acceleration responded to changes in the force, in order to facilitate your analysis of data, plot a graph of force *vs.* acceleration.

3. If the relationship between force and acceleration appears to be linear, fit a straight line to your data. If possible, print a copy of your data table and graph.

4. Write the equation that represents the relationship between the force, F, acting on the cart and the cart's acceleration, a. Be sure to record the value of the mass of the cart, sensor and any additional masses you used.

5. Write a statement that describes the relationship between the force acting on the cart and the cart's acceleration.

6. Compare your results to those obtained by others in class. What relationship appears to exist between the slope of the graph of F *vs.* a and the mass that was accelerated?

7. Assuming that the conclusion you reached in Step 6 is correct, express the SI unit of force, N, in terms of the fundamental units of mass, length and time.

8. Write a general equation that expresses the relationship between the force, mass and acceleration.

EXTENSION

Examine the force *vs.* time graph for one of your trials. Explain why the force reading decreases as the cart is released and accelerates.

Suppose that you had kept the net force acting on the cart the same, but varied the mass instead. Predict the shape of the graph of acceleration *vs.* mass. Your answer to the question above should suggest why it would be very difficult to perform an experiment to test your prediction. Explain.

Experiment

4

INSTRUCTOR INFORMATION

Newton's Second Law

This is the second in a series of three experiments designed to help students develop an understanding of Newton's laws of motion. It is written with the assumption that the instructor will engage the students in discussions at critical junctures. These discussions can take place with the entire class or with individual lab groups. The icon at left indicates where these discussions should occur.

The Microsoft Word files for the student pages can be found on the CD that accompanies this book. See *Appendix A* for more information.

OBJECTIVES

In this experiment, the student objectives include

- Identify the forces acting on an object both when its change in velocity, Δv, is zero and when it is accelerating.
- Collect force, velocity and time data as a cart is accelerated on a track.
- Use graphical methods to determine the acceleration of the cart.
- Determine the relationship between the cart's acceleration and the net force applied to it.
- Determine the effect of the mass on the relationship between acceleration and net force.

During this experiment, you will help the students

- Recognize that the acceleration of an object is the slope of a velocity *vs.* time graph.
- Recognize that the slope of a graph of two variables in a physical system is usually a function of variables(s) that are held constant.
- Derive an equation for Newton's second law: $F_{net} = ma$ or $\sum F = ma$.

REQUIRED SKILLS

In order for student success in this experiment, it is expected that students know how to

- Zero a Motion Detector. This is addressed in Activity 2.
- Perform linear and curve fits to graphs. This is addressed in Activity 1.
- Zero a Force Sensor.

EQUIPMENT TIPS

There are a number of ways one could collect suitable data for the analysis called for in this experiment. Three variations are described in the student version of the experiment. A fourth, using the Wireless Dynamic Sensor System (WDSS) is described in this document. Depending on the equipment available to students, you may wish to allow them to use different approaches to collect the data.

The first method suggested for obtaining velocity-time data–using a photogate and cart picket fence–is intended to build on student experience with the larger picket fence in Experiment 2. Students should recognize that, as the cart accelerates through the photogate, the time from blocked state to blocked state decreases. It follows that the fixed distance (1.0 cm) divided by smaller intervals of time yields increasing values of velocity (see Figure 1).

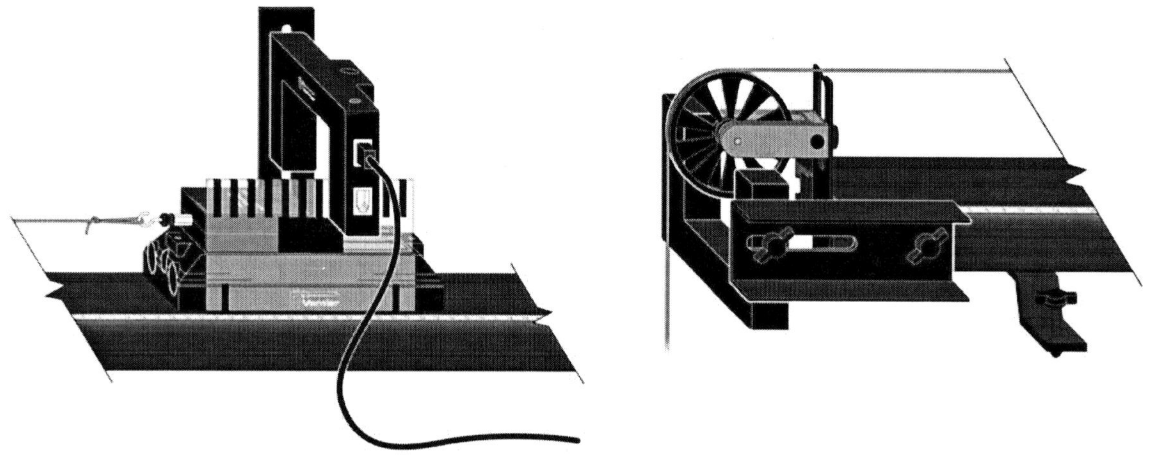

Figure 1 *Figure 2*

In the second method, the photogate is attached to the end of the track with a bracket. However, it may require some discussion to convince students that the spokes on the ultra pulley serve the same purpose as the dark bands on the picket fence (see Figure 2).

The third method, use of a Motion Detector, follows naturally from Experiment 3 and yields good data as well. Make sure that the Motion Detector is set to Track mode. If you do not have enough Dynamics Tracks for all the groups, this experiment can be performed on a lab table top. You will have to attach the pulley to the edge of the table.

The Wireless Dynamic Sensor System has the advantage over the DFS of not worrying about wire to the Force Sensor impeding the motion of the cart. While it provides acceleration as well as force data, we suggest using the other timing methods because the acceleration *vs.* time data for this experiment are sufficiently noisy that students would find it difficult to interpret the data.

Slotted or drilled lab masses can be suspended from the string using a lightweight hanger, or even a large paper clip. Use of a lightweight mass hanger allows the opportunity to examine accelerations for a wider range of values of force. The sample data were obtained using hanging masses ranging from 25–125 g.

The mass of the standard cart and DFS is approximately 630 g. If you instruct different lab groups to place additional masses on the cart (100 g increments work well), they can investigate the effect of mass on their force *vs.* acceleration graph when they compare their results.

PRE-LAB DISCUSSION

This experiment should be performed only after a thorough discussion of Newton's first law. Students should recognize that if the sum of the forces acting on an object is not zero, the object accelerates. The purpose of this experiment is to determine the relationship between the net force acting on an object and its acceleration.

Demonstrate to the students the modified Atwood's apparatus. Remind the students that the Force Sensor measures only the force applied by the hanging weights. In order to accurately determine the relationship between force and acceleration, it is important to reduce the effect of frictional forces (which are not easily measured) on the cart. To achieve this, the end of the track opposite the pulley is elevated slightly to introduce a small component of the force of gravity. When these two forces cancel, the cart should move with nearly constant velocity when given a gentle shove.

Once this condition has been met, connect the hanging weight to the cart, release the cart and allow it to accelerate toward the pulley. Make sure that the string is long enough so that the weights reach the floor before the cart reaches the pulley. Ask the students if they could observe a change in the motion of the cart once a net force no longer acted on the cart. You may have to use a low towing weight for students to notice. Now, ask students to predict the shape of the graph of acceleration *vs.* force. Once they have done this, they can begin the experiment.

LAB PERFORMANCE NOTES

Whichever way students collect data, they should record force and acceleration data for multiple trials for each hanging mass in their lab notebooks. When they feel that they have three reasonably consistent values of force and acceleration, they can increment the hanging mass. They should obtain force and acceleration data for at least five different hanging masses.

Computer Data Collection

Students should connect the Dual-Range Force Sensor (DFS)[1] to an analog port and a Photogate (or Motion Detector) to a digital port on the interface; then start the Logger *Pro*. They will need only graphs of force *vs.* time and velocity *vs.* time, so they should delete others and then re-size graphs to make them easier to see. More explicit instructions than usual are provided in the student pages because there are multiple timing options. If students are using a version of Logger *Pro* prior to version 3.8.3, they will need to use the following steps to set up the Cart Picket Fence or Ultra Pulley.

- Choose Set Up Sensors ▶ Show All Interfaces from the Experiment menu.
- Click the image of the photogate, select Set Distance or Length, then set the appropriate conditions (User Defined ▶ 0.01 m with the cart picket fence, or Ultra Pulley (10 Spoke) Inside Edge).
- If you are using the cart picket fence, choose Data Collection from the Experiment menu. In the Mode list, click Digital Events. End collection after 20 Events.

LabQuest App Data Collection

Students should connect the Dual-Range Force Sensor (DFS) to an analog port and a Photogate (or Motion Detector) to a digital port on the interface. Force and Gate State (or Position) fields

[1] A Wireless Dynamic Sensor System can be used in place of the DFS.

appear in the Sensor window. They will need only graphs of force *vs.* time and velocity *vs.* time, so they should select Graph Options in the Graph tab and choose velocity for the Graph 2 *y*-axis. Explicit instructions are provided in the student pages for each of these timing options.

With either method of data collection, after each trial, students need to find statistics for the portion of the force *vs.* time graph during which the cart was accelerating; they should record the mean force for this interval. They should perform a linear fit on the portion of the velocity-time graph during which the velocity was changing smoothly, and record the slope as the acceleration. An example is shown in Figure 3.

This analysis is more easily done in Logger *Pro*, but can be accomplished without great difficulty in LabQuest App.

Note: Students might wonder why the force reading decreases as the cart is released and accelerates. Suggest that they will address this phenomenon in the post-lab extension.

SAMPLE RESULTS AND POST-LAB DISCUSSION

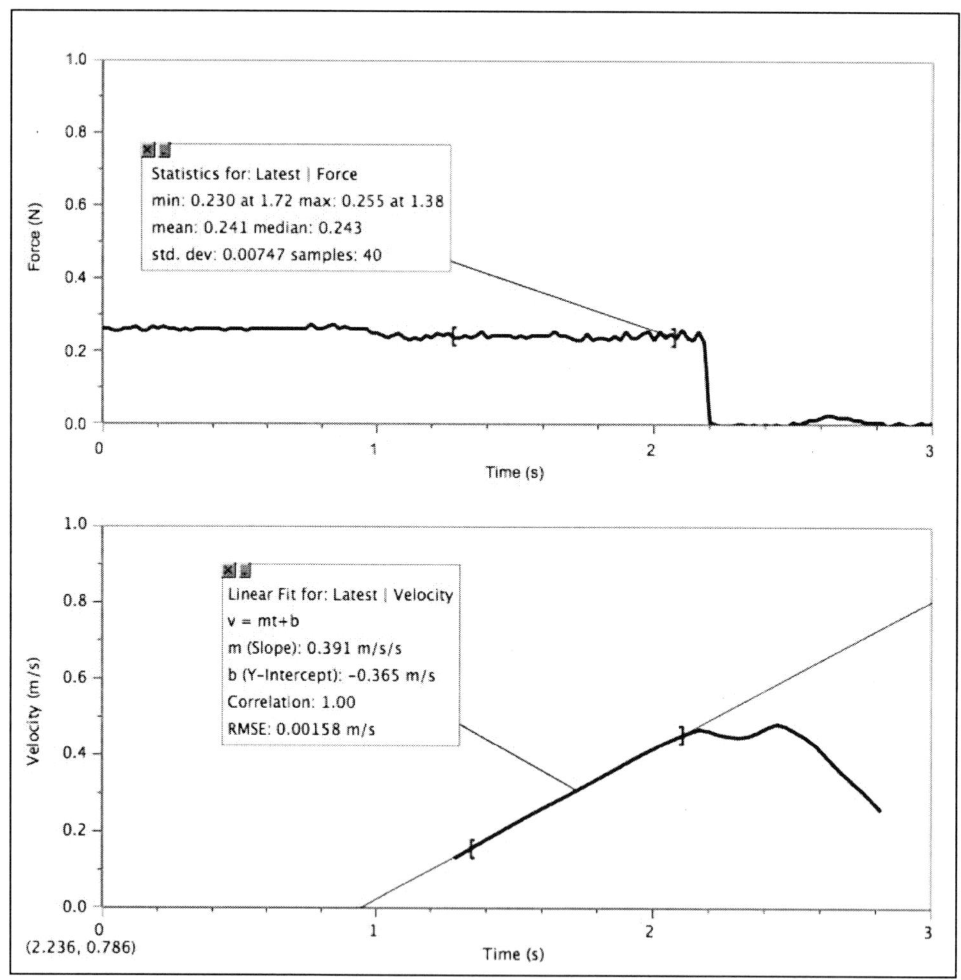

Figure 3 Timing with the photogate and ultra pulley

Steps 1–3

Suggest to the students that while we ordinarily place the independent variable on the *x*-axis, sometimes, to facilitate the analysis of the graph, we choose to ignore that convention. The graph in Figure 4 shows the plot of force *vs*. acceleration using the photogate with the cart picket fence and Logger *Pro* to collect data. Similar results were obtained with the other methods of data collection.

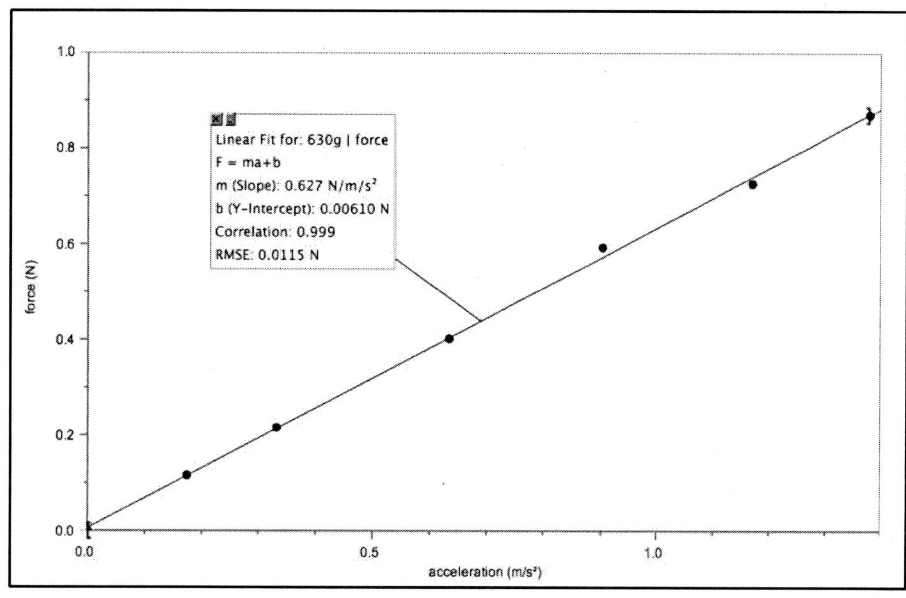

Figure 4

Steps 4–5

For the graph above, the equation describing the relationship between force and mass is

$$F_{net} = (0.627 \frac{\text{N}}{\text{m/s}^2})a$$

Students should be able to state that the force is directly proportional to the acceleration of the cart and masses.

Step 6

Note that the slope of the graph (0.627 N/(m/s^2)) is within 1% of the mass of the cart and the sensor. With increasing mass, the agreement is slightly worse, but still within 3%. This is an appropriate time to suggest that the slope of a graph of variables in a physical system is usually related, in some way, to a variable that was held constant during the experiment. In this case, students held mass constant.

Step 7

If the value of the slope of the graph is equivalent to the mass of the object undergoing acceleration, then the units of the slope must simplify to kilograms. This suggests that the fundamental units of the newton are $\frac{\text{kg}}{\text{m/s}^2}$. Encourage students to make this substitution for N in the numerator and cancel out m/s^2, leaving kg, the unit for mass.

Step 8

The analysis in Step 7 leads to the conclusion that the general equation that describes the relationship between net force and acceleration is $F_{net} = ma$.

EXTENSION

Students might wonder why the force decreases slightly when the cart is released (see Figure 3). To answer this question, they should sketch free-body diagrams for the hanging mass and the cart when they are both motionless. The tension force in the string has the same magnitude as the weight of the hanging mass. This same force acts on the cart. When the cart is released, the tension force must be smaller than the weight of the hanging mass; otherwise, the hanging mass (and cart) would not accelerate.

Some portion of the force of gravity acting on the hanging mass accelerates this mass as well as cart/sensor/mass system. As the hanging mass increases, the difference between the force of tension in the static and accelerating conditions increases because the hanging mass represents an increasing fraction of the overall mass of the *system* that is accelerating.

If you were to keep the hanging mass constant, but vary the mass of the cart, the force of tension in the static condition would remain the same. However, when the cart was released, the tension acting on the cart would decrease slightly for each increase in mass.

Experiment

5

Newton's Third Law

INTRODUCTION

In your discussion of the force concept you may have heard that a force is a push or pull exerted by one object on another. You may also have heard a popular expression of Newton's third law: "For every action, there is an equal, but opposite reaction." Unfortunately, this expression leads to some naïve conceptions about forces. Unlike Newton's first two laws, which deal with the effects of forces on a single object, the third law describes the interaction between two objects. In this experiment, you will examine this interaction in a variety of situations so that you might develop a better understanding of the forces involved when two objects interact.

OBJECTIVES

In this experiment, you will

- Observe the magnitude and direction of forces exerted by interacting objects.
- Observe the time variation of these forces.
- Develop a more robust expression of Newton's third law.

MATERIALS

Vernier data-collection interface
Logger *Pro* or LabQuest App
two Vernier Dual-Range Force Sensors, **or** two Wireless Dynamic Sensor Systems (WDSS)
Vernier Dynamics Track
two standard carts
string
rubber band
(optional) two magnetic bumpers **or** two hoop springs from the Bumper and Launcher Kit

PRE-LAB INVESTIGATION

Before you begin your experiment, consider the following situations involving an interaction between two objects. Sketch a graph of the forces exerted by the objects as a function of time. Recall that force is a vector–it has direction as well as magnitude.

1. You and your little brother are engaged in a tug of war. You are bigger and stronger, but you choose to not pull your brother over during the contest. Assume that your force is in the positive direction. Graph both forces on one set of axes during the contest.

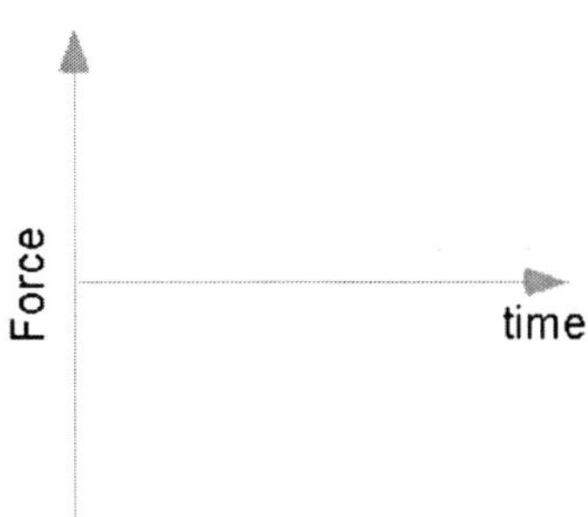

2. You are pushing on a wall. Graph the forces exerted by you and by the wall (if any). Assume that your force is in the positive direction.

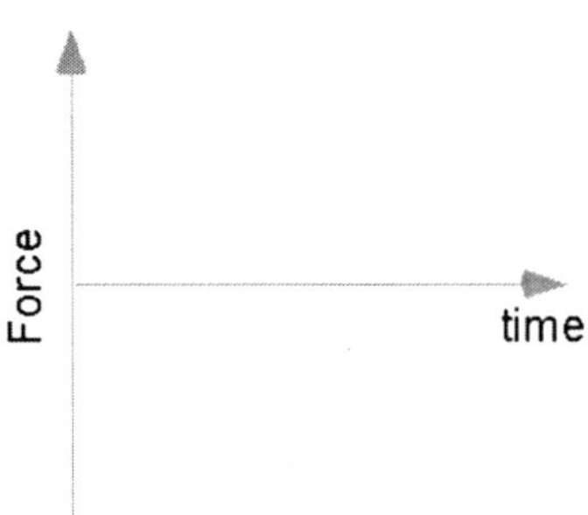

PROCEDURE

In this experiment, you will collect force-time data in a number of ways. In your lab notebook, record details about the way you collected the data for each trial.

1. Set the range switch on both sensors to 10 N. Connect both Force Sensors to the interface and start the data-collection program. The default data-collection mode works well for this experiment.

2. Make a loop with the string so that you can connect the hook ends of the two sensors. See Figure 1.

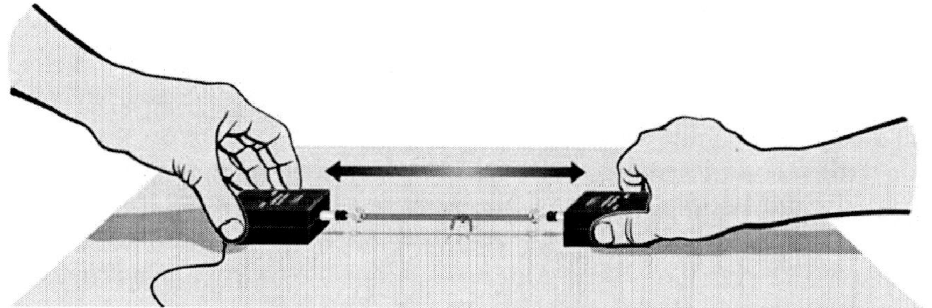

Figure 1

3. Since force is a vector quantity, you must reverse the direction of one of the sensors in the software. Decide which one you should adjust.

4. Before any force is applied to the force sensors, zero both sensors. Be sure to re-zero the sensors after each change you make in the configuration of the apparatus.

5. Begin data collection, wait a second, then you and your partner should pull on your sensor, being careful not to exceed the 10 N limit for data collection. During the 10 s data collection interval, vary the amount of force each of you applies to the other.

6. Observe the plots of force *vs*. time for both sensors, then store this run.

7. Collect data again, but this time, one of you should simply hold onto your force sensor while the other applies the force. Observe the plots of force *vs*. time for both sensors, then store this run.

8. Replace the loop of string with a rubber band. Collect force-time data as you did in Step 5. Observe the plots of force *vs*. time for both sensors, then store this run.

9. Replace the hooks on the force sensors with bumpers (see Figure 2). Before any force is applied to the force sensors, zero both sensors.

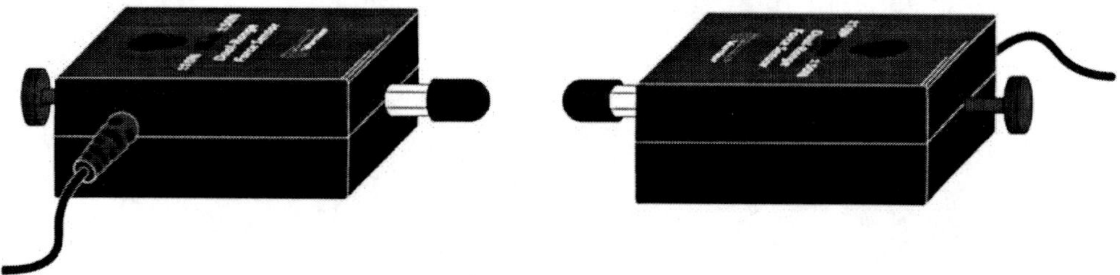

Figure 2

10. Collect force-time data while both of you push on your force sensor. Observe the plots of force *vs*. time for both sensors, then store this run.

11. To help you test your prediction posed in the second situation in the pre-lab, attach one force sensor to a support rod that is clamped to the table so that it cannot move. Position the other sensor so that the bumpers on the two sensors line up. Begin collecting data, then push on the free sensor so that both sensors are in contact. Observe the plots of force *vs*. time for both sensors, then store this run.

12. Be sure to save your data file. You will refer to it for the Evaluation of Data.

EVALUATION OF DATA

1. Examine your data and graph for your first trial. Compare the magnitude and direction of the forces that you and your partner applied.

2. Were you able to observe any time delay between the appearance of either force; i.e., does Force 2 appear to be a *reaction* to Force 1, or vice-versa?

3. Could you tell from the graph whether you or your partner was pulling harder than the other at any time during the run?

4. Examine the data and graph for your second trial. From the graph, can you tell which of you pulled on your force sensor and which one simply held on?

5. Examine your data and graph for your third trial. What effect, if any, did replacing the string with the rubber band have on your results?

6. Examine the data and graph for your fourth trial. Did it make a difference if you were *pushing* on the sensors instead of pulling on them?

7. Examine the data and graph for your fifth trial. Is it true that only one object was applying a force? Explain.

8. Reflect on the graphs for your various trials. Can you tell, without the help of your lab notes, how the forces were applied?

9. Re-state Newton's third law in your own words without using the words "action" and "reaction." If you said in Step 1 that the forces were *both* equal *and* opposite, find a more precise way to state the relationship between the forces.

EXTENSION

If the accessories in a Bumper and Launcher Kit (magnetic bumpers, hoop springs) are available, replace the regular bumpers with these and repeat data collection.

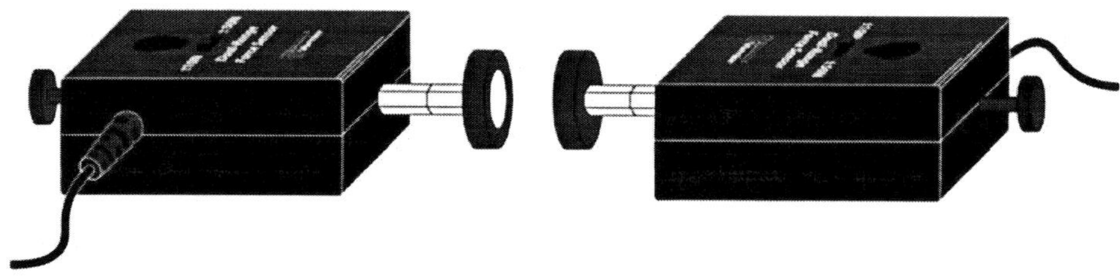

Figure 3

ANIMATED DISPLAY

If you're using Logger *Pro*, inserting an animated display allows you to represent both force vectors during the experiment. Your instructor will show you how to set up the point display options for such a display.

Experiment
5

INSTRUCTOR INFORMATION

Newton's Third Law

This is the last in a series of three experiments designed to help students develop an understanding of Newton's laws of motion. It is written with the assumption that the instructor will engage the students in discussions at critical junctures. These discussions can take place with the entire class or with individual lab groups. The icon at left indicates where these discussions should occur.

The Microsoft Word files for the student pages can be found on the CD that accompanies this book. See *Appendix A* for more information.

Considerable evidence[1,2] exists that Newton's third law of motion is the least well understood aspect of the force concept. Students hold a number of naïve conceptions about forces, including the association of force with human action, and the view that an interaction is a struggle between competing forces, where the stronger agent prevails. These naïve conceptions are unlikely to be dislodged by simply asserting Newton's third law. This experiment is designed to provide students with experiences that will help them address their own naïve conceptions in class discussion about the nature of forces.

OBJECTIVES

In this experiment, the student objectives include

- Observe the magnitude and direction of forces exerted by interacting objects.
- Observe the time variation of these forces.
- Develop a more robust expression of Newton's third law.

During this experiment, you will help the students

- Recognize that paired forces are part of an interaction, rather than an action-reaction pair.
- Recognize that forces can be exerted by other than animate agents.
- Recognize that the magnitudes of the paired forces are equal, regardless of the exertion of the various agents involved in an interaction.

REQUIRED SKILLS

In order for student success in this experiment, it is expected that students know how to zero and reverse the direction of a Dual-Range Force Sensor or WDSS.

EQUIPMENT TIPS

While students can collect suitable data on pulling forces using the Force Sensors configured with hooks, they can examine a variety of situations involving pushing forces using the bumpers or the accessories in the Bumper and Launcher Kit.

---

1 D. Hestenes, M. Well, G. Swackhamer, "Force Concept Inventory", *The Physics Teacher*, 30: 141-158, 1992.
2 D. Hestenes, "Modeling Methodology for Physics Teachers", 1996.

Remind the students that it is important to re-zero the Force Sensors whenever they change attachments.

PRE-LAB DISCUSSION

Begin the discussion by asking the students what they know about the forces involved in a variety of common interactions: a tug-of-war, a baseball contacting a bat, a person leaning on a wall. If the common expression, “For every action there is an equal but opposite reaction.” does not arise, bring it up and ask students to discuss its implications. Don’t attempt to correct any of the students’ naïve conceptions at this point. The opportunity to help students come to a Newtonian view of forces will take place at the end of the experiment.

Ask the students to make predictions about the plots of force vs. time for the situations in their pre-lab investigation. You may need to remind students that force is a vector, so that the sign of force B should be negative. The first situation is designed to probe the notion that the dominant agent in a pair exerts a stronger force than does the other. The second addresses the issue of whether an inanimate object can exert a force.

LAB PERFORMANCE NOTES

When the Dual-Range Force Sensors are connected to the data-collection interface and Logger *Pro* or LabQuest App is started, the default digital meters and graphs can be used without modification.

Student are expected to know how to zero the sensors and reverse the direction of one of the sensors. Nevertheless, students may need some assistance.

SAMPLE RESULTS AND POST-LAB DISCUSSION

The following graphs were obtained using a LabQuest as a standalone device and importing the data into Logger *Pro*. The plots of force *vs.* time appear in difference colors on the screen of the computer or LabQuest to help distinguish them.

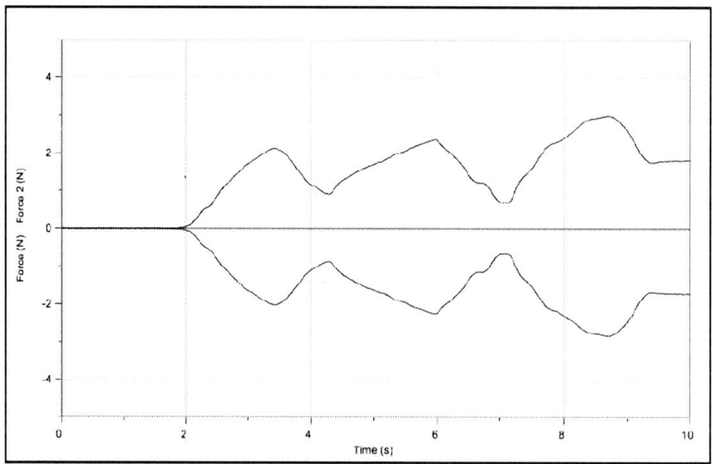

Figure 1 Two persons pulling

Steps 1–4

The symmetry of the plots is striking. It is apparent that the magnitudes of the paired forces are equal. Therefore it is incorrect to claim that one agent exerts a stronger pull than the other. The negative sign on values of force 2 imply that this force acts in the opposite direction. Furthermore, changes in the forces occur simultaneously; one force does not lag behind the other.

Step 5

Apparently, the stretching of the rubber band has no effect on the paired forces in the interaction.

Steps 6–8

Figure 2 shows that a plot of force *vs.* time when the interaction involves pushing is indistinguishable from that when agents pull on one another. It is not possible to tell, just by looking at the graph, how the forces were applied. What's more, it is apparent that forces come in pairs even when only one animate agent is involved (Step 7).

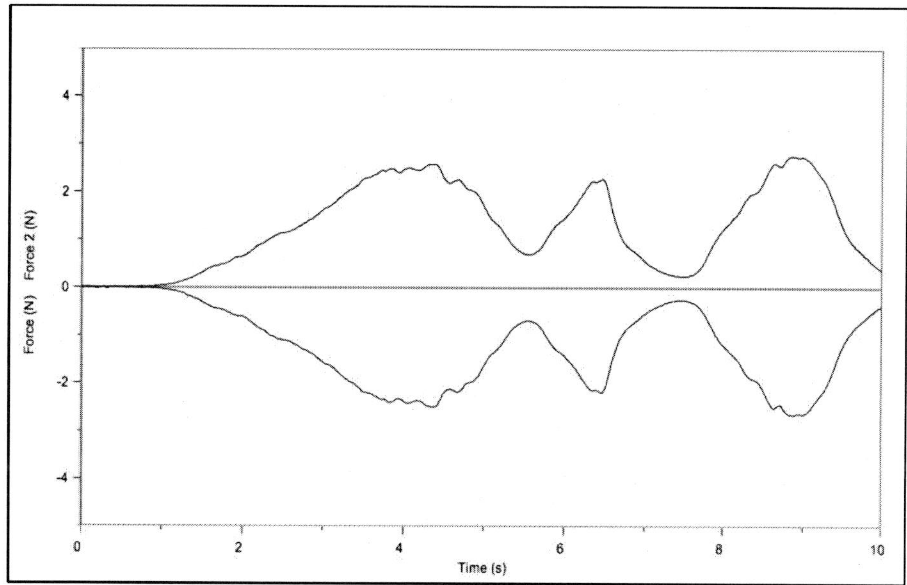

Figure 2 Two persons pushing

Step 9

Students should conclude that the forces between objects are paired because they are part of an *interaction* between the objects. These paired forces are always equal in *magnitude*, but, because they act in opposite directions, these forces have different signs. This expression can be expressed algebraically, using agent-object notation, where A/B is read as "that A exerts on B."

$$\vec{F}_{A/B} = -\vec{F}_{B/A}$$

This experiment alone will not clear up all student misconceptions regarding the force concept. In fact, after students accept this aspect of forces, they are likely to have difficulty explaining, for example, how a horse can pull a cart, if the force the horse exerts on a cart has the same magnitude as the force the cart exerts on the horse. The resolution of this dilemma involves focusing one's attention on one object at a time, and examining all the forces acting on that single object. The paired forces in an interaction always act on different objects.

EXTENSION

Students should obtain results similar to those with the bumpers or hooks. However, if students use the magnetic bumpers, they will have to make an effort to keep them aligned as they push the sensors towards one another.

ANIMATED DISPLAY

Use of the Animated Display tool in Logger *Pro* allows students to represent the force vectors with their origin at the point of contact of the two bumpers during this experiment. Suggest that students first view the example given in the file: 1 Using Animated Displays following the path: Experiments ▶ Sample Data ▶ Physics ▶ Animated Display Vectors.

They should then choose Meter ▶ Animated Display from the Insert menu to insert an Animated Display into their Logger *Pro* experiment file. Stretch the window horizontally, then double-click it to bring up the Animated Display Options window. Some sample settings are provided below.

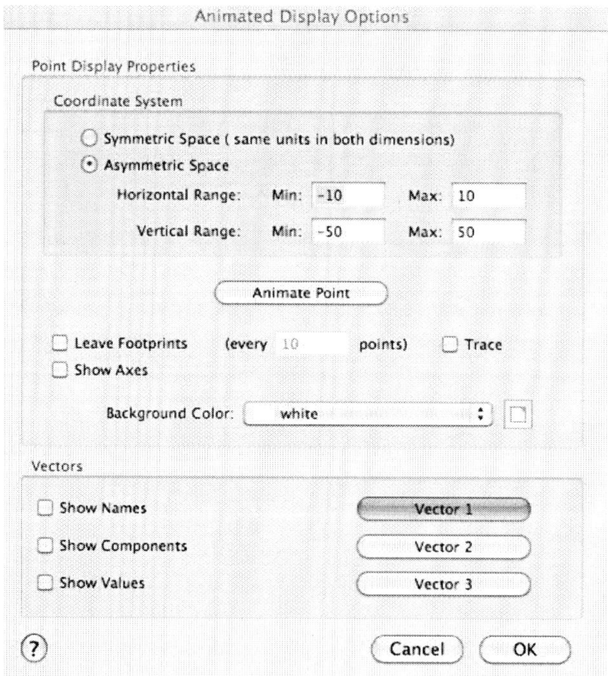

Figure 3

After setting the horizontal range, define vectors 1 and 2 to correspond to the readings from the force sensors.

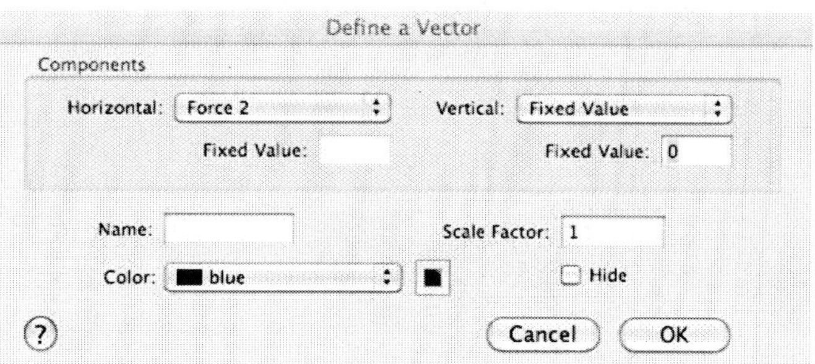

Figure 4

Choosing Replay from the Analyze menu brings up the Replay control window. This allows one to view the trace of the graph at the same time the force vectors are displayed. The Logger *Pro* file, 05 Newtons third sample data.cmbl, with the animated display is provided on the CD that accompanies this book. See *Appendix A* for more information.

Experiment

6

Projectile Motion

INTRODUCTION

Up to this point it is likely that you have examined the motion of an object in one dimension only – either on a horizontal or inclined surface, or falling vertically under the influence of the force of earth's gravity.

In this experiment, you will examine the behavior of a projectile – an object moving in space due to the exertion of some launching force. Such an object can undergo motion in two dimensions simultaneously. Using the video analysis features of Logger *Pro*, you will compare features of the position-time and velocity-time graphs with those you have studied earlier.

OBJECTIVES

In this experiment, you will

- Use video analysis techniques to obtain position, velocity, and time data for a projectile.
- Analyze the position *vs.* time and velocity *vs.* time graphs for both the horizontal and vertical components of the projectile's motion.
- Determine the best fit equations for the position *vs.* time and velocity *vs.* time graphs for both the horizontal and vertical components of the projectile's motion.
- Relate the parameters in the best-fit equations for position *vs.* time and velocity *vs.* time graphs to their physical counterparts in the system.
- Relate the horizontal and vertical components of the projectile's motion to any forces acting on the object while it is moving.
- Produce a movie of an object undergoing projectile motion.

MATERIALS

digital video camera or digital still camera capable of shooting in movie mode
tripod
means to transfer a movie file in the camera to a computer
meter stick or some other object to provide scale
projectile –this could be something as simple as a ball (point particle) or an extended body
video editing software such as iMovie or QuickTime Pro may be helpful

PRE-LAB INVESTIGATION

Your instructor will launch a projectile. Observe its motion carefully, then discuss its position-time and velocity-time behavior.

PART 1 – ANALYSIS OF AN EXISTING MOVIE

PROCEDURE

1. Start Logger *Pro*. Choose Movie from the Insert menu. Insert the Basketball Shot movie from the Sample Movies folder in Logger *Pro*. Your instructor will help you if you cannot locate this movie.

2. Make the movie window large enough to easily see the projectile.

3. Enable Video Analysis by clicking on the button in the lower-right corner. This brings up a toolbar with a number of buttons (see Figure 1).

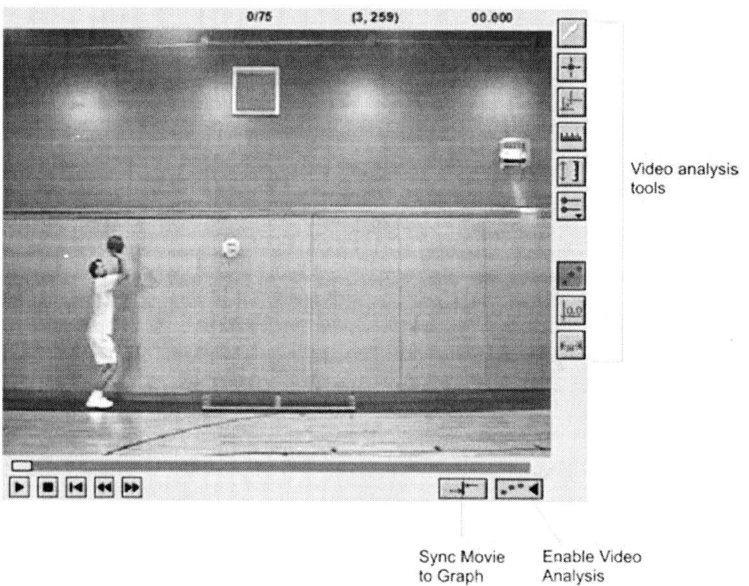

Figure 1

3. Click the **Set Origin** button (third from top), then click in the movie frame to set the location of the origin. If needed, this coordinate system can be rotated by dragging the yellow dot on the horizontal axis.

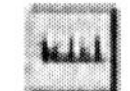

4. Click the **Set Scale** button (fourth from top), then drag across an object of known length in the movie. In this movie, the object of known length is the 2 m stick on the floor. When you release the mouse button, enter the length of the object; be sure the units are correct.

5. Use the forward and back movie buttons to advance the movie until the ball is released from the shooter's hands. Next to the button you used to enable analysis is the **Sync Movie to Graph** button. Click this button, then enter **0** in the graph time window. Select Use This Synchronization in Video Capture.

6. Now click the **Add Point** button (second from the top). Decide where on the object you will mark its location (center, top, other) and then click the object in the movie. **Important**: Be consistent in your marking.

   Each time you mark the object's location, the movie advances one frame. Depending on the frame rate, you may choose to mark the position every other frame. Notice that data are being plotted on the graph.

7. Continue this process as long as is desired. Should you wish to edit a point, click the **Select Point** button (top). This allows you to move or delete a mismarked point.

8. Select the graph window. Logger *Pro* defaults to display both the x and y positions of the object as a function of time. You may find it easier to examine the position-time behavior of just one of these components at a time.

EVALUATION OF DATA

1. Examine the graph of x-position *vs.* time. If it appears to be linear, fit a straight line to your data. If the slope of the graph appears to change abruptly, fit separate straight lines to each portion of the graph that appears to be linear.

2. Write the equation that describes the x-position *vs.* time behavior of the ball in each segment; be sure to include units.

3. Based on your previous experiments, describe the horizontal component of the motion of the projectile. Note when any change in the horizontal component of the motion occurs.

4. Now, examine the graph of y-position *vs.* time. Fit an appropriate curve to this graph (or to each portion of the graph). Write the equation that describes the y-position *vs.* time behavior of the ball in the first segment; be sure to include units.

5. Based on what you have learned in previous experiments, describe the vertical component of the position of the projectile.

6. Now, to test your analysis in Step 5, examine the graph of y-velocity *vs.* time. Fit a straight line to the first portion of the graph.

7. What can you say about the *rate of change* of the y-velocity as a function of time? How does the value of the slope of the linear fits compare to the acceleration of a freely falling object?

8. Compare the A and B parameters (values and units) to the curve fits you performed in Step 4 to the slope and intercept of the linear fits you performed in Step 6.

9. Explain the differences in the horizontal and vertical components of the motion of the projectile in terms of the force(s) acting on it after it was launched.

PART 2 – PRODUCTION AND ANALYSIS OF YOUR OWN MOVIE

PROCEDURE

You will need either a digital video camera or a digital still camera set to "movie mode". Keep the following tips in mind when you shoot your movie.

1. It is best to have a plain background that provides sufficient contrast with the projectile. Good lighting is essential.
2. Set up the camera on a tripod so that it is looking square at the background, and so that the plane of motion is perpendicular to the view
3. Position the camera as far from the plane of motion as is practical to reduce problems with scaling and parallax. Use the zoom feature to fill the screen with the motion.
4. The object used for scaling must be in the same plane as the motion of the projectile (see Figure 2).

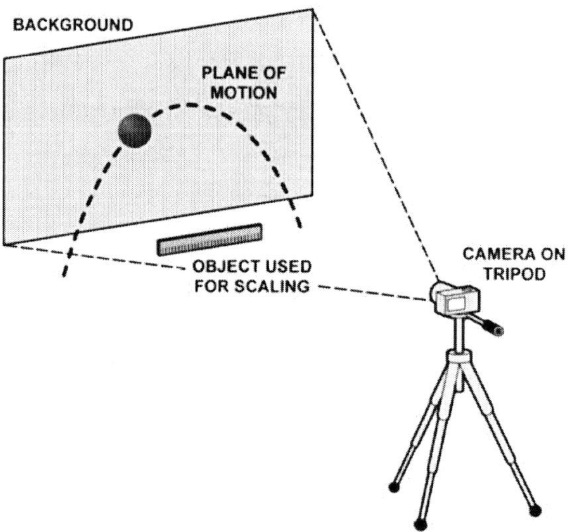

Figure 2

Once you have shot your movie, use the directions that accompany your camera to transfer the video clip to the computer you will use for the analysis. If you have captured more video than you need and your movie is much too large, you can use video editing software (e.g. QuickTime Pro or iMovie) to edit the clip down to a more manageable length.

EVALUATION OF DATA

Perform the evaluation of data as you did with the movie clip provided to you in Part 1.

EXTENSIONS

1. Suppose that, in the shooting of your movie, you placed the meter stick used for scaling against the wall you used for your background. However, the plane of the ball's motion was 0.50 m in front of the wall. The distance between the camera and the wall was 5.0 m. Would

this error result in a value for a_g in your analysis of the *y*-velocity *vs.* time graph that was smaller or larger than the accepted value? By what factor would this value differ from the expected value? Explain, using a diagram.

2. Repeat the production and video analysis of a projectile, but this time use an extended body; i.e., an object that cannot be readily modeled by a point-particle. Consider carefully how best to mark the position of such an object during its motion. Interpret your position-time and velocity-time graphs as you did before.

ANIMATED DISPLAY

Inserting an animated display in Logger *Pro* gives you another tool to represent the velocity of the projectile at a number of instants during the experiment. Your instructor will show you how to set up the point display options for such a display.

INSTRUCTOR INFORMATION

Experiment 6

Projectile Motion

This lab is written with the assumption that the instructor will engage the students in discussions at critical junctures. These discussions can take place with the entire class or with individual lab groups. The icon at left indicates where these discussions should occur.

The Microsoft Word files for the student pages can be found on the CD that accompanies this book. See *Appendix A* for more information.

OBJECTIVES

In this experiment, the student objectives include

- Use video analysis techniques to obtain position, velocity, and time data for a projectile.
- Analyze the position *vs.* time and velocity *vs.* time graphs for both the horizontal and vertical components of the projectile's motion.
- Determine the best fit equations for the position *vs.* time and velocity *vs.* time graphs for both the horizontal and vertical components of the projectile's motion.
- Relate the parameters in the best-fit equations for position *vs.* time and velocity *vs.* time graphs to their physical counterparts in the system.
- Relate the horizontal and vertical components of the projectile's motion to any forces acting on the object while it is moving.
- Produce a movie of an object undergoing projectile motion.

During this experiment, you will help the students

- Edit a video clip to a manageable length or perform video capture using Logger *Pro*.
- Recognize that the horizontal component of a projectile's motion is best described as an object moving with constant velocity.
- Recognize that the vertical component of a projectile's motion is best described as an object undergoing uniform acceleration.
- Recognize that, neglecting air resistance, the acceleration of the object very nearly equals a_g.

REQUIRED SKILLS

In order for student success in this experiment, it is expected that students know how to

- Perform linear and curve fits to graphs. This is addressed in Activity 1.
- Manipulate graphs and pages in Logger *Pro*. This is addressed in Activity 1.

EQUIPMENT TIPS

What makes this experiment interesting to students is the possibility of examining movies they have made, rather than one provided in Logger *Pro* or on the CD that accompanies this manual. To do this, students will need either a digital video camera or a digital still camera that shoots in movie mode (most do), a tripod and some object they can use for scaling. The first object they should use for a projectile should be a small ball that is massive enough not to experience significant slowing from air resistance. One that bounces will provide an opportunity to examine how the horizontal component of the motion changes after it bounces. However, students are likely to want to examine the behavior of more "interesting" objects. Encourage them to save those for the Extension activity.

PRE-LAB DISCUSSION

This experiment should be performed only after students have had the opportunity to explore the behavior of an object moving with constant velocity as well as one that undergoes uniform acceleration. Demonstrate a ball toss, then ask the students to describe the position-time behavior and velocity-time behavior of the ball. They are likely to describe the ball's path as parabolic but may have more difficulty describing how its velocity changes with time. Suggest that separate analyses of the horizontal and vertical components of the ball's motion could be revealing. To do this they will need to use Logger *Pro* to analyze a video clip of the ball's motion.

LAB PERFORMANCE NOTES

Part 1

We advise that students first learn how to do video analysis on the Basketball Shot movie provided in Logger *Pro*. Analysis of this movie is likely to yield results that are easier to interpret. Other movies may suffer from poor scaling, perspective problems, or blurred images. This movie can be found in the following locations in Logger *Pro* 3.8.2 or later:

On a Mac: /Library/Applications Support/Logger Pro/Experiments/Sample Movies

For computers running Windows refer to the information in the Tech Info Library: http://www.vernier.com/til/2247.html

You might find it more convenient to place this movie (on the CD that accompanies this book) in a location students can easily access.

Part 2

After students have become familiar with performing video analysis on a sample movie clip, they can move on to making one of their own. If they do so, they need to keep the following tips in mind.[1]

1. It is best to have a plain background that provides sufficient contrast with the projectile. Good lighting is essential.

2. Set up the camera on a tripod so that it is looking square at the background and the motion is in a plane perpendicular to the view.

3. Position the camera as far from the plane of motion as is practical (and zooming in) to reduce problems with scaling and parallax.

[1] More detail is provided at the Vernier web site, http://www.vernier.com/til/1464.html.

4. The object used for scaling must be in the same plane as the motion of the projectile (see Figure 1).

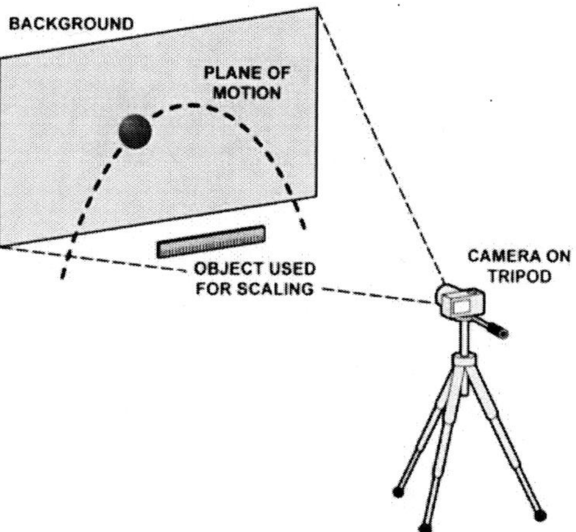

Figure 1

If students choose to use an extended body that undergoes rotation during the motion, they should mark its center of gravity (a piece of contrasting tape works well) so they can mark the position of its center of mass.

SAMPLE RESULTS AND EVALUATION OF DATA

Steps 1–3

Make sure that students recognize that the *x*-position *vs.* time graph has two distinct linear segments, so two best-fit lines are appropriate (see Figure 2). Each shows that the horizontal component of the motion can be described as constant velocity. The lower slope in the second segment is due to the fact that the ball slowed after it bounced.

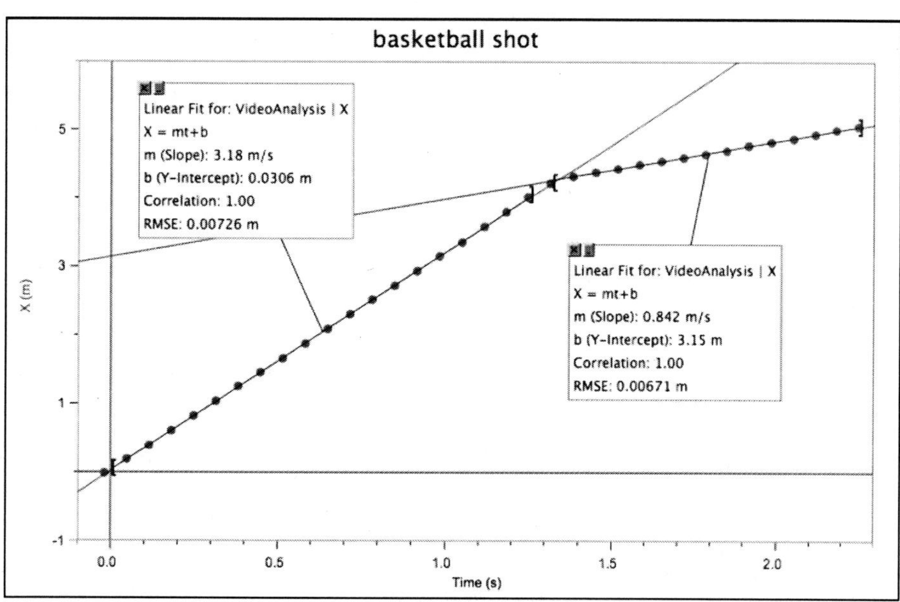

Figure 2 x-t graph

The equations for the graph in Figure 2 are x=(3.18 m/s)t + 0.031 m and x=(0.842 m/s)t + 3.15 m respectively.

Steps 4–5

Based on their experience in Experiment 1, students should recognize that a quadratic fit is appropriate for the y-position *vs.* time graph. They should perform a separate quadratic fit for each parabolic segment (see Figure 3).

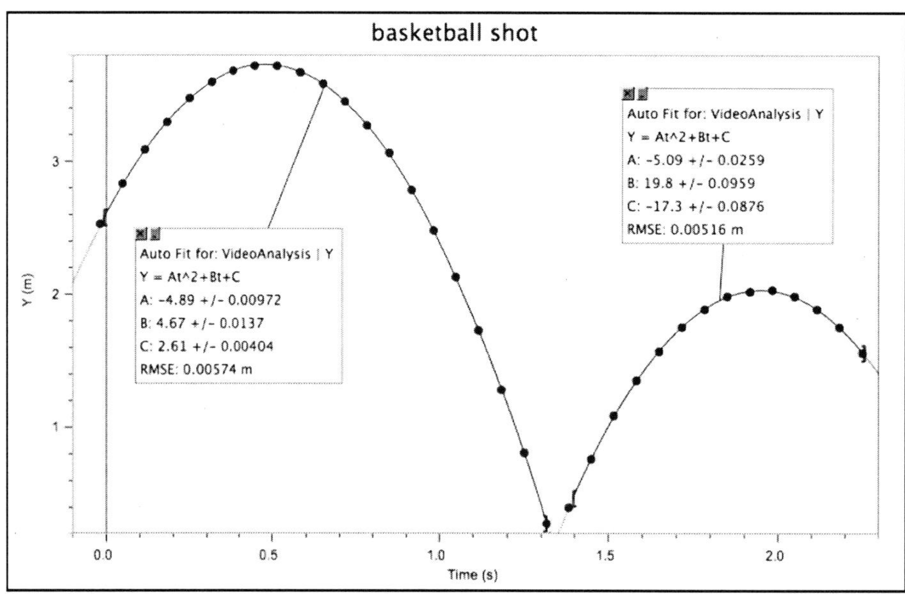

Figure 3 y-position vs. *t graph*

The equation for the first segment is y=(–4.9 m/s^2)t^2 + (4.67 m/s)t + 2.61 m. The fact that a quadratic fits the y-position *vs.* time graph so nicely indicates that the ball is accelerating. The A parameter for the quadratic is very close to the expected value. However, after the bounce the path of the ball was no longer quite in the same plane as the 2-meter stick. As a result, the A parameter for the second parabola is nearly 4% greater than the expected value. This shows that a slight difference in the distance between the camera and the plane of motion and the object used for scaling can result in a non-trivial error. This error can be reduced by increasing the distance between the camera and the projectile.

Steps 6–7

The interval used for performing the linear fit should not include the first few and last few points.[2] The y-velocity *vs.* time graph shows that vertical component of the ball's velocity changed at a constant rate (see Figure 4). Students should recognize the *rate of change* of the velocity as the acceleration of the ball. If the object used for scaling and the plane of the ball's path coincide, the value of the acceleration should be nearly –9.8 m/s^2, a_g.

[2] This is due to the way that Logger *Pro* calculates the velocity of the object at each clock reading.

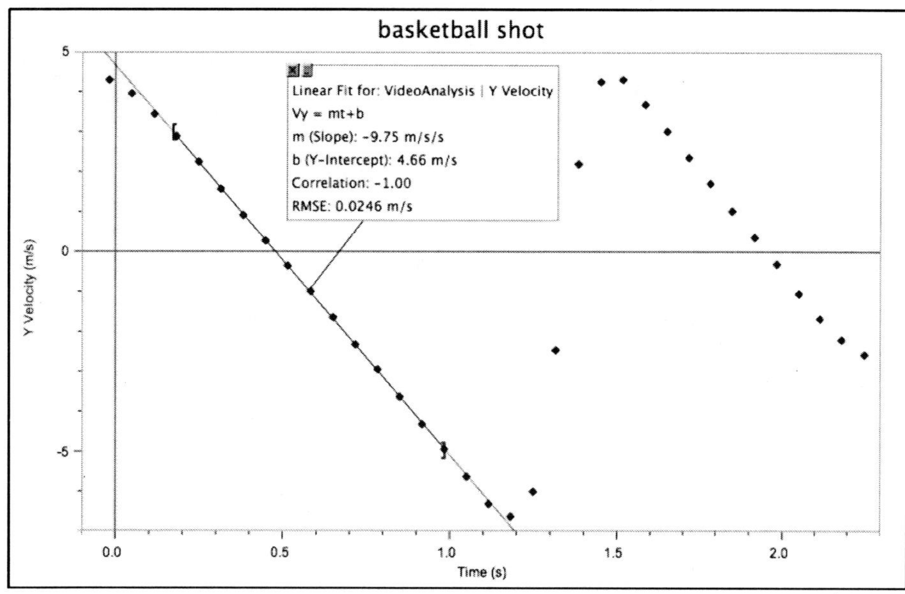

Figure 4 y-velocity vs. *time graph*

Step 8

By comparing the value of A in the quadratic fit to the y-position vs. time graph to the slope of the linear fit to the y-velocity vs. time graph, students should be able to conclude that A is half of the acceleration. Furthermore, note that for the parabola before the bounce, the B parameter is the same as the intercept of the line of best fit to the corresponding y-velocity vs. time graph. This intercept is the y-velocity for the first frame used in the analysis. Once these connections have been made, students should recognize that the general equation for the position of an object moving with constant acceleration, in this case, $y = \frac{1}{2}at^2 + v_0t$ describes the vertical component of the motion of the projectile.

Step 9

From their understanding of the relationship between motion and forces, students should conclude that no net force acts on the projectile in the horizontal direction, except when it makes contact with the ground. Note that the *x*-velocity decreases after the bounce. The projectile accelerates in the vertical direction because of the force exerted on it by the earth. The sign of the acceleration is consistent with the direction of gravitational force.

EXTENSIONS

1. If the object used for scaling were farther from the camera than the plane of the motion, the video analysis will report values of position and velocity that are larger than they actually are. This will result in a value of the slope of the y-velocity *vs.* time graph that is 5.0/4.5 = 1.1 times larger than the accepted value (a_g would appear to be –10.9 m/s^2).

2. For the analysis of the motion of an extended body that undergoes rotation during the motion, students should mark its center of gravity (a piece of contrasting tape works well) so they can mark the position of its center of mass. Analysis of the motion of the center of mass of the extended body should yield results much like those of a point particle.

ANIMATED DISPLAY

Use of the Animated Display tool in Logger *Pro* allows students to display vectors representing the velocity of the projectile at various points in its flight. If your students are not familiar with Animated Displays they will need to study the example given in the file: 1 Using Animated Displays following the path: Experiments ► Sample Data ► Physics ► Animated Display Vectors.

To begin this process, students should choose Meter ►Animated Display from the Insert menu into their Logger *Pro* experiment file, then double-click the window to bring up the Animated Display Options window. Some sample settings are provided below. Choosing Replay from the Analyze menu brings up the Replay control window. When the movie is played at reduced speed, one can see the points and vectors appear on the movie window during the projectile's flight.

The following settings were used to produce the display in Figure 7.

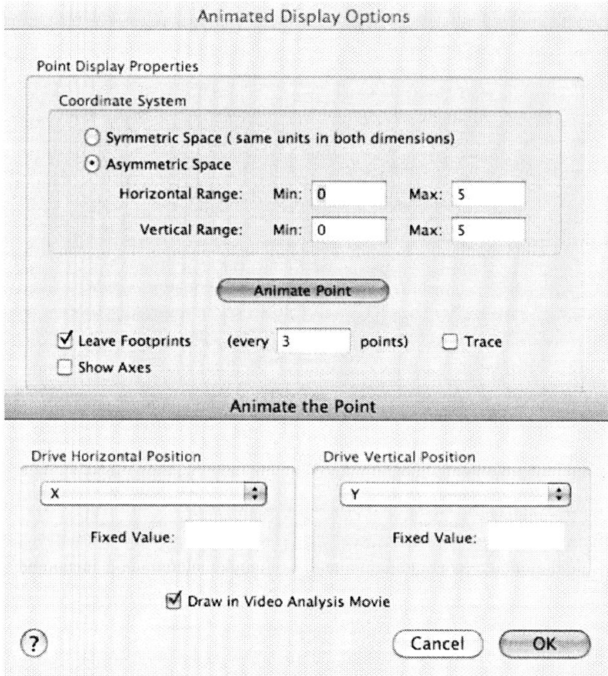

Figure 5

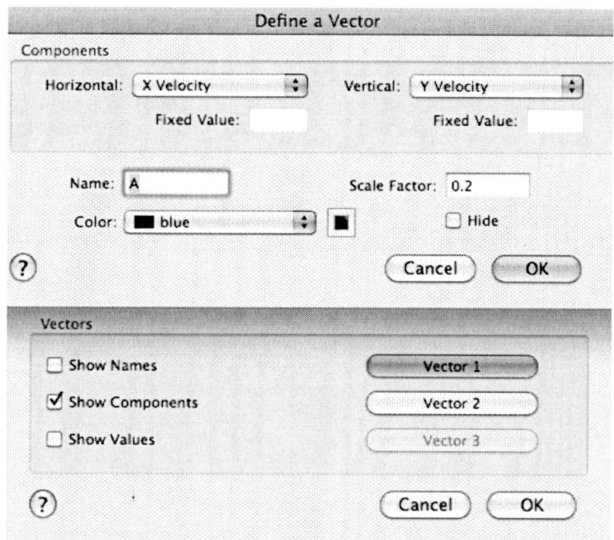

Figure 6

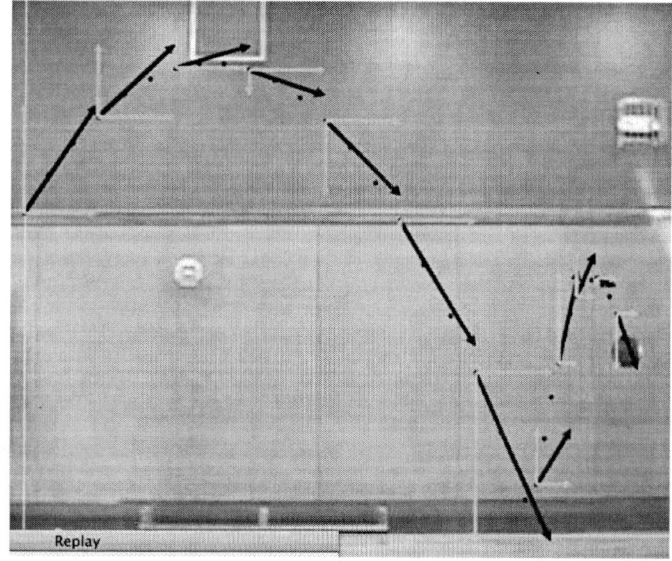

Figure 7 Animated display representing the velocity of the projectile

Note that the horizontal velocity vectors show that the x-velocity is constant (in each segment of the motion). The vertical velocity vectors are consistent with an object that is accelerating in the y-direction.

Experiment

7

Energy Storage and Transfer: Elastic Energy

PART 1 – ELASTIC ENERGY

As Richard Feynman described it, energy is the currency of the universe. If you want to speed it up, slow it down, change its position, make it hotter or colder, bend it, break it, whatever, you'll have to pay for it (or be paid to do it). This is the first of a series of experiments in which you will investigate the role of energy in changes in a system.

If you grasp one end of a rubber band and pull on the other, you realize that the stretched rubber band differs from the same band when it is relaxed. Similarly, if you compress the spring in a toy dart gun by exerting a force on it, you know that the state of the compressed spring is different from that of the relaxed spring. By exerting a force on the object through some distance you have changed the energy state of the object. We say that the stretched rubber band or compressed spring stores *elastic energy* – the energy account used to describe how an object stores energy when it undergoes a reversible deformation. This energy can be transferred to another object to produce a change – for example, when the spring is released, it can launch a dart. It seems reasonable that the more the spring is compressed, the greater the change in speed it can impart to the toy dart.

If we want to *quantify* the amount of energy stored by a spring when it is deformed, we must first study the relationship between the force applied and the extent to which the length of the spring is changed.

OBJECTIVES

- Determine the relationship between the applied force and the deformation of an elastic object (spring or rubber band).
- Determine an expression for the elastic energy stored in spring or rubber band that has been compressed or stretched.

MATERIALS

Vernier data-collection interface
Logger *Pro* or LabQuest App
Vernier Dual-Range Force Sensor
Vernier Dynamics Track
standard cart
Vernier Bumper and Launcher Kit (recommended) **or**
lightweight (3–4 N/m) extensible spring **or**
heavy rubber band

PRE-LAB INVESTIGATION

Examine whatever elastic systems (hoop, extensible spring, or rubber band) are available to you. Apply varying force to the system and note the deformation.

On the axes to the right, sketch a graph of the applied force *vs*. the change in length, based on what you felt.

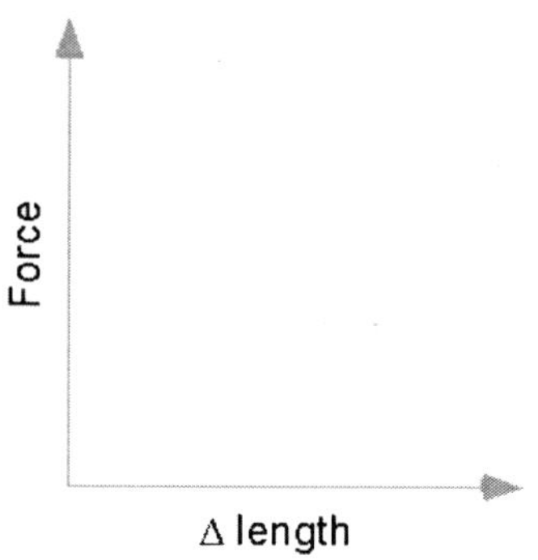

PROCEDURE

1. Connect the Dual-Range Force Sensor to the interface and start the data-collection program.

2. Check to see if the sensor is working by applying a gentle force to the hook.

3. Devise a quick test to determine which range of the Force Sensor you should use.

4. For this experiment, you will collect data in the Events with Entry mode. Enter **Change in Length** as the Name. While there are a number of variables one could use to represent the change in length, physicists have adopted the convention of using x for this quantity. Enter **x** as the Short Name and **m** as the Units. This is the distance that the hoop spring is compressed or the extensible spring or rubber band is stretched.

5. Replace the hook end of the Dual-Range Force Sensor (DFS) with the hoop spring bumper.[1] Attach the DFS to the bumper launcher assembly as shown.

Note: Shown inverted for assembly.

6. Attach the bumper launcher assembly near one end of a Vernier dynamics track. Place the cart on the track and position it so that the cart bumper just makes contact with the hoop spring bumper. Note the position of the rear of the cart; this represents the 0 value of spring compression. It may help to adjust the track adaptor so that 0 position falls on a "convenient" value on the scale. It is important that you sight the scale from a position directly above the cart so as to avoid parallax error.

7. Push gently on the hoop with the cart. If the force is negative, reverse the direction in the data-collection program.

8. With the cart just touching the hoop spring bumper, zero the force sensor.

9. Begin data collection. Without compressing the spring, enter **0** for the value of x.

[1] If this type spring is not available, your instructor will show you how to use an alternate arrangement to collect data for this experiment.

10. Now, compress the spring in equal increments, collecting a data point each time, until you have sufficient data points to determine a relationship. Be careful not to compress the hoop spring more than half of its diameter; doing so will damage the spring.

EVALUATION OF DATA

1. If the relationship between force and the change in length appears to be linear, fit a straight line to your data. If possible, print a copy of your data table and graph.
2. Write the equation that represents the relationship between the force, F, applied to the spring and its change in length, x.
3. Write a statement that describes the relationship between the force you applied to the spring and the extent to which it was compressed (stretched).
4. If you were to double the change in length of the spring, what effect would that have on the force required to produce this change?
5. Examine the slope of the graph (units as well as numerical value). Write a statement describing what the slope tells you about the spring. In your class discussion, you will give a name to this quantity.
6. Now write the general equation describing the relationship between the applied force and the change in the length of the spring.
7. As you learned in kinematics experiments, the area under a curve can also have physical significance. In this case, the area represents the work that was done on the spring as you applied a force parallel to the change in the spring's length. This work you did increased the *elastic energy* stored in the spring. Noting the shape of the area, write an equation relating the elastic energy to the applied force, F, and the change in length of the spring, x.
8. Now, replace the variable F with an equivalent expression from the general equation you wrote in Step 4. After you simplify the equation you will have derived a general equation for the elastic energy stored in a spring.
9. Determine the energy stored by the spring when it was compressed 0.020 m. Do this both algebraically, using the equation you derived in Step 7, and graphically, by finding the area under this portion of the curve. How do these values compare?
10. If you were to double the change in length of the spring, what effect would this have on the energy stored by the spring? Explain.

Experiment
7

INSTRUCTOR INFORMATION

Energy Storage and Transfer: Elastic Energy

PART 1 – ELASTIC ENERGY

This is the first in a series of experiments developing the concept of energy storage and transfer. It is written with the assumption that the instructor will engage the students in discussions at critical junctures. These discussions can take place with the entire class or with individual lab groups. The icon at left indicates where these discussions should occur.

The Microsoft Word files for the student pages can be found on the CD that accompanies this book. See *Appendix A* for more information.

OBJECTIVES

In this experiment, the student objectives include

- Determine the relationship between the applied force and the deformation of an elastic object (spring or rubber band).
- Determine an expression for the elastic energy stored in spring or rubber band that has been compressed or stretched.

During this experiment, you will help the students

- Recognize that the force required to compress a spring is proportional to the deformation.
- Recognize that the slope of a graph of force *vs.* compression (stretch) is the spring constant, k, for the object undergoing deformation.
- Recognize that the area under the graph of force *vs.* change in length is the elastic energy stored in the spring.

REQUIRED SKILLS

In order for student success in this experiment, it is expected that students know how to

- Set up the data-collection mode in Events with Entry.
- Reverse the direction of the Dual-Range Force Sensor in the software.
- Zero the Force Sensor.
- Interpret the area under the curve on a graph. This is addressed in Activity 3.

EQUIPMENT TIPS

It is best to use a spring for this lab. The problem with rubber bands is that their effective spring constant, k, tends to change with use. The Vernier Bumper and Launcher Kit (BLK) comes with two hoop springs (~26 N/m and 75 N/m). You can eliminate problems with sensor cables by mounting the Dual-Range Force Sensor (DFS) upside down to the Dynamics Track Bracket.

Then attach the assembly to the track as directed by the instructions that accompany the BLK as shown in Figures 1 and 2.

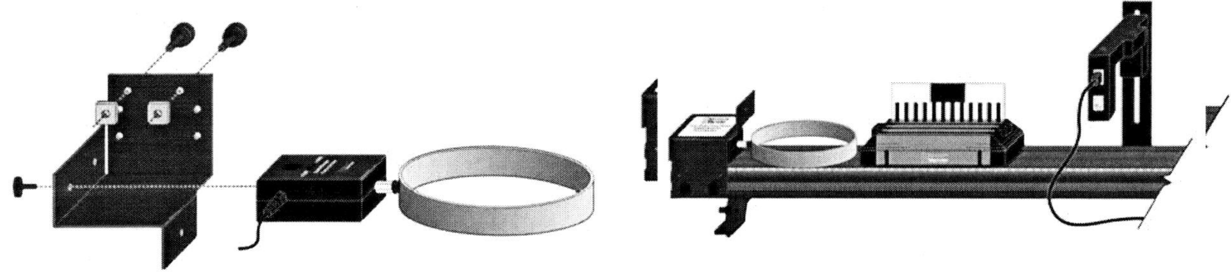

Figure 1 *Figure 2*

Note: Shown inverted for assembly.

Once the general equation for elastic energy is derived, either of the hoop springs could be used in the subsequent labs in which students investigate the transfer of energy from the elastic account to kinetic and then gravitational accounts.

One can also connect the hoop spring to the DFS and affix the sensor to the Dynamics Track using the aluminum rod that comes with the DFS and a 1/4" hex nut, as shown in Figure 3.

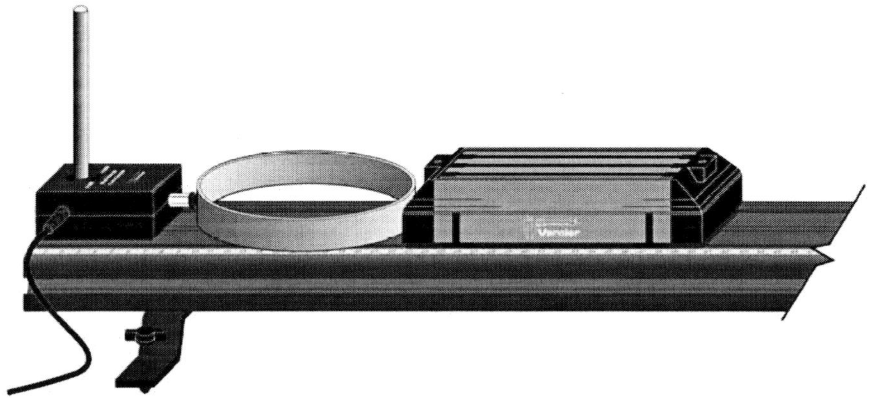

Figure 3

If the hoop springs are not available, a lightweight (~ 3 N/m) extensible spring could be used instead. Replace Steps 5–10 in the procedure with the following:

5. Tie a short (~30 cm) length of string to one end of the spring. Attach the other end to the dynamics cart. Attach the free end of the string to the track endstop.

6. Attach another short length of string to the other end of the dynamics cart. Tie a loop in the free end of the string; attach this to the hook on the Dual-Range Force Sensor.

7. Place the sensor on the track and gently pull on it to take any slack out of the strings and barely begin to stretch the spring. Now, zero the force sensor (see Figure 4).

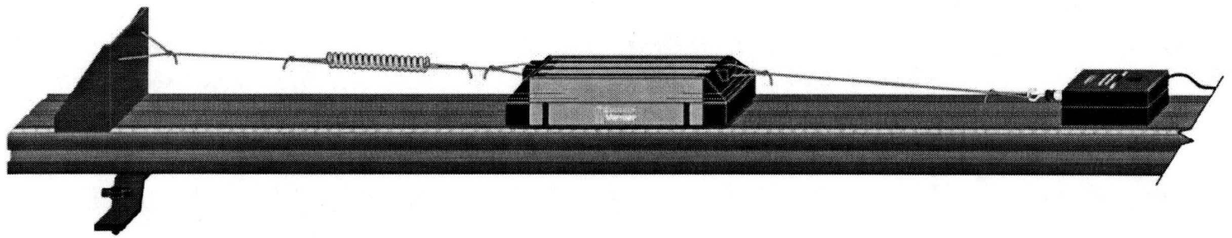

Figure 4

8. Begin data collection. Enter **0** for the value of stretch, *x*.

9. Now, stretch the spring, collecting data points until you have sufficient data points to determine a relationship. Be careful not to stretch the spring too far; doing so will damage the spring.

If neither of these options is available, one can obtain less ideal, but acceptable results by using a strong rubber band stretched between posts (ring stands clamped to the table) on either side of the track as a substitute for the bumper band spring. The hook on the force sensor is replaced with the bumper. The student places the DFS on the track and pushes on the cart. The sensor is zeroed when the applied force causes the portion of the band that contacts the cart to just touch the other portion of the rubber band.

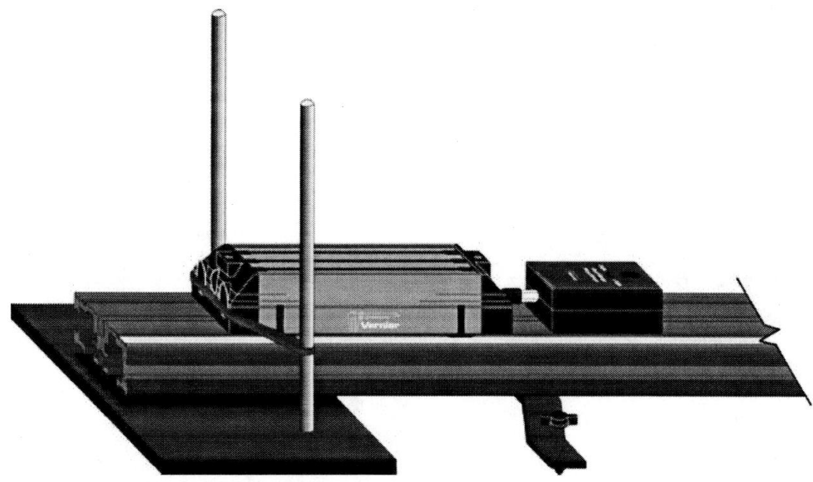

Figure 5

The difficulty with this arrangement is that the rubber band does not consistently exhibit a linear relationship between the applied force and stretch. The directions in the student version of the lab can be used with the exception that you will have to determine appropriate increments of the stretch to use. The analysis of the data for the rubber band is found in the extension at the end of the lab.

PRE-LAB DISCUSSION

This series of experiments best follows a brief discussion about the role energy plays in the changes we observe in matter. Rather than use the "forms of energy" locution, it is more straightforward to describe energy using a unitary concept – there is only one kind of energy, but there are many different ways it can be stored in a system. Like money in a bank or credit union, energy can be stored in a system in different "accounts" and can be transferred between system and surroundings in various ways (e.g., via work or heat) without changing its identity. Avoid identifying all the various accounts or the expressions for elastic, gravitational, and kinetic energy; these will be developed naturally by the analysis of the data the students collect in this series of experiments.

Prior to distributing the lab to the students, initiate a discussion with the students about energy along the lines of the introduction in the student version of the experiment. Explain to them that they are about to engage in a series of experiments in which they will work to derive quantitative expressions for the various ways energy is stored in a system. It would be helpful to have a rubber band (and, if possible, a dart gun) to help students see that an object capable of a reversible deformation can store elastic energy. Inform the students that they first need to find a relationship between the force applied to the spring or rubber band and the change in length (the distance the spring is compressed or stretched). Show students the experimental apparatus and ask them to predict the relationship between the force and the change in length. Then, ask the students how they might go about collecting the data that would allow them to test their prediction. They should be able to suggest a procedure similar to the one provided in the student version of the lab.

LAB PERFORMANCE NOTES

Advise the students to take care when they read the position of the cart to determine the change in length of the spring. They should not compress the hoop spring more than half of its diameter, nor should they stretch the extensible spring too far; doing so will damage the spring.

After students have stopped collecting data, ask them to work on Evaluation Questions 1–5, then pause for in-class discussion.

SAMPLE RESULTS

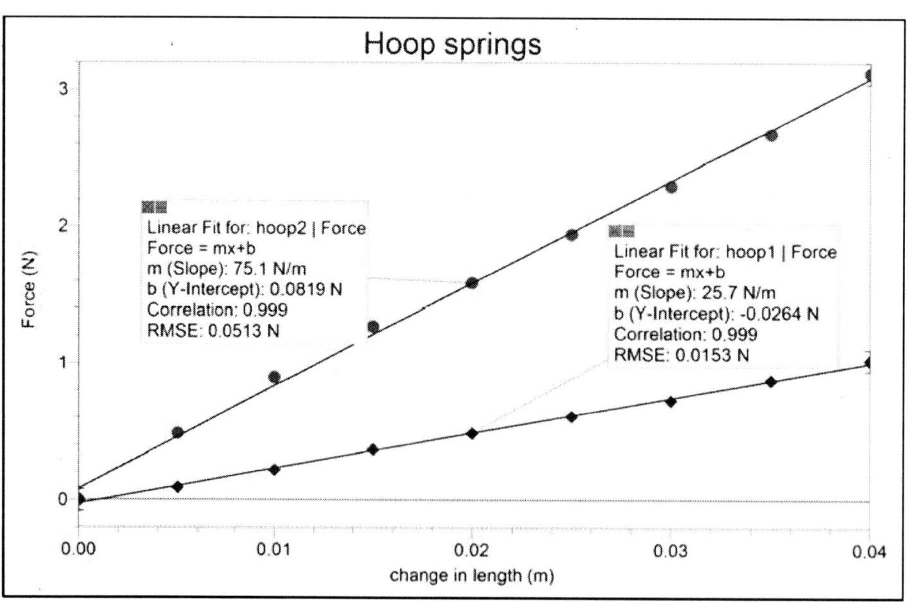

Figure 6 With hoop springs from the Bumper and Launcher Kit

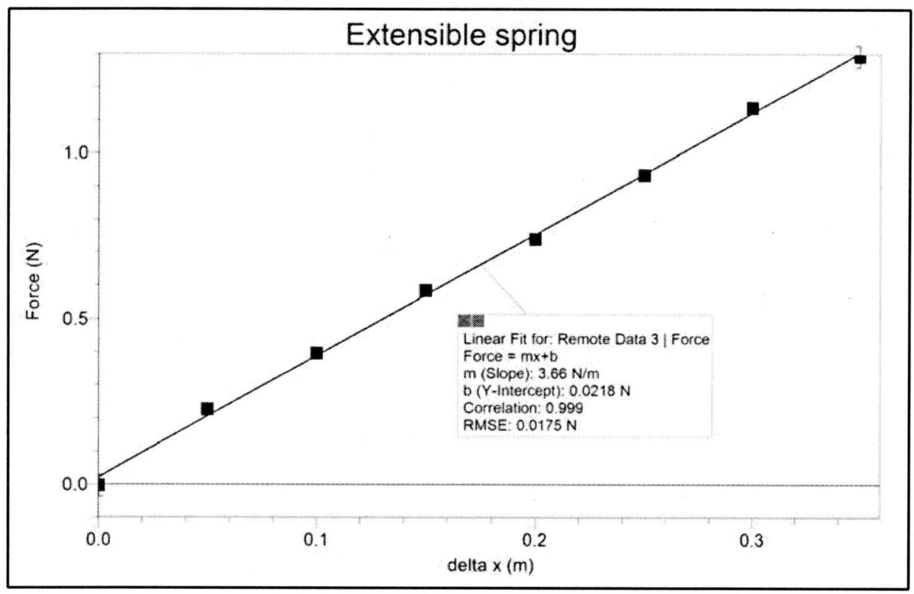

Figure 7 With an extensible spring

POST-LAB DISCUSSION

Steps 1–4

Have the students prepare presentations of their findings (whiteboard, chart paper, etc). These should include a sketch of the graph and the equation of their line of best fit. Suggest to the students that the vertical intercept is negligible and can be dropped from their equation of the line. Make sure that they have included the units on their value of slope. For the line in Figure 7, the equation is $F = (3.66\,\text{N/m})x$.

In the discussion of their findings, try to get the students to say that the force applied to the spring is proportional to the change in its length. Therefore, if the change in length were doubled, the force applied would also have to be doubled.

Steps 5–6

When asked about the meaning of the slope, students might say that it is "the number of newtons per meter." Push them to give a more meaningful statement, such as "It is the force required to compress (stretch) the spring one meter." or "For every meter of compression (stretch), this much force must be applied." After they have a clear idea of what the slope means, tell them that this quantity is given the name *–spring constant*, and is represented by the variable, *k*.

For Step 6 students should be able to write the general equation $F = kx$ to describe the relationship between the applied force and the change in length of the spring.

Steps 7–10

After students have worked out the general equation $F = kx$, display the graph (or simply sketch it on the board), shade in the area under the curve and point out that the area under the curve represents the work they did compressing (or stretching) the spring. The work they did was stored as elastic energy in the spring. This quantity can be found by

$$W = \Delta E_{el} = \tfrac{1}{2}F \cdot x$$

Next, suggest to students that they can substitute for F using the general equation they found in Question 5. Hopefully, they will conclude that

$$E_{el} = \frac{1}{2}kx^2$$

Answers to Step 9 will vary. Sample calculations for the stiffer hoop spring are shown below.

Algebraic solution	Graphical solution
$E_{el} = \frac{1}{2}(75.1\ \text{N/m})(0.02\ \text{m})^2$ $E_{el} = 0.0150\ \text{N}\cdot\text{m} = 0.0150\ \text{J}$	$E_{el} = \frac{1}{2}(1.58\ \text{N})(0.020\ \text{m})$ $E_{el} = 0.0158\ \text{N}\cdot\text{m} = 0.0158\ \text{J}$

In Step 10, students should recognize that since the energy stored is proportional to the *square* of the change in length, doubling x results in 4x the energy stored.

EXTENSION-RUBBER BAND

Below is a graph of force *vs.* stretch for a 7 inch rubber band stretched between the support rods of two ring stands.

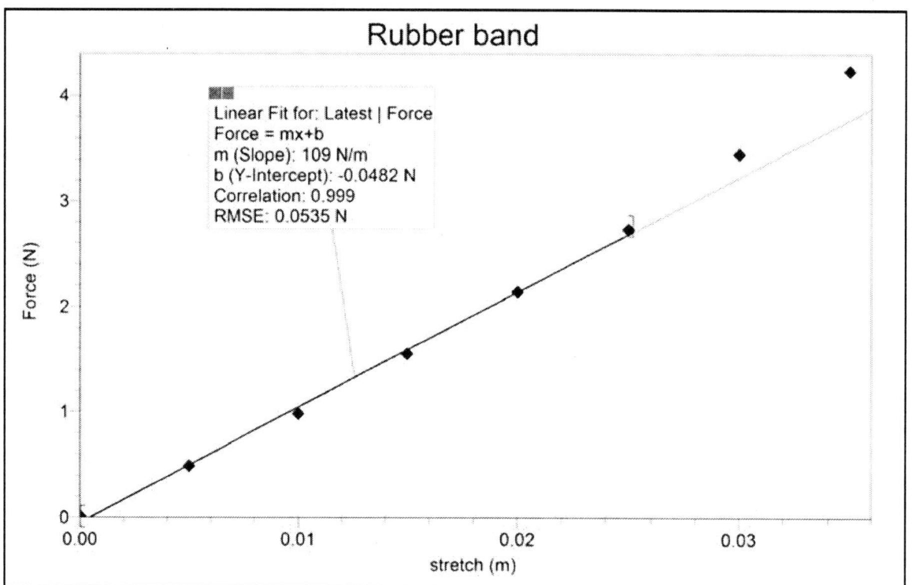

Figure 8 With a rubber band

Note that for $x > 2.5$ cm, the data begin to deviate from a linear relationship. For the purposes of this lab, the linear fit did not include the last two points. If students were to do this investigation as an extension, or if a group used this apparatus instead of a spring, the analysis of data would provide an opportunity to discuss the limitations of the linear relationship used to fit the data.

You should also keep in mind that changing the configuration of the support rods that hold the band would most likely affect the value of the effective spring constant, k. Before performing the second and third parts of this suite of labs, students should quickly collect F *vs.* x data again and use the most recent value of the slope of the graph in their subsequent analysis.

Energy Storage and Transfer: Kinetic Energy

PART 2 – KINETIC ENERGY

In the first of this series of labs exploring the role of energy in change, you found that the energy stored in an elastic system was proportional to the square of the change in the length of the spring or rubber band deformed by the applied force. We called the energy stored in this way *elastic energy*.

This energy can be transferred to another object to produce a change–for example, when the spring is released, it can launch a toy dart. It seems reasonable that the more the spring is compressed, the greater the change in speed it can impart to the object. As the spring or band returns to its original shape, it transfers energy to the moving object. We say that the moving object stores energy in an account called *kinetic energy*. It seems reasonable that an object's kinetic energy is a function of its mass and velocity. It would be useful to determine a quantitative relationship between the kinetic energy and its velocity for a given mass.

OBJECTIVES

In this experiment, you will

- Recognize that the energy stored in an elastic system (spring, rubber band) can be transferred to another object, resulting in a change in the state of that object.
- Determine an expression for the kinetic energy stored in a moving body.

MATERIALS

Vernier data-collection interface
Logger *Pro*
Vernier Photogate
cart with Cart Picket Fence
Vernier Bumper and Launcher
Kit (recommended) **or**
lightweight (34 N/m) extensible spring **or**
heavy rubber band

PRE-LAB INVESTIGATION

1. Examine the same elastic system (hoop or extensible spring, rubber band) that you used in the previous experiment. Change the length of the spring varying amounts, allow it to launch the dynamics cart, and note the speed of the cart.

2. On the axes at right, sketch a graph of kinetic energy *vs*. the velocity of the cart, assuming that the energy in the spring is completely transferred to the cart.

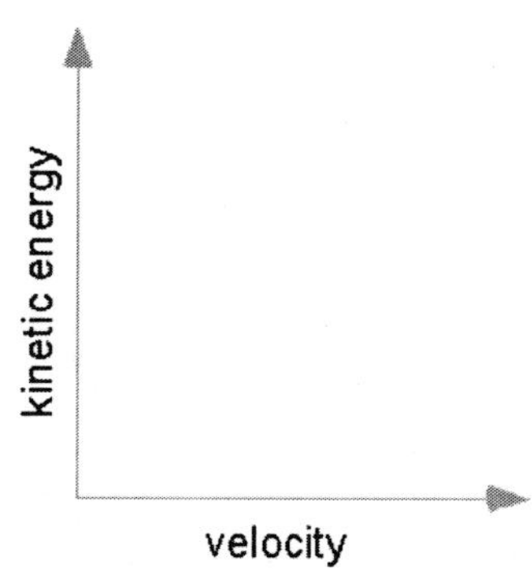

PROCEDURE

1. Attach the same spring that you used in Experiment 7 to the Dynamics Track Bracket, then mount the bracket on the end of a Dynamics Track.[1]

2. Obtain the value of the spring constant, *k*, for the spring you used in the previous experiment.[2]

3. In your pre-lab discussion, you agreed on a way to adjust the level of the dynamics track to minimize the effect of friction on the transfer of energy from elastic to kinetic accounts. Make the necessary adjustment now.

4. Attach the Cart Picket Fence to the dynamics cart. Set up a photogate near the dynamics cart so that the flag on the fence interrupts the sensor shortly after the cart leaves the spring, as shown in Figure 1.

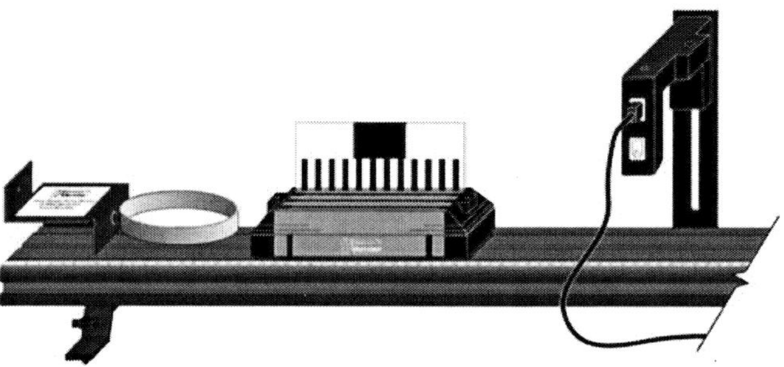

Figure 1

5. Determine the mass of your cart, fence, and any additional masses your instructor may have assigned you to use.

6. Start the data-collection program. Set up the photogate for Gate Timing. The length of the flag passing through the photogate is 0.05 m. The Gate State should read **Blocked** when the flag interrupts the sensor and **Unblocked** when it is moved beyond the sensor.

7. For this experiment, you will first collect velocity data for varying changes in spring length, *x*, and record these manually in your lab notebook. The change in length, *x*, is the distance that the hoop spring is compressed (or the extensible spring is stretched). Later, you will construct a Logger *Pro* file (or use one your instructor provides) to analyze the data you have collected.

8. Consult with your instructor about appropriate values of change in length, *x*, for your apparatus. Begin data collection. Perform several trials for each change in length, *x*, and keep the three most consistent velocity values. Be sure to stop data collection after each trial and then begin again before the next trial. As before, it is important that you sight the scale from a position directly above the cart so as to avoid parallax error.

---

1 If you used an extensible spring or a rubber band in the previous lab, your instructor will show you an alternative way to set up the apparatus.

2 If you used a rubber band to store energy in the previous lab, you will need to quickly collect *F* vs. *x* data again to make sure that the band's effective "spring constant" has not changed since the last use.

EVALUATION OF DATA

1. To evaluate the relationship between energy and velocity, disconnect the photogate and choose New from the File menu. You will need to set up Logger *Pro* with the necessary columns of **x, v1, v2, v3**, where **x** is the change in length of the spring and **v1**, **v2**, and **v3** are the three consistent velocities of the cart as it passes through the photogate. You will enter these data manually. You also need two calculated columns, **ave v** (the average of the three velocities) and $\mathbf{E_{el}}$ (elastic energy). Your instructor may guide you in the design of this file.
2. Enter the data for change in length, *x*, and the three velocity values you obtained during the experiment. Be sure to include (0,0) in your table.
3. Discuss how the system energy is stored once the spring returns to its original shape. Make a new manual column for $\mathbf{E_k}$ (kinetic energy) and then fill in the $\mathbf{E_k}$ column with the appropriate values.
4. Create a graph of kinetic energy *vs.* average velocity. Write a statement that describes the relationship between the kinetic energy of the cart system and its average velocity.
5. If your graph of E_k vs. *average v* is not linear, then you need to take steps to modify a column so as to produce a linear relationship. When you have done so, save your file and (if possible) print a copy of your original and then linearized graph.
6. Write the equation of the line that best fits your linearized graph. Take special care to get the units of the slope of this line correct.
7. In the previous lab, the SI unit of energy, joule, was defined as a N m. From your knowledge of the relationship between force and acceleration, express the joule in terms of its fundamental units (kg, m, s). Simplify the units of your slope as much as possible.
8. Prepare a summary of your analysis of data (whiteboard or chart paper). Include the original and linearized graph and the equation of the line of best fit. Also report the system mass. In your class discussion you will compare your findings with those of other groups.
9. When a quantity (in this case, kinetic energy) is a function of more than one variable, it is usually the case that the slope of the graph is related to the variable that was held constant during the experiment. Write a statement describing the relationship between the slope of the line and the mass of the system.
10. Write a general equation describing the relationship between an object's kinetic energy and its velocity.

EXTENSIONS

1. Suppose that you used a spring with a *k* value of 15.0 N/m to launch a lighter (250 g) dynamic cart. If the spring were compressed 0.0200 m, what should be the velocity of the cart when it left the spring?
2. If you were to double the compression of the spring in the previous question, what effect would that have on the cart's velocity? Explain.
3. Suppose you had not adjusted the track for friction as you did in the lab, and 20% of the elastic energy in Extension 1 went into internal energy instead. Determine the velocity of the cart with the reduced amount of kinetic energy.

INSTRUCTOR INFORMATION

Energy Storage and Transfer: Kinetic Energy

PART 2 – KINETIC ENERGY

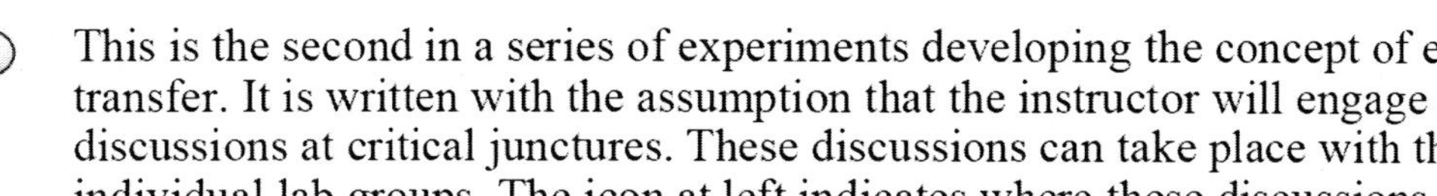

This is the second in a series of experiments developing the concept of energy storage and transfer. It is written with the assumption that the instructor will engage the students in discussions at critical junctures. These discussions can take place with the entire class or with individual lab groups. The icon at left indicates where these discussions should occur.

The Microsoft Word files for the student pages can be found on the CD that accompanies this book. See *Appendix A* for more information.

OBJECTIVES

In this experiment, the student objectives include

- Recognize that the energy stored in an elastic system (spring, rubber band) can be transferred to another object, resulting in a change in the state of that object.
- Determine an expression for the kinetic energy stored in a moving body.

During this experiment, you will help the students

- Recognize that the slope of a graph of kinetic energy *vs.* the square of velocity is proportional to the mass of the moving object.
- Recognize that the kinetic energy of a moving body is proportional to the square of the velocity.

REQUIRED SKILLS

In order for student success in this experiment, it is expected that students know how to

- Set up a photogate in Gate mode. This is addressed in Activity 4.
- Create manual and calculated columns in the software. This is addressed in Activity 1
- Linearize data. This is addressed in Activity 1.

EQUIPMENT TIPS

Students should use the same spring for this lab as they did in Experiment 7 when dealing with elastic energy. The hoop springs or extensible spring recommended for use in the previous lab have spring constants that are quite stable. The problem with rubber bands is that their effective spring constant, *k*, tends to change with use. If a rubber band is the only option available, then students should repeat the measurement of force *vs.* stretch with the Dual-Range Force Sensor (DFS) to make sure that they have the current value of the effective spring constant for the rubber band. This can be accomplished in a few minutes.

The hoop spring can be left connected to the DFS as shown in the previous lab. However, one can simply attach the hoop spring directly to the Dynamics Track Bracket, as shown in Figure 1.

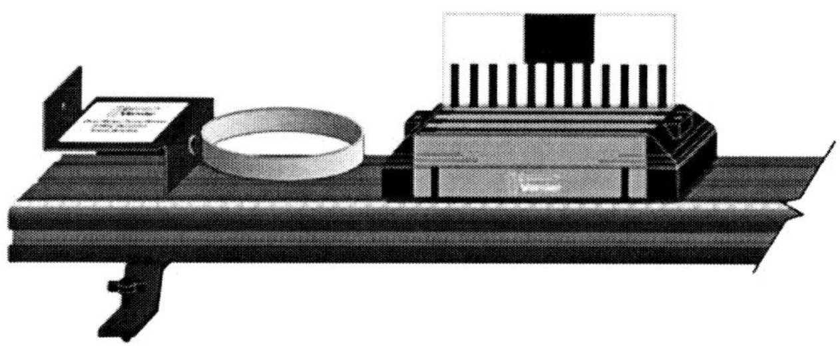

Figure 1

To determine the velocity of the cart, place the cart picket fence on the cart and position a photogate so that the 5 cm flag interrupts the photogate shortly after the cart no longer contacts the spring, as shown in Figure 2.

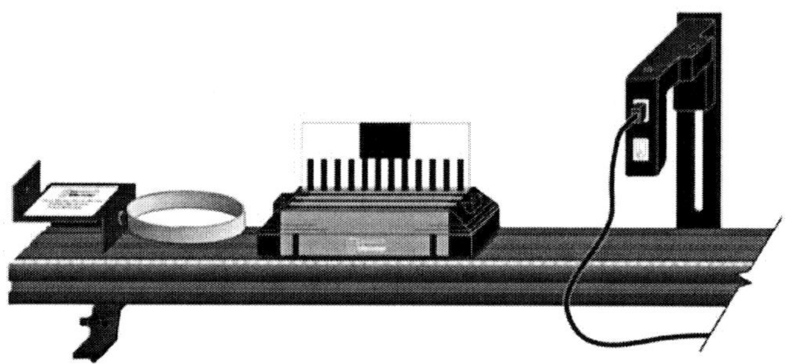

Figure 2

Note: An adjustment to this apparatus is necessary to insure good results. In order to make the assumption that all the elastic energy stored in the spring is transferred to the kinetic energy of the cart, one must minimize the energy "lost" due to frictional interactions. To accomplish this, the end of the track with the spring should be raised enough so that the cart, given a gentle push, will travel to the far end of the track at nearly constant velocity. The leveling screws are sufficient for this task.

PRE-LAB DISCUSSION

This is the second in the series of experiments investigating energy storage and transfer. Prior to distributing the lab to the students, demonstrate how the compressed spring can launch the dynamics cart when released. In this initial discussion with the students, try to get them to describe the transfer of energy from the compressed spring (elastic) to the moving cart (kinetic). Explain to them that they are about to derive a quantitative expression of the kinetic energy of the moving cart. If students say that they already know that this expression is $E_k = 1/2\,mv^2$, ask them "How do you know?" Inform them that this experiment affords them the opportunity to determine this expression experimentally.

Elicit from the students what system variables might be involved in the cart's kinetic energy. They should be able to suggest that the mass and speed (velocity) are the relevant variables. Inform the students that they will be investigating the relationship between kinetic energy and cart velocity. Furthermore, by varying the mass in subsequent trials, they can determine the effect of mass in this relationship.

With the track level, give the cart a gentle push so that the students can see the cart slow down. Ask for a possible explanation. Students are likely to suggest that energy is "lost" due to friction. Re-direct their thinking to another view: as the cart slows, its kinetic energy is transferred to another account, which we call "internal energy". While the effects of this account are not easy to detect, (the energy is now stored in the increased random thermal motion of the particles that make up the cart and track) this description is superior to stating that energy is "lost."

Now ask students to suggest a way to compensate for this transfer from kinetic to internal energy so that they can make the assumption that the cart's kinetic energy is equivalent to the elastic energy in the spring. You might need to show them the trick of elevating the end of the track with the spring to the point that the cart, gently launched by the spring, moves down the track with constant velocity. You can also suggest that the students should position the photogate relatively close to the point where the cart leaves the spring. At this point, you should distribute the lab instructions.

LAB PERFORMANCE NOTES

Students should record the change in spring length, x, (used to determine the elastic energy that is transferred to the cart) and the velocities for multiple trials in their lab notebooks. Multiple trials are recommended because it is somewhat tricky to release the cart cleanly.

Computer Data Collection

Students should connect the photogate to a digital port on the interface and start Logger *Pro*. They should click on the Gate State in the status bar for the sensor; this brings up the sensor information window. Students should click on the icon of the photogate, choose "Gate Timing," then close the window. The default distance value of 0.05 m should be used.

LabQuest App Data Collection

Students should connect the photogate to a digital port on the LabQuest and change the mode from Photogate Timing to Gate. The default distance value of 0.05 m should be used.

With either method, the Gate State should read [Blocked] when the flag interrupts the sensor and [Unblocked] when it has moved beyond the sensor. When they start data collection, they can compress the spring, release the cart, then read the velocity.

Remind the students that they should stop data collection after each trial and then begin again before the next trial. We recommend that students perform several trials for every value of x and keep the three most consistent values of velocity. Students should start with a compression of 1.0 cm (hoop springs) and increment x by 0.5 cm. For the extensible spring, the initial stretch should be 10 cm with increments of 5 cm thereafter. The data for seven values of x can be collected in 10–15 minutes; this number is recommended to insure that students see the relationship between energy and velocity is not linear. Depending on the time you wish to devote to the experiment, you can have students repeat the experiment, adding masses to the cart, or you can assign different masses to groups in the class. In this way, during the reporting of lab results, students can see the effect of the cart mass on the relationship between kinetic energy and velocity.

EVALUATION OF DATA

Now, the students need to construct a Logger *Pro* data analysis file with the interface disconnected from the computer. There should be manual columns for **x**, **v1, v2 and v3,** where **x** is the change in length of the spring and **v1**, **v2**, and **v3** are the three consistent velocities of the cart as it passes through the photogate. Students also need calculated columns for average velocity, **ave v**, and elastic energy, $\mathbf{E_{el}}$. Later, students will be prompted to conclude that the energy stored in the spring was transferred completely to kinetic energy. This experiment provides an opportunity for students to use a defined user parameter **spring constant**, although they can simply enter that value directly into the formula for elastic energy. A sample Logger *Pro* file (Ek vs v.cmbl) is provided on the CD that accompanies this book for guidance. To save time, you may provide this file for students to use directly. If you exercise this option, make sure that students examine the equations for the calculated columns. In any event, once students have concluded that the system energy is now stored as kinetic energy, they should create a new manual column for kinetic energy and fill in the values.

SAMPLE RESULTS

Below are sample data using the lighter hoop spring.

500 g						
x (m)	v1 (m/s)	v2 (m/s)	v3 (m/s)	ave v (m/s)	E_{el} (J)	E_k (J)
0	0	0	0	0	0	0
0.0100	0.0621	0.0538	0.0568	0.0576	0.00129	0.00129
0.0150	0.0921	0.953	0.0977	0.950	0.00289	0.00289
0.0200	0.135	0.134	0.134	0.134	0.00514	0.00514
0.0250	0.170	0.173	0.176	0.173	0.00803	0.00803
0.0300	0.208	0.212	0.206	0.209	0.0116	0.0116
0.0350	0.246	0.243	0.243	0.244	0.0157	0.0157
0.0400	0.279	0.277	0.279	0.278	0.0206	0.0206

Below is a graph of sample data collected using the lighter hoop spring and a standard dynamics cart.

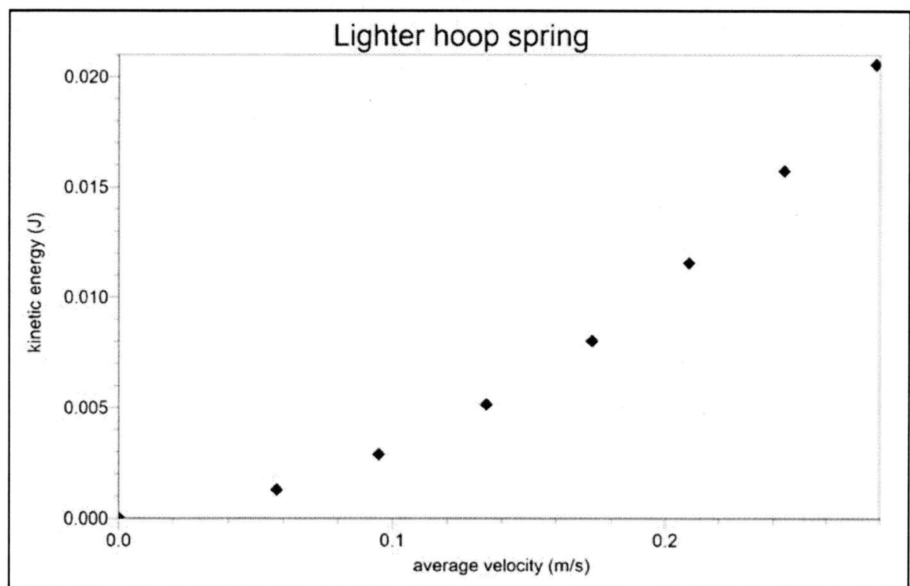

Figure 3 Lighter hoop spring and dynamics cart

This plot of kinetic energy *vs*. velocity appears to be a top-opening parabola. By this point, students should be familiar with "linearizing" a graph. If not, or if they need a reminder, see 10-1 Linearization.cmbl in the Tutorial folder in Logger *Pro*.

Students should add a new calculated column for velocity squared and display the graph of kinetic energy *vs*. velocity squared on a new page (see 08 Ek v velocity.cmbl). The linearized graph is displayed below.

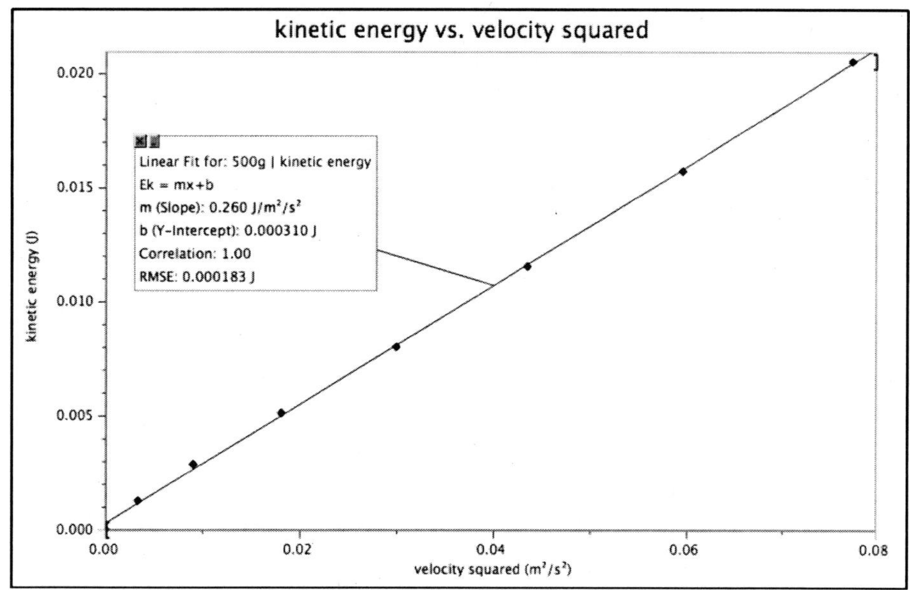

Figure 4 Linearized graph for the cart

A linearized graph for the cart plus 200 g is shown below.

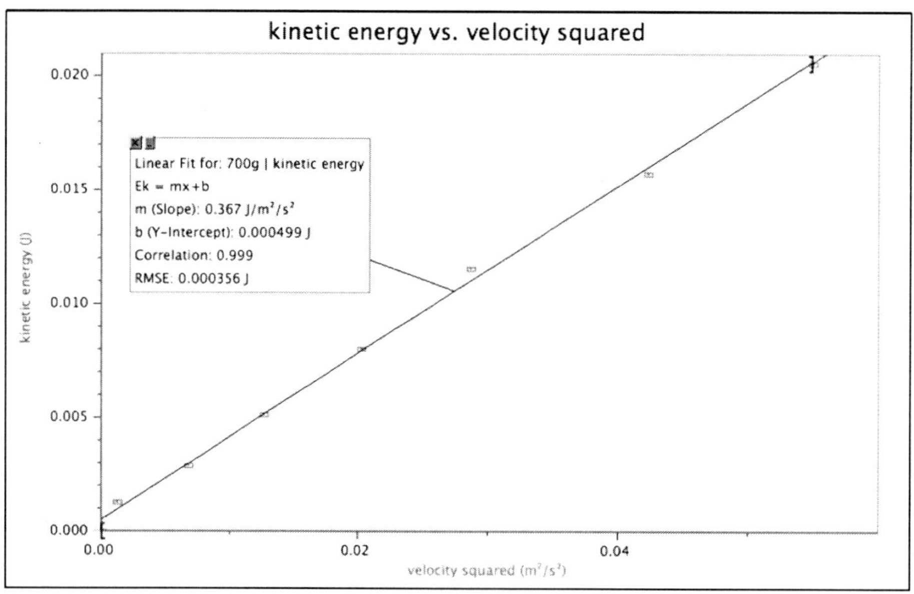

Figure 5

POST-LAB DISCUSSION

Steps 1–6

Guide students in step 3 to the conclusion that the elastic energy once stored in the spring is now stored as the kinetic energy of the cart. During Step 4, students should recognize that the graph of kinetic energy *vs.* average velocity appears to be a top-opening parabola, suggesting that kinetic energy is proportional to the square of the velocity ($E_k \propto v^2$). Their prediction is borne out in the linearized graph (see Figure 4). If the students collected data carefully, the vertical intercept is negligible and can be dropped from their equation of the line.

The equation of the line in Figure 4 is $E_k = (0.260 \frac{\text{J}}{\text{m}^2/\text{s}^2})v^2$. Make sure that students have included the units on their value of the slope.

Step 7

Point out that in the previous lab, the joule was introduced as a unit of energy defined as a N· m. From their previous experience with Newton's Second law, students should be able to simplify the units of slope by replacing J with the fundamental units from which it is derived.

$$\frac{\text{J}}{\text{m}^2/\text{s}^2} \Rightarrow \frac{\text{N}\cdot\text{m}}{\text{m}^2/\text{s}^2} \Rightarrow \frac{(\text{kg}\cdot\text{m/s}^2)\cdot\text{m}}{\text{m}^2/\text{s}^2} \Rightarrow \text{kg}$$

Step 8

Have the students prepare presentations of their findings (whiteboard, chart paper, etc). These should include a sketch of both the original and linearized graph and the equation of their line of best fit.

Steps 9–10

Remind students that the slope of a graph is frequently related to a system variable that is held constant during the experiment. During the subsequent discussion, students should note that the plot of kinetic energy *vs*. velocity squared is linear, regardless of the mass of the cart. Students might notice that the slope of the line increases with the mass. With some guidance they should conclude that slope of the graph is approximately equal to half the mass of the cart (in kg). If they have not worked out the units before, now is the appropriate time to show that the units of the slope simplify to kilograms. So, students should be able to write an expression for the kinetic energy as a function of mass and velocity: kinetic energy equals half the mass times the square of the velocity, or

$$E_k = \tfrac{1}{2}mv^2$$

EXTENSIONS

1. For a compression of 0.0200 m, the energy stored should be

$$E_{el} = \tfrac{1}{2}kx^2$$

$$E_{el} = \tfrac{1}{2}(15.0\ \text{N/m})(0.0200\ \text{m})^2 = 0.00300\ \text{J}$$

Assuming all of this is transferred to kinetic energy, the velocity of the cart can be found by

$$E_k = \tfrac{1}{2}mv^2 \Rightarrow v = \sqrt{\frac{2E_k}{m}}$$

$$v = \sqrt{\frac{2(0.00300\ \text{J})}{0.250\ \text{kg}}} = 0.155\ \text{m/s}$$

2. See if you can get your students to reason as follows rather than substituting 0.0400 m into the equation for elastic energy and solving for velocity as if it were a totally new problem. Since the elastic energy is proportional to the square of the compression $E_{el} \propto x^2$ and the kinetic energy is proportional to the square of the velocity $E_k \propto v^2$, it follows that doubling the compression should double the cart's velocity.

3. If 20% of the elastic energy went to internal energy, then the kinetic energy of the cart would be only 80% of the value found in Extension 1. The new velocity would be

$$v' = \sqrt{\frac{2(0.00240\ \text{J})}{0.250\ \text{kg}}} = 0.139\ \text{m/s}$$

Energy Storage and Transfer: Kinetic Energy

PART 2 – KINETIC ENERGY

In the first of this series of labs exploring the role of energy in change, you found that the energy stored in an elastic system was proportional to the square of the change in the length of the spring or rubber band deformed by the applied force. We called the energy stored in this way *elastic energy*.

This energy can be transferred to another object to produce a change—for example, when the spring is released, it can launch a toy dart. It seems reasonable that the more the spring is compressed, the greater the change in speed it can impart to the object. As the spring or band returns to its original shape, it transfers energy to the moving object. We say that the moving object stores energy in an account called *kinetic energy*. It seems reasonable that an object's kinetic energy is a function of its mass and velocity. It would be useful to determine a quantitative relationship between the kinetic energy and its velocity for a given mass.

OBJECTIVES

In this experiment, you will

- Recognize that the energy stored in an elastic system (spring, rubber band) can be transferred to another object, resulting in a change in the state of that object.
- Determine an expression for the kinetic energy stored in a moving body.

MATERIALS

Vernier data-collection interface
LabQuest App
Vernier Photogate
cart with Cart Picket Fence

Vernier Bumper and Launcher Kit (recommended) **or**
lightweight (34 N/m) extensible spring **or**
heavy rubber band

PRE-LAB INVESTIGATION

1. Examine the same elastic system (hoop or extensible spring, rubber band) that you used in the previous experiment. Change the length of the spring varying amounts, allow it to launch the dynamics cart, and note the speed of the cart.

2. On the axes on the right, sketch a graph of kinetic energy *vs.* the velocity of the cart, assuming that the energy in the spring is completely transferred to the cart.

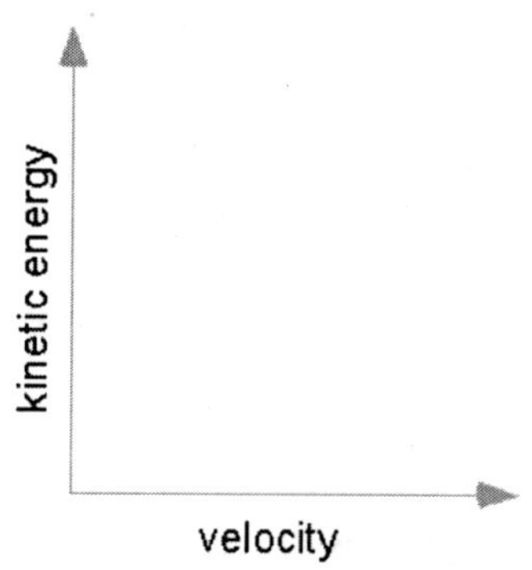

PROCEDURE

1. Attach the same spring that you used in Experiment 7 to the Dynamics Track Bracket, then mount the bracket on the end of a Dynamics Track.[1]

2. Obtain the value of the spring constant, k, for the spring you used in the previous experiment.[2]

3. In your pre-lab discussion, you agreed on a way to adjust the level of the dynamics track to minimize the effect of friction on the transfer of energy from elastic to kinetic accounts. Make the necessary adjustment now.

4. Attach the Cart Picket Fence to the dynamics cart. Set up a photogate near the dynamics cart so that the flag on the fence interrupts the sensor shortly after the cart leaves the spring, as shown in Figure 1.

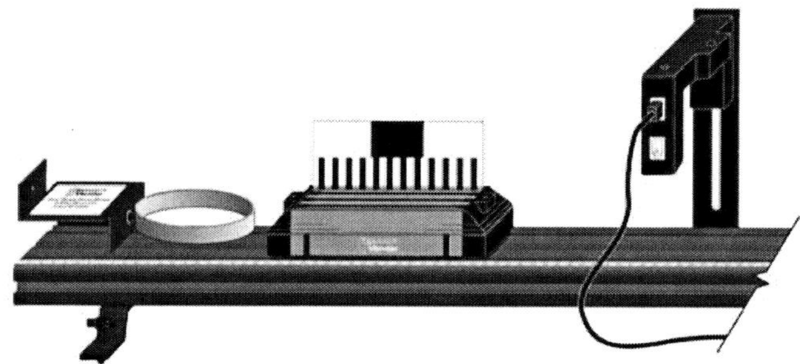

Figure 1

5. Determine the mass of your cart, fence, and any additional masses your instructor may have assigned you to use.

6. Start the data-collection program. Set up the photogate for Gate Timing. The length of the flag passing through the photogate is 0.05 m. The Gate State should read **Blocked** when the flag interrupts the sensor and **Unblocked** when it is moved beyond the sensor.

7. For this experiment, you will first collect velocity data for varying changes in spring length, x, and record these manually in your lab notebook. The change in length, x, is the distance that the hoop spring is compressed (or the extensible spring is stretched).

8. Consult with your instructor about appropriate values of change in length, x, for your apparatus. Begin data collection. Perform several trials for each change in length, x, and keep the three most consistent velocity values. Be sure to stop data collection after each trial and then begin again before the next trial. As before, it is important that you sight the scale from a position directly above the cart so as to avoid parallax error.

EVALUATION OF DATA

1. To evaluate the relationship between energy and velocity, disconnect the photogate and choose New from the File menu. Tap the Table tab to set up the columns **x** (the change in

---

[1] If you used an extensible spring or a rubber band in the previous lab, your instructor will show you an alternative way to set up the apparatus.

[2] If you used a rubber band to store energy in the previous lab, you will need to quickly collect F vs. x data again to make sure that the band's effective "spring constant" has not changed since the last use.

length of the spring) and **v** (the average of three consistent velocities of the cart as it passes through the photogate). You will enter these data manually. You also need a calculated column $\mathbf{E_{el}}$ (elastic energy). Your instructor may guide you in the design of this file.

2. Enter the data for change in length, *x*, and the average of the velocity values you obtained for each value of *x* during the experiment. Be sure to include (0,0) in your table.
3. Discuss how the system energy is stored once the spring returns to its original shape. Then you can change the header for the column $\mathbf{E_{el}}$ (elastic energy) to the appropriate form of energy.
4. Tap the Graph tab; from the Graph menu choose to show one graph with kinetic energy on the vertical axis and velocity on the horizontal. Write a statement that describes the relationship between the kinetic energy of the cart system and its average velocity.
5. If your graph of E_k vs. v is not linear, then you need to take steps to modify a column so as to produce a linear relationship. When you have done so, save your file and (if possible) print a copy of your original and then linearized graph.
6. Write the equation of the line that best fits your linearized graph. Take special care to get the units of the slope of this line correct.
7. In the previous lab, the SI unit of energy, joule, was defined as a N m. From your knowledge of the relationship between force and acceleration, express the joule in terms of its fundamental units (kg, m, s). Simplify the units of your slope as much as possible.
8. Prepare a summary of your analysis of data (whiteboard or chart paper). Include the original and linearized graph and the equation of the line of best fit. Also report the system mass. In your class discussion you will compare your findings with those of other groups.
9. When a quantity (in this case, kinetic energy) is a function of more than one variable, it is usually the case that the slope of the graph is related to the variable that was held constant during the experiment. Write a statement describing the relationship between the slope of the line and the mass of the system.
10. Write a general equation describing the relationship between an object's kinetic energy and its velocity.

EXTENSIONS

1. Suppose that you used a spring with a *k* value of 15.0 N/m to launch a lighter (250 g) dynamic cart. If the spring were compressed 0.0200 m, what should be the velocity of the cart when it left the spring?
2. If you were to double the compression of the spring in the previous question, what effect would that have on the cart's velocity? Explain.
3. Suppose you had not adjusted the track for friction as you did in the lab, and 20% of the elastic energy in Extension 1 went into internal energy instead. Determine the velocity of the cart with the reduced amount of kinetic energy.

INSTRUCTOR INFORMATION

Energy Storage and Transfer: Kinetic Energy

PART 2 – KINETIC ENERGY

This is the second in a series of experiments developing the concept of energy storage and transfer. It is written with the assumption that the instructor will engage the students in discussions at critical junctures. These discussions can take place with the entire class or with individual lab groups. The icon at left indicates where these discussions should occur.

The Microsoft Word files for the student pages can be found on the CD that accompanies this book. See *Appendix A* for more information.

OBJECTIVES

In this experiment, the student objectives include

- Recognize that the energy stored in an elastic system (spring, rubber band) can be transferred to another object, resulting in a change in the state of that object.
- Determine an expression for the kinetic energy stored in a moving body.

During this experiment, you will help the students

- Recognize that the slope of a graph of kinetic energy *vs*. the square of velocity is proportional to the mass of the moving object.
- Recognize that the kinetic energy of a moving body is proportional to the square of the velocity.

REQUIRED SKILLS

In order for student success in this experiment, it is expected that students know how to

- Set up a photogate in Gate mode. This is addressed in Activity 4.
- Create manual and calculated columns in the software. This is addressed in Activity 1
- Linearize data. This is addressed in Activity 1.

EQUIPMENT TIPS

Students should use the same spring for this lab as they did in Experiment 7 when dealing with elastic energy. The hoop springs or extensible spring recommended for use in the previous lab have spring constants that are quite stable. The problem with rubber bands is that their effective spring constant, *k*, tends to change with use. If a rubber band is the only option available, then students should repeat the measurement of force *vs*. stretch with the Dual-Range Force Sensor (DFS) to make sure that they have the current value of the effective spring constant for the rubber band. This can be accomplished in a few minutes.

The hoop spring can be left connected to the DFS as shown in the previous lab. However, one can simply attach the hoop spring directly to the Dynamics Track Bracket, as shown in Figure 1.

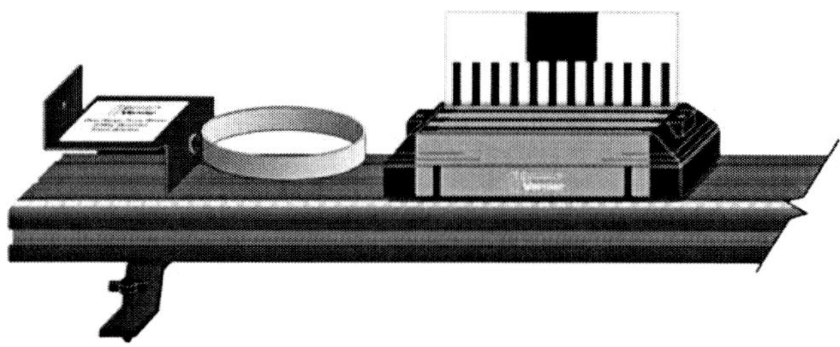

Figure 1

To determine the velocity of the cart, place the cart picket fence on the cart and position a photogate so that the 5 cm flag interrupts the photogate shortly after the cart no longer contacts the spring, as shown in Figure 2.

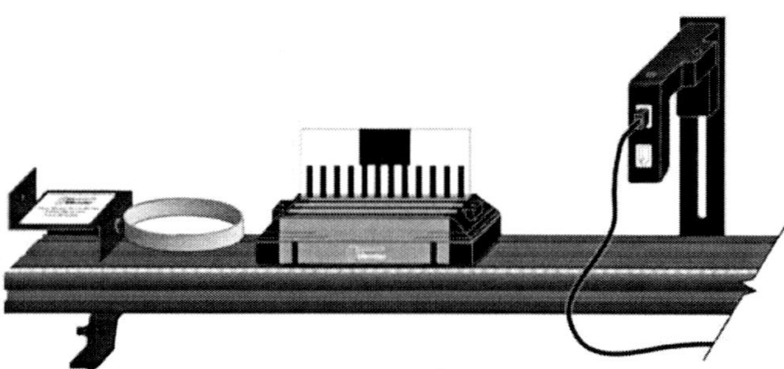

Figure 2

Note: An adjustment to this apparatus is necessary to insure good results. In order to make the assumption that all the elastic energy stored in the spring is transferred to the kinetic energy of the cart, one must minimize the energy "lost" due to frictional interactions. To accomplish this, the end of the track with the spring should be raised enough so that the cart, given a gentle push, will travel to the far end of the track at nearly constant velocity. The leveling screws are sufficient for this task.

PRE-LAB DISCUSSION

This is the second in the series of experiments investigating energy storage and transfer. Prior to distributing the lab to the students, demonstrate how the compressed spring can launch the dynamics cart when released. In this initial discussion with the students, try to get them to describe the transfer of energy from the compressed spring (elastic) to the moving cart (kinetic). Explain to them that they are about to derive a quantitative expression of the kinetic energy of the moving cart. If students say that they already know that this expression is $E_k = 1/2\,mv^2$, ask them "How do you know?" Inform them that this experiment affords them the opportunity to determine this expression experimentally.

Elicit from the students what system variables might be involved in the cart's kinetic energy. They should be able to suggest that the mass and speed (velocity) are the relevant variables. Inform the students that they will be investigating the relationship between kinetic energy and cart velocity. Furthermore, by varying the mass in subsequent trials, they can determine the effect of mass in this relationship.

With the track level, give the cart a gentle push so that the students can see the cart slow down. Ask for a possible explanation. Students are likely to suggest that energy is "lost" due to friction. Re-direct their thinking to another view: as the cart slows, its kinetic energy is transferred to another account, which we call "internal energy". While the effects of this account are not easy to detect, (the energy is now stored in the increased random thermal motion of the particles that make up the cart and track) this description is superior to stating that energy is "lost."

Now ask students to suggest a way to compensate for this transfer from kinetic to internal energy so that they can make the assumption that the cart's kinetic energy is equivalent to the elastic energy in the spring. You might need to show them the trick of elevating the end of the track with the spring to the point that the cart, gently launched by the spring, moves down the track with constant velocity. You can also suggest that the students should position the photogate relatively close to the point where the cart leaves the spring. At this point, you should distribute the lab instructions.

LAB PERFORMANCE NOTES

Students should record the change in spring length, x, (used to determine the elastic energy that is transferred to the cart) and the velocities for multiple trials in their lab notebooks. Multiple trials are recommended because it is somewhat tricky to release the cart cleanly.

Students should connect the photogate to a digital port on the LabQuest and change the mode from Photogate Timing to Gate. The default distance value of 0.05 m should be used.

The Gate State should read [Blocked] when the flag interrupts the sensor and [Unblocked] when it has moved beyond the sensor. When they start data collection, they can compress the spring, release the cart, then read the velocity.

Remind the students that they should stop data collection after each trial and then begin again before the next trial. We recommend that students perform several trials for every value of x and keep the three most consistent values of velocity. Students should start with a compression of 1.0 cm (hoop springs) and increment x by 0.5 cm. For the extensible spring, the initial stretch should be 10 cm with increments of 5 cm thereafter. The data for seven values of x can be collected in 10–15 minutes; this number is recommended to insure that students see the relationship between energy and velocity is not linear. Depending on the time you wish to devote to the experiment, you can have students repeat the experiment, adding masses to the cart, or you can assign different masses to groups in the class. In this way, during the reporting of lab results, students can see the effect of the cart mass on the relationship between kinetic energy and velocity.

EVALUATION OF DATA

We recommend using Logger *Pro* for the Evaluation of Data, whether you use computers or LabQuest to collect data. To evaluate the data using LabQuest App, students need to perform some of the calculations on their own to simplify the process of entering data.

Students need to disconnect the photogate and select New from the File menu to construct a LabQuest App data analysis file. There should be manual columns for **x** (change in length) and **v** (average of the three consistent velocities of the cart as it passes through the photogate). Students also need to create a calculated column for $\mathbf{E_{el}}$. When they do so, they should choose AX^B as the equation type, with x as the Column for X, half the value of the spring constant, k, for A and 2 for B. Later, students will be prompted to conclude that the energy stored in the spring was transferred completely to kinetic energy. A sample LabQuest App file (08 Ek v velocity.qmbl) is provided on the CD that accompanies this book for guidance. To save time, you may provide this file for students to use directly. If you exercise this option, make sure that students examine the equations for the calculated columns. In any event, once students have concluded that the system energy is now stored as kinetic energy, they should change the column header for elastic energy to kinetic energy, $\mathbf{E_k}$.

SAMPLE RESULTS

Below are sample data using the lighter hoop spring. **Note:** students need not enter values for $v1$, $v2$ and $v3$ in their LabQuest App file.

500 g						
x (m)	v1 (m/s)	v2 (m/s)	v3 (m/s)	ave v (m/s)	E_{el} (J)	E_k (J)
0	0	0	0	0	0	0
0.0100	0.0621	0.0538	0.0568	0.0576	0.00129	0.00129
0.0150	0.0921	0.953	0.0977	0.950	0.00289	0.00289
0.0200	0.135	0.134	0.134	0.134	0.00514	0.00514
0.0250	0.170	0.173	0.176	0.173	0.00803	0.00803
0.0300	0.208	0.212	0.206	0.209	0.0116	0.0116
0.0350	0.246	0.243	0.243	0.244	0.0157	0.0157
0.0400	0.279	0.277	0.279	0.278	0.0206	0.0206

Below is a graph of sample data collected using the lighter hoop spring and a standard dynamics cart.

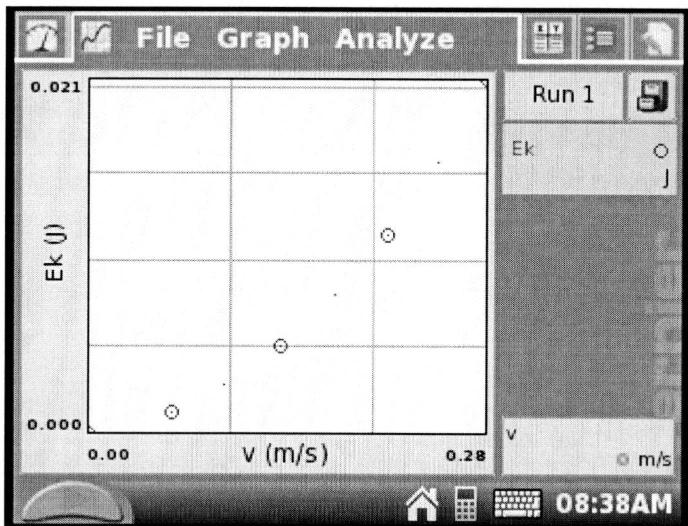

Figure 3 Lighter hoop spring and dynamics cart

This plot of kinetic energy *vs.* velocity appears to be a top-opening parabola. By this point, students should be familiar with "linearizing" a graph.

In much the same way as they did for $\mathbf{E}_{el}$, students should add a new calculated column for velocity squared and display the graph of kinetic energy *vs.* velocity squared. They should then choose to perform a linear fit of the values as shown in Figure 4.

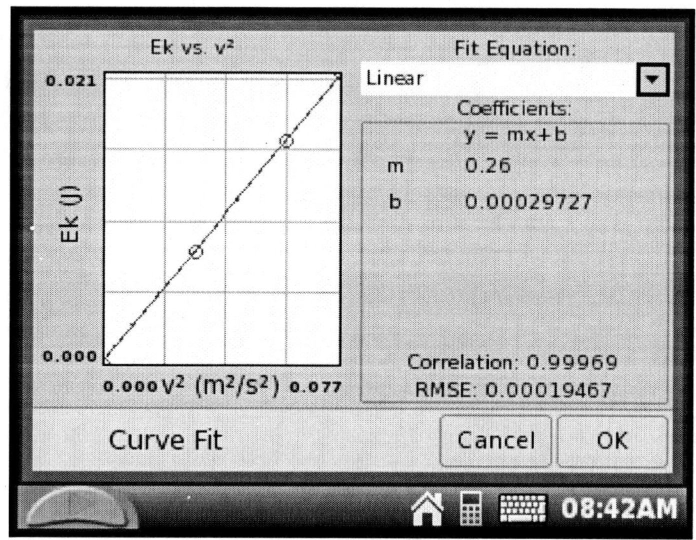

Figure 4 Linearized graph for the cart

A linearized graph for the 700 g (cart plus 200 g) is shown below. Note that the slope is very nearly half of the mass.

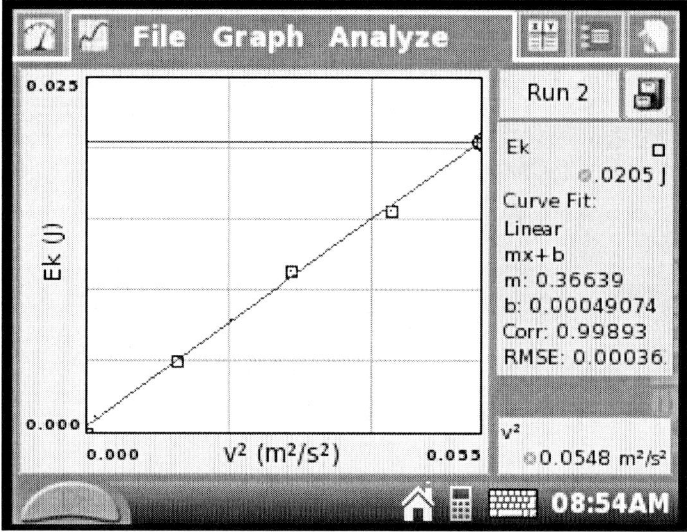

Figure 5

POST-LAB DISCUSSION

Steps 1–6

Guide students in Step 3 to the conclusion that the elastic energy once stored in the spring is now stored as the kinetic energy of the cart. During Step 4, students should recognize that the graph of kinetic energy *vs.* average velocity appears to be a top-opening parabola, suggesting that kinetic energy is proportional to the square of the velocity ($E_k \propto v^2$). Their prediction is borne out in the linearized graph (see Figure 4). If the students collected data carefully, the vertical intercept is negligible and can be dropped from their equation of the line.

The equation of the line in Figure 4 is $E_k = (0.260 \frac{\text{J}}{\text{m}^2/\text{s}^2})v^2$. Make sure that students have included the units on their value of the slope.

Step 7

Point out that in the previous lab, the joule was introduced as a unit of energy defined as a N· m. From their previous experience with Newton's Second law, students should be able to simplify the units of slope by replacing J with the fundamental units from which it is derived.

$$\frac{\text{J}}{\text{m}^2/\text{s}^2} \Rightarrow \frac{\text{N}\cdot\text{m}}{\text{m}^2/\text{s}^2} \Rightarrow \frac{(\text{kg}\cdot\text{m/s}^2)\cdot\text{m}}{\text{m}^2/\text{s}^2} \Rightarrow \text{kg}$$

Step 8

Have the students prepare presentations of their findings (whiteboard, chart paper, etc). These should include a sketch of both the original and linearized graph and the equation of their line of best fit.

Steps 9–10

Remind students that the slope of a graph is frequently related to a system variable that is held constant during the experiment. During the subsequent discussion, students should note that the plot of kinetic energy *vs*. velocity squared is linear, regardless of the mass of the cart. Students might notice that the slope of the line increases with the mass (see Figure 5). With some guidance they should conclude that slope of the graph is approximately equal to half the mass of the cart (in kg). If they have not worked out the units before, now is the appropriate time to show that the units of the slope simplify to kilograms. So, students should be able to write an expression for the kinetic energy as a function of mass and velocity: kinetic energy equals half the mass times the square of the velocity, or

$$E_k = \tfrac{1}{2}mv^2$$

EXTENSIONS

1. For a compression of 0.0200 m, the energy stored should be

$$E_{el} = \tfrac{1}{2}kx^2$$

$$E_{el} = \tfrac{1}{2}(15.0\ \text{N/m})(0.0200\ \text{m})^2 = 0.00300\ \text{J}$$

Assuming all of this is transferred to kinetic energy, the velocity of the cart can be found by

$$E_k = \tfrac{1}{2}mv^2 \Rightarrow v = \sqrt{\frac{2E_k}{m}}$$

$$v = \sqrt{\frac{2(0.00300\ \text{J})}{0.250\ \text{kg}}} = 0.155\ \text{m/s}$$

or, algebraically, $\tfrac{1}{2}kx^2 = \tfrac{1}{2}mv^2$, so $v = \sqrt{\frac{k}{m}} \cdot x$

2. See if you can get your students to reason as follows rather than substituting 0.0400 m into the equation for elastic energy and solving for velocity as if it were a totally new problem. Since the elastic energy is proportional to the square of the compression $E_{el} \propto x^2$ and the kinetic energy is proportional to the square of the velocity $E_k \propto v^2$, it follows that doubling the compression should double the cart's velocity.

3. If 20% of the elastic energy went to internal energy, then the kinetic energy of the cart would be only 80% of the value found in Extension 1. The new velocity would be

$$v' = \sqrt{\frac{2(0.00240\ \text{J})}{0.250\ \text{kg}}} = 0.139\ \text{m/s}$$

Experiment

9

Energy Storage and Transfer: Gravitational Energy

PART 3 – GRAVITATIONAL ENERGY

In the first of this series of labs exploring the role of energy in change, you found that the energy stored in an elastic system was proportional to the square of the change in the length of the spring or rubber band deformed by the applied force. We called the energy stored in this way *elastic energy*.

In the previous experiment you found that this energy could be transferred to a cart to produce a change in its speed. We said that the moving cart stored energy in an account called *kinetic energy*. Suppose that, instead of moving horizontally, the cart were to move up an incline. Gradually, the cart would come to a stop before it began to roll back down the incline. Let's examine for a moment the energy of the system when the object reaches its maximum height and its velocity is zero. While kinetic energy has diminished to zero, the energy of the system isn't "lost." It must be stored in some other account, which we call *gravitational energy*. This is the energy stored in the Earth-cart system as a function of its new height. Consider for a moment what system variables might affect the gravitational energy of the Earth-cart system.

While it is not a simple matter to measure this quantity directly, determining the *change* in the gravitational energy is straightforward. We can simplify this discussion if we arbitrarily assign a value of *zero* to both the gravitational energy of the system and the height of the object when it is as close to the Earth as it will get during the course of our investigation. Your goal is to determine a quantitative relationship between the gravitational energy and the height of an object above the zero-reference position.

OBJECTIVES

In this experiment, you will

- Recognize that the energy stored in an elastic system (spring, rubber band) can be transferred to another object, resulting in a change in the state of that object.
- Determine an expression for the gravitational energy as a function of the height of an object above the Earth.

MATERIALS

Vernier data-collection interface
Logger *Pro* or LabQuest App
Vernier Dynamics Track
standard cart
standard lab masses (100 g increments)
Vernier Bumper and Launcher Kit (recommended) **or** heavy rubber band
Vernier Photogate (Extension only)
Cart Picket Fence (Extension only)

PRE-LAB INVESTIGATION

Examine the apparatus for this experiment. Assume that after you released the cart, all the elastic energy was transferred completely, first to kinetic energy, and then to gravitational energy as the cart reached its maximum height.

On the axes to the right, predict what a graph of the gravitational energy *vs*. the height would look like. Compare your graph to those sketched by other students in your class.

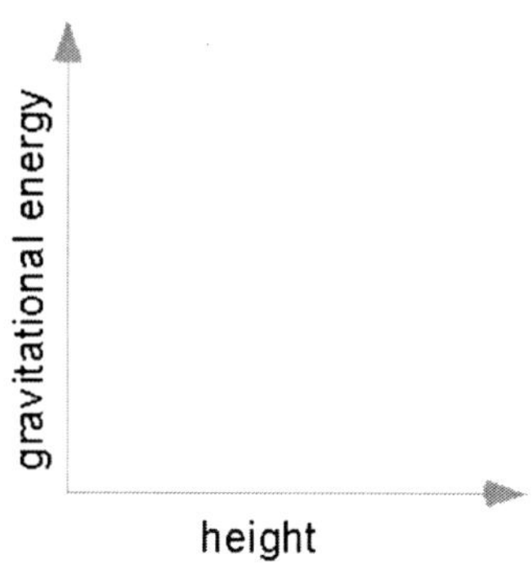

PROCEDURE

1. Attach the same spring you used in the last experiment to the Dynamics Track Bracket, then mount the bracket on the end of a Dynamics Track.
2. Obtain the value of the spring constant, *k*, for the spring you used in the previous experiment.[1]
3. Let *D* represent the distance between the leveling feet and *H* the height you have elevated one end of the track. By measuring the distance, *d*, the cart moves up the track, you can use similar triangles to determine the cart's height, *h*, above its zero position.

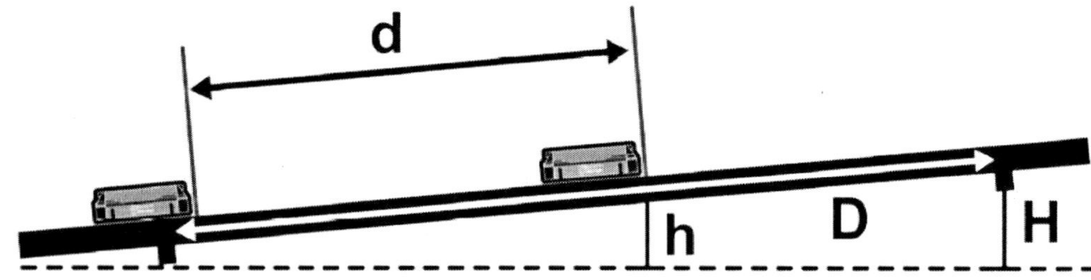

Figure 1

4. Before you elevate the track, recall the need to minimize the effect of friction on the transfer of energy from elastic to gravitational accounts. Use the leveling screws on one of the feet to raise that end of the track slightly. After you have made the necessary adjustment, place a block of height, *H*, under the other foot. Be sure to record these parameters.

[1] If you used a rubber band to store energy in the previous lab, you will need to quickly collect *F* vs. *x* data again to make sure that the band's effective "spring constant" has not changed since the last use.

5. Position the cart so that it just touches the hoop spring without deforming it. It is helpful to adjust the position of the bracket so that one end of the cart falls on a "convenient" value on the scale. Note this value as the zero reference position, x_0, for the spring.

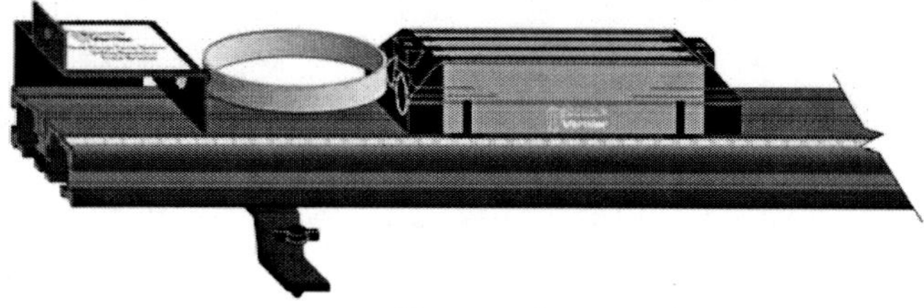

Figure 2

6. For this experiment, you will measure the distance, *d*, the cart travels before stopping as you vary the compression of the spring. This value, along with *H* and *D* (see Figure 1) will help you determine the height the cart reaches. Later, you will create a file (or use one your instructor provides) to analyze the data you have collected.

7. Consult with your instructor about appropriate values of change in length, *x*, for your apparatus. Begin data collection. Perform several trials for each change in length, *x*, and average the three most consistent values of distance. As before, it is important that you sight the scale from a position directly above the cart so as to avoid parallax error. Keep in mind that the initial position of the cart changes as you increase the extent to which you compress the spring

EVALUATION OF DATA

1. To evaluate the relationship between gravitational energy, E_g *vs.* height, set up a Logger *Pro* or LabQuest App file with the columns of **x** and **d,** where **x** is the change in length of the spring, and **d** is the average distance the cart moved up the inclined track. You will enter these data manually. You also need two calculated columns, **h** (the change in height of the cart) and $\mathbf{E_{el}}$ (elastic energy). Your instructor may guide you in the design of this file.

2. Discuss how the system energy is stored once the spring returns to its original shape. Make a new manual column for $\mathbf{E_g}$ (gravitational energy) and then fill in the $\mathbf{E_g}$ column with the appropriate values.

3. Create a graph of gravitational energy *vs.* change in height. If the relationship between gravitational energy, E_g, and the change in height, *h*, appears to be linear, fit a straight line to your data. If possible, print a copy of your data table and graph.

4. Write a statement that describes the relationship between the gravitational energy and the height of the cart. Keep in mind any assumptions made in the design of the file for this experiment.

5. Examine the slope of the graph (units as well as numerical value). Recall that the SI unit of energy, joule, is defined as a N m. Simplify the units of the slope, then compare the value that you obtained with that obtained by other groups.

6. Now write the general equation describing the relationship between the gravitational energy and the height of an object (above the zero-reference position).

EXTENSIONS

1. Suppose that you used a spring with a *k* value half as great as the one you used in your experiment to launch the dynamics cart. How would this change have affected the values of distance, *d*, you would have obtained? Describe the effect of this change on the graph of E_g *vs*. *h*. Explain.

2. Suppose that you had used a more massive dynamics cart (mass = 750g) in your experiment. How would this change have affected the values of distance, *d*, you would have obtained? Describe the effect of this change on the graph of E_g *vs*. *h*. Explain.

3. Suppose that the block that you used to elevate your track were 50% higher than the one that you used in your experiment. How would this change have affected the values of distance, *d*, you would have obtained? Describe the effect of this change on the graph of E_g *vs*. *h*. Explain.

4. Suppose that you had been able to perform your experiment on Mars where the acceleration due to gravity is about one third that on Earth. How would this change have affected the values of distance, *d*, you would have obtained? Describe the effect of this change on the graph of E_g *vs*. *h*. Explain.

EXTENSION ACTIVITY

Suppose you released a cart on the Dynamics Track at some height, *h*, above a zero-reference position. Assuming that you made an adjustment to minimize frictional losses, the principle of Conservation of Energy should lead you to predict that the kinetic energy (E_k) at the zero reference position would equal the gravitational energy (E_g) at the point of release. The following activity will enable you to test this prediction.

1. As you did in the experiment, raise the left end of the track slightly by adjusting the leveling screws on the left foot so as to minimize frictional losses. After you have made the necessary adjustment, place a block of height, *H*, under the right foot. Use the same value of *D*, (distance between feet) as you did in the experiment. Be sure to record these parameters.

2. Place the cart picket fence on the cart, then position a photogate near the lower end of the track so that the leading edge of the flag interrupts the beam when the front end of the cart falls on some "convenient" value on the scale, as shown in Figure 3.

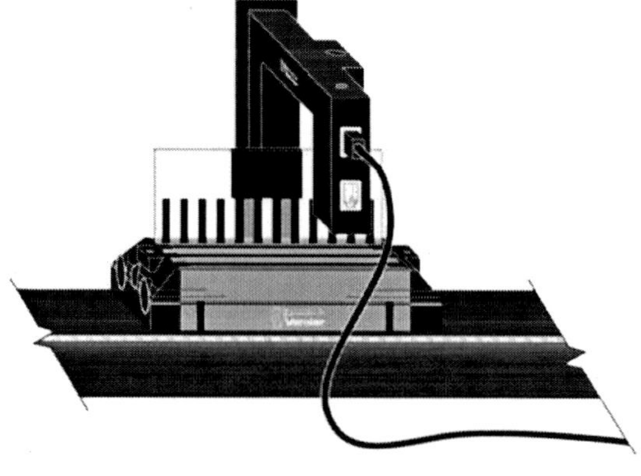

Figure 3

3. Start the data-collection program. Set up the photogate for Gate Timing. The length of the flag passing through the photogate is 0.05 m. The GateState should read **Blocked** when the flag interrupts the sensor and **Unblocked** when it has moved beyond the sensor.

4. You will first collect velocity data for 5–6 distances, *d*, and record these manually in your lab notebook. Later, you will create a file (or use one your instructor provides) to analyze the data you have collected. From the distance, *d*, the cart travels before the flag blocks the beam, you can determine the value of *h*. Using this, you can determine the value of the gravitational energy, E_g, in the system before you released the cart. With the cart mass and velocity, you can determine the kinetic energy, $E_{k,}$, of the cart at the zero-reference position.

5. Produce a graph of kinetic energy *vs*. gravitational energy for your data. If the relationship appears linear, fit a straight line to the data. What can you conclude from the value of the slope?

INSTRUCTOR INFORMATION

Energy Storage and Transfer: Gravitational Energy

PART 3 – GRAVITATIONAL ENERGY

This is the third in a series of experiments developing the concept of energy storage and transfer.

The Microsoft Word files for the student pages can be found on the CD that accompanies this book. See *Appendix A* for more information.

OBJECTIVES

In this experiment, the student objectives include

- Recognize that the energy stored in an elastic system (spring, rubber band) can be transferred to another object, resulting in a change in the state of that object.
- Determine an expression for the gravitational energy stored in a planet-object system.

During this experiment, you will help the students

- Recognize that the gravitational energy stored in the planet-object system is proportional to the height (h) of the object above the Earth.
- Recognize that the slope of a graph of gravitational energy *vs.* height is proportional to the mass of the object as well as to the gravitational field strength (g) for the planet.

REQUIRED SKILLS

In order for student success in this experiment, it is expected that students know how to

- Set up a photogate for Gate mode. This is addressed in Activity 4.
- Create manual columns in the software. This is addressed in Activity 1.
- Create calculated columns in the software. This is addressed in Activities 1 and 3.

EQUIPMENT TIPS

If students used a hoop spring or rubber band in the previous experiments, they should use it for this one. Unfortunately, the extensible spring does not yield acceptable results when it is used to pull the cart up an inclined track. In a pinch, students could use the apparatus described in the extension activity by making the assumption that the cart's kinetic energy at the bottom is equivalent to the gravitational energy at the height from which it was released. However, this approach is not as conceptually straightforward as the one employed in this experiment.

Keep in mind that the effective spring constant, k, of rubber bands tends to change with use. If a rubber band is used, then students should repeat the measurement of force *vs.* stretch with the Dual-Range Force Sensor to make sure that they have the current value of the effective spring constant for the rubber band. This can be accomplished in a few minutes.

Attach the hoop spring directly to the Dynamics Track Bracket, as shown in Figure 1.

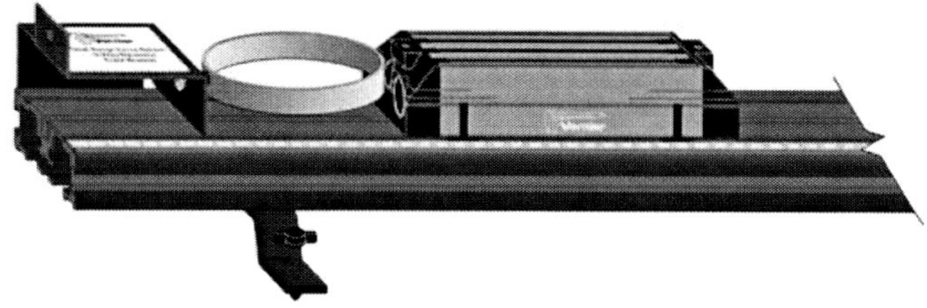

Figure 1

PRE-LAB DISCUSSION

This is the third in a series of experiments investigating energy storage and transfer. Prior to distributing the lab to the students, demonstrate how the compressed spring can launch the dynamics cart when released. Then elevate the far end of the track a few centimeters and repeat the demonstration. Ask the students what the cart's velocity is at its maximum distance from the spring. When students answer, "It is zero," ask them to explain what happened to the kinetic energy of the cart. This is the appropriate time to introduce another way a system can store energy – *gravitational energy*. This is the energy stored in the Earth-cart system as a function of its position above the Earth. Elicit from the students what system variables might affect the gravitational energy of the Earth-cart system. Students should note that the cart is higher than it was when it was in contact with the spring. If necessary, remind them that the $F_{earth/cart}$ depends on the cart's mass, so mass, too, should play a role.

Be careful to avoid stating that gravitational energy is stored by the cart. This account is a bit trickier than the two studied previously where the energy was stored in an object – whether it was due to reversible deformation (elastic) or its motion (kinetic). An object, by itself, cannot store gravitational energy. Instead, this energy is stored in the Earth-cart system due to the cart's position in the Earth's gravitational field. A change in the position of an object in the Earth's gravitational field results in a transfer of energy either into or out of this account. This is why we need to define a zero-reference position where we arbitrarily assign a value of zero to the gravitational energy of the earth-object system. We refer to the location of an object above this zero-reference position as its height.

Inform the students that they will be investigating the relationship between gravitational energy and the height. Their goal is to derive a quantitative expression for the gravitational energy of the Earth-cart system. Try to get students to recognize that we can examine the role of mass by varying the mass of the cart used by different lab groups. At this point, you should distribute the lab instructions.

As in the second experiment, an adjustment to this apparatus is necessary to insure good results. In order to make the assumption that all the elastic energy stored in the spring is transferred to the gravitational energy of the system, one must minimize the energy "lost" due to frictional interactions. To accomplish this, students should use the leveling screws on the foot at the end of the track with the spring to raise that end just enough so that the cart, given a gentle push, will travel to the other end of the track at nearly constant velocity. Make sure that students do this *before* they raise the other end of the track.

Point out that direct measurement of the cart's height at the point where its velocity (and E_k) is zero would very difficult. Guide them to recognize that, due to similar triangle (see illustration in student version):

$$\frac{h}{d} = \frac{H}{D}, \text{ rearranging yields}$$

$$h = d \cdot \frac{H}{D}$$

So, by measuring the distance, *d*, the cart travels up the track, they can calculate the cart's height.

LAB PERFORMANCE NOTES

Students should record the change in length, *x* (used to determine the elastic energy that will be transferred to the gravitational account), and the distance, *d*, the cart travels when it reaches its maximum height in their lab notebooks. Students need to keep in mind that the initial position changes as they compress the spring. Multiple trials are recommended for each value of *x*, because it is somewhat tricky to release the cart cleanly, and students have to judge the final position of the cart. With practice, fairly consistent results can be obtained.

EVALUATION OF DATA

Students will need to construct a Logger *Pro* or LabQuest App file in which they perform the analysis of the data they have collected. There should be manual columns for **x** and **d**, where **x** is the change in length of the spring, and **d** is the average distance the cart moved up the inclined track. There should be calculated columns for **h**, and elastic energy $\mathbf{E_{el}}$. This experiment provides an opportunity for students to use defined user parameters **spring constant**, **H** and **D** in Logger *Pro*, though these are not necessary if students simply incorporate these fixed values into their equation for *h* and E_{el}. The sample Logger *Pro* file (09 Eg v height.cmbl) is provided on the CD that accompanies this book for guidance. The LabQuest App equivalent (09 Eg v height.qmbl) is also provided. Students will have to adjust the value of *A* in the calculations for height and elastic energy, E_{el} to match the parameters in their experiment. To save time, you may provide the appropriate file for students to use directly. If you choose to do so, make sure that students examine the equations for the calculated columns. In any event, once students have concluded that the system energy at the maximum height is now stored as gravitational energy, they should create a new manual column for gravitational energy and fill in the values.

SAMPLE RESULTS

Below are sample data obtained with the heavier hoop spring in the Bumper and Launcher Kit.

Δx (m)	d (m)	height (m)	E_{el} (J)	E_g (J)
0.000	0.00	0	0	0
0.0100	0.044	0.00123	0.0037	0.0037
0.0150	0.081	0.00227	0.0084	0.0084
0.0200	0.134	0.00375	0.0150	0.0150
0.0250	0.193	0.00540	0.0235	0.0235
0.0300	0.274	0.00767	0.0338	0.0338
0.0350	0.349	0.00977	0.0460	0.0460
0.0400	0.449	0.0126	0.0601	0.0601

Figure 2 BLK hoop spring and 500 g cart

The graph of gravitational energy *vs.* height is shown in Figure 3.

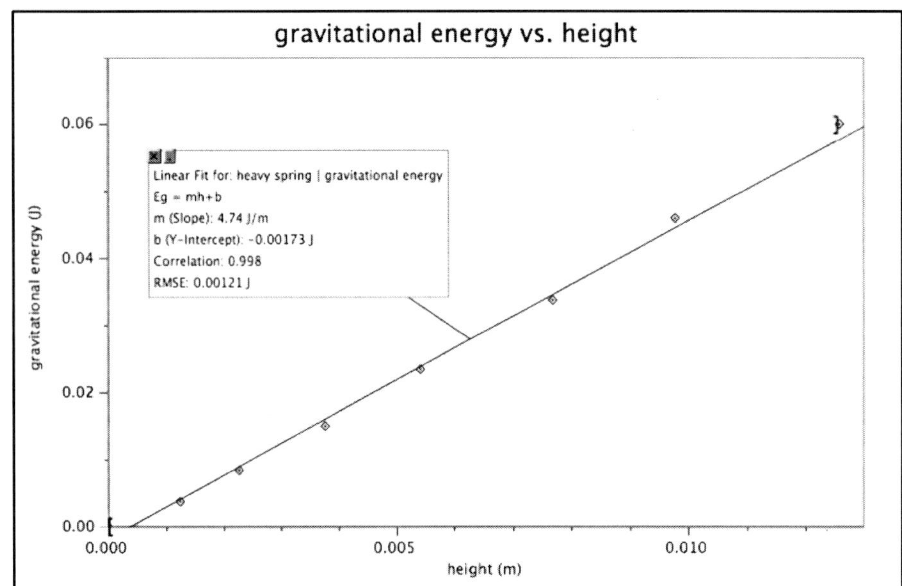

Figure 3 BLK hoop spring and 500 g cart

Note that the slope of the graph, 4.74 J/m, is within 3.3% of the expected value for *mg* for this system since *the mass* was 500 g.

POST-LAB DISCUSSION

Steps 1–3

As they did in Experiment 8, students should conclude that the system energy when the cart reaches its maximum height is stored as gravitational energy. After they have plotted gravitational energy *vs.* height, they should prepare presentations of their findings (whiteboard, chart paper, etc). These should include a sketch of the graph and the equation of their line of best fit. Suggest to the students that the vertical intercept is negligible and can be dropped from their equation of the line. Make sure that they have included the units (J/m) on their value of slope. They should also have recorded the mass of their cart.

Students' results will vary. The equation of the graph in Figure 3 is $E_g = (4.87\,\text{J/m})h$.

Step 4

Students should find that their graph of E_g vs. h is linear. In the discussion of their findings, try to get the students to say that the gravitational energy stored in the system is proportional to the height of the cart above the zero-reference position. Therefore, if the height were doubled, the energy stored would also be doubled.

Step 5

When asked about the meaning of the slope, students should note that the units, J/m, reduce to newtons, a unit of force. Have the students compare the numerical value of their slope with those obtained by other groups. Remind students that the slope of a graph is frequently related to a system variable that is held constant during the experiment. They will note that slope of their graph is close to the weight of their cart in newtons.

Step 6

Students should now be able to write a general expression for gravitational energy as a function of height: $E_g = mgh$.

ANSWERS TO THE EXTENSION QUESTIONS

For all of these questions, try to get students to use proportional reasoning.

1. If the value of the spring constant, k, had been half as great, the energy stored by the spring, and eventually transferred to gravitational energy would also have been half as great for each change in length, x. However, with only half the energy, the distance the cart would have traveled up the track would also have been half as great. Since the height is proportional to the distance, it, too, would also be halved. As a result, the graph would have smaller values of E_g and h, but there would be no change in the slope.

2. If the cart used had been more massive, the distance it would have traveled up the ramp would have been smaller for each value of x. This would have the effect of reducing the value of h, while the value of E_g would remain unchanged. Since the slope of the graph is:

$$m = \frac{\Delta E_g}{\Delta h}$$

reducing h would have the effect of increasing the slope of the graph.

3. If the block used to elevate the track were higher, the cart would not travel as far up the steeper ramp, reducing the value of *d*. However, the value of h was calculated by using the equation:

$$h = d \cdot \frac{H}{D}$$

The increase in *H* would offset the increase in *d* and the value for *h* would remain unchanged for each value of *x*. The higher block would have no effect on E_g so the graph would be unchanged.

4. If students are uncertain how this would matter, ask them to see if they can find the value of the acceleration due to gravity on Mars. A Google search will result in their finding that *g* for Mars is 3.73 m/s^2. With a weaker gravitational force slowing the cart as it moves up the ramp, the value of *d*, and thus *h* would increase. Nothing about the spring has changed, so the values of E_g would remain unchanged. Increasing *h* would reduce the value of the slope.

EXTENSION ACTIVITY

Thus far, we have made the assumption that the elastic energy stored in the spring (or band) was transferred to other accounts (kinetic or gravitational) with negligible losses due to friction. This would be a good opportunity to test that assumption further by examining a different type of transfer. Suggest to the students that they make a minor modification to the apparatus used for this experiment to check the transfer from gravitational to kinetic accounts. To do this, add a photogate near the bottom end of the track and a flag to the cart (the cart picket fence works nicely). They can release the cart from various distances from the photogate and use the relationship for determining the height from the lab to determine the gravitational energy at each release point.

Using Gate Timing students can determine the kinetic energy from the velocity of the cart at the zero-reference position. Then they can plot kinetic energy *vs.* gravitational energy – if these values are equal, then the slope of this graph will be one.

To determine the zero-reference position, students should position the photogate so that the flag interrupts the beam when the front end of the cart falls on a "convenient" value on the scale, as shown in Figure 4.

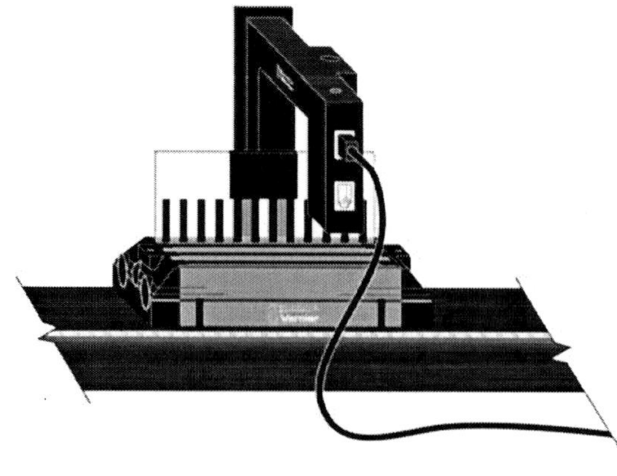

Figure 4

Computer Data Collection

Students should connect the photogate to a digital port on the interface and start Logger *Pro*. They should click on the Gate State in the status bar for the sensor; this brings up the sensor information window. Students should click on the icon of the photogate, choose "Gate Timing", then close the window. The default distance value of 0.05 m should be used.

LabQuest App Data Collection

Students should connect the photogate to a digital port on the LabQuest and change the mode from Photogate Timing to Gate. The default distance value of 0.05 m should be used.

With either method, the Gate State should read [Blocked] when the flag interrupts the sensor and [Unblocked] when it has moved beyond the sensor. When they click Collect, they can release the cart, then read the velocity.

Once the students have obtained the velocity data, they should open a new file to perform the analysis of data. Students might need some guidance to set up this file; they should recall what they did in the previous experiments to design the file for this extension. A sample file (09 Ek v Eg.cmbl) is provided on the CD for your use to guide the students to set up this file. The LabQuest App equivalent (09 Eg v Eg.qmbl) is also provided.

The following data were obtained setting the photogate such that the front edge of the cart was at the 30 cm mark on the track for the zero-reference position.

d (m)	height (m)	Eg (J)	v (m/s)	E_k (J)
0	0.000	0.000	0.000	0.000
0.20	0.006	0.027	0.344	0.030
0.30	0.008	0.041	0.413	0.043
0.40	0.011	0.055	0.474	0.056
0.50	0.014	0.069	0.529	0.070
0.60	0.017	0.082	0.578	0.084
0.70	0.020	0.096	0.625	0.098

Figure 5

In this table, E_g is calculated using $E_g = mgh$. E_k is calculated using $E_k = \frac{1}{2}mv^2$.

From these data the following graph was obtained.

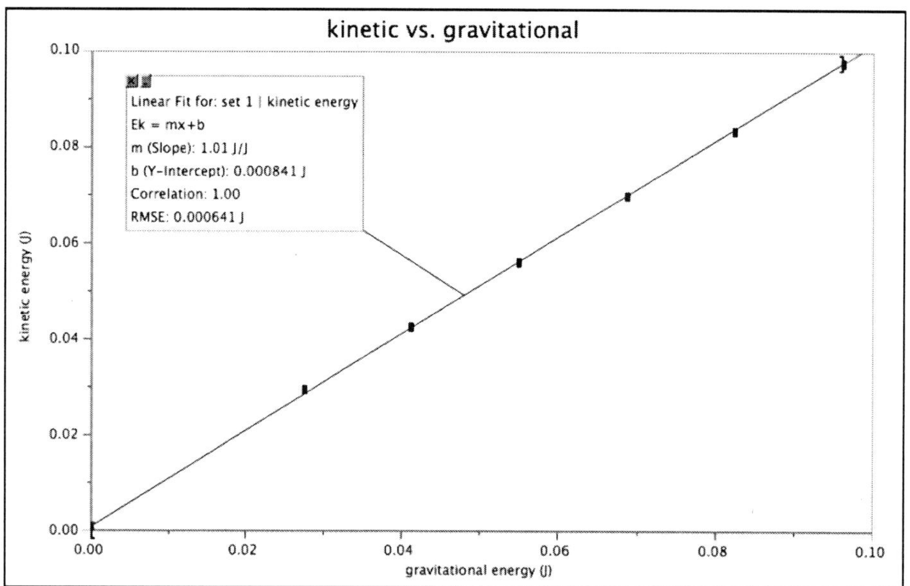

Figure 6

Ask students to draw a conclusion from the value of the slope and the fact that the units cancel. Since the slope of the graph is very nearly 1, it is reasonable to conclude that the kinetic energy when the cart is at the bottom of the track is equal to the gravitational energy when the cart is at the top.

Experiment

10A

Impulse and Momentum

INTRODUCTION

You are no doubt familiar with everyday uses of the term *momentum*; e.g., a sports team that has begun to exert superiority over an opponent is said to have gained "momentum." However, in physics, this term has a precise definition: momentum, p, is the product of the mass and velocity of an object, $p = mv$.

You have learned that a net force is required to change the *velocity* of an object. In this experiment you will examine how the *momentum* of a cart changes as a force acts on it. This will enable you to determine the relationship between force, the length of time the force is applied, and the change in the momentum of the cart.

OBJECTIVES

In this experiment, you will

- Collect force, velocity, and time data as a cart experiences different types of collisions.
- Determine an expression for the change in momentum, Δp, in terms of the force and duration of a collision.

MATERIALS

Vernier data-collection interface
Logger *Pro* or LabQuest App
Vernier Motion Detector
Vernier Dual-Range Force Sensor
Vernier Dynamics Track
Motion Detector bracket
Vernier Bumper and Launcher Kit
standard cart
string
elastic cord (optional)

PRE-LAB QUESTIONS

1. In a car collision, the driver's body must change speed from a high value to zero. This is true whether or not an airbag is used, so why use an airbag? How does it reduce injuries?
2. Suppose airbags were not vented to allow the gas inside to escape, but remained inflated (like a balloon). Would they be as effective in protecting a passenger in a collision?

PROCEDURE

1. Attach the Motion Detector to the bracket that will allow you to position it near one end of the Dynamics Track.

2. If your motion detector has a switch, set it to Track.

3. Adjust the leveling screws on the feet as needed to level the track. To make sure the track is level, give the cart a gentle push. It should reach the opposite end of the track without a noticeable change in velocity.

4. Connect the motion detector and the Dual-Range Force Sensor (DFS) to the interface and start the data-collection program. Increase the data-collection rate to 500 samples/second[1]. The duration of the experiment can be reduced to 5 seconds.

5. Make the necessary adjustments so that two graphs: force *vs*. time and velocity *vs*. time appear in the graph window.

Part 1 Elastic collisions

6. Replace the hook end of the force sensor with the hoop spring bumper.[2] Attach the force sensor to the bumper launcher assembly as shown. Then attach the bumper launcher assembly to the end of the track opposite the motion detector.

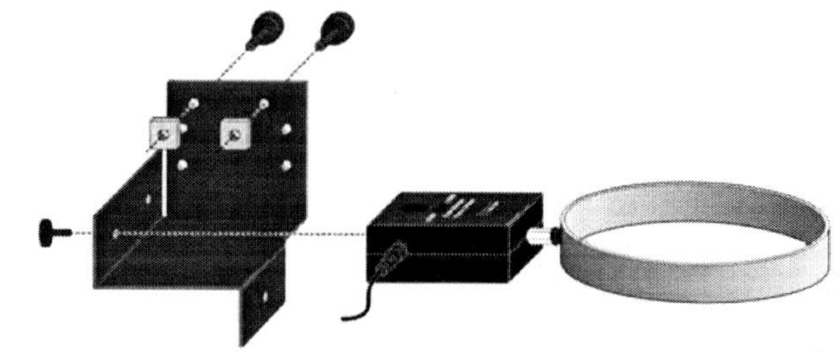

Figure 1

Note: *Shown inverted for assembly.*

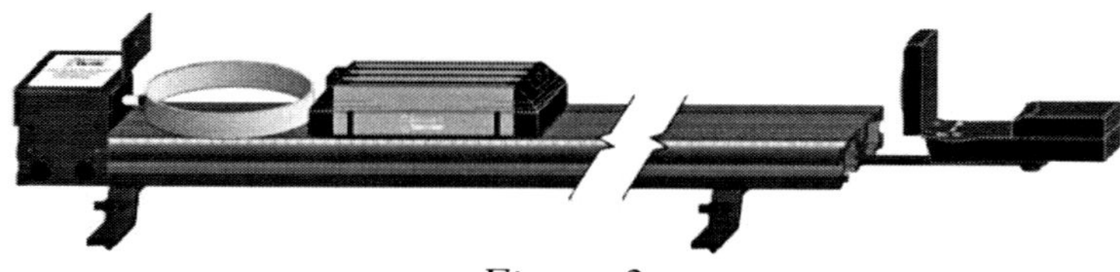

Figure 2

7. Practice launching the cart with your finger so that when it collides with the hoop spring bumper, it slows to a stop and reverses direction smoothly. An abrupt collision will not yield satisfactory data.

---

[1] 250 samples/second for LabPro

[2] If this type spring is not available, your instructor will show you how to use an alternate arrangement to collect the data for this experiment.

8. Position the cart at least 20 cm from the motion detector[3], then zero both the motion detector and the force sensor.

9. Start data collection. When you hear the motion detector clicking, launch the cart toward the hoop spring bumper. Be sure to catch the cart once it has returned to its starting position. Because both force and velocity are vector quantities, check to see if the signs of force and velocity match your experimental setup. If necessary, reverse the direction of one or both sensors.

10. Collect data for at least three elastic collisions, varying the mass of the cart. Be sure to store the data for each run.

Part 2 Inelastic collisions

1. Replace the hoop spring bumper with one of the clay holders from the Bumper and Launcher Kit. Attach cone-shaped pieces of clay to both the clay holder and to the front of the cart, as shown in Figure 3.

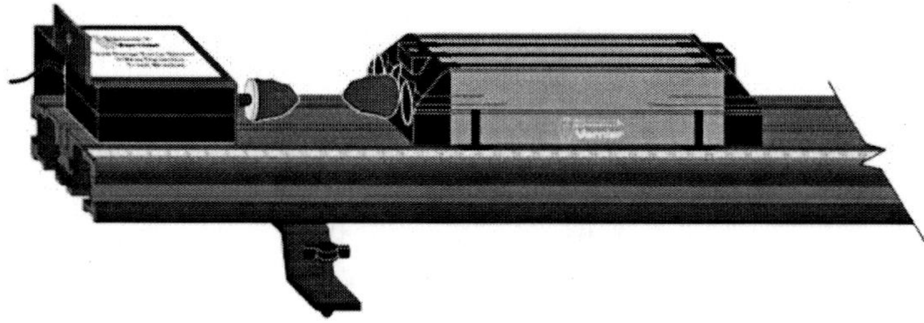

Figure 3

2. Set the switch on the force sensor to the 50 N position; reset the data-collection parameters as you did in Step 4.

3. Practice launching the cart with your finger so that when the clay "nose" on the front of the cart collides with the clay on the force sensor, the cart comes to a stop without bouncing. A collision that is too jarring will not yield satisfactory data.

4. Position the cart at least 20 cm from the motion detector, then zero both the motion detector and the force sensor.

5. Collect data as before for at least three inelastic collisions, varying the mass of the cart. Be sure to store the data for each run.

EVALUATION OF DATA

Part 1 Elastic collisions

1. On the velocity *vs.* time graph, select the interval corresponding to the period of time when the spring was acting on the cart. LabQuest App automatically selects the same interval on both graphs when you drag the stylus across an interval on one of the graphs. However, in Logger *Pro*, you first have to select both graphs and group them (x-axes). Next, turn on the Examine tool for each graph. Then, make the v-t graph the active window and drag the cursor across the appropriate interval.

---

[3] If you are using an older motion detector without a switch, the cart needs to be at least 45 cm from the detector.

In either program, when you choose Statistics from the Analyze menu, you will have the velocity of the cart just before and just after the collision with the spring.

2. From the mass of the cart and its change in velocity, $v_f - v_i$, determine the change in momentum, Δp, of the cart.

3. As you learned in kinematics, the area under a curve often has physical significance. In the case of the *F-t* graph, the area of the interval you selected is the product of the average force and the time during which the spring was interacting with the cart. You can determine this area by choosing Integral from the Analyze menu. In your class discussion you will give a name to this quantity.

4. Compare the value (both magnitude and sign) of the quantity you determined in Step 3 with the change in momentum of the cart.

5. Perform similar analyses for your remaining elastic collisions. Determine the % difference between the impulse, $F\Delta t$, and the change in momentum, Δp, for each of the collisions. Compare your findings to those of others in your class. What can you conclude about these quantities?

Part 2 Inelastic collisions

1. As you did in Step 1 of Part 1, select the interval corresponding to the period of time from slightly before to slightly after the collision. Due to the shorter duration of this type of collision, you should zoom in on this portion of both graphs. Note any differences in the shape of the *F-t* graph for this type of collision. Try to account for this difference.

2. Because some bouncing is unavoidable, you should discuss how to select an appropriate interval of the *F-t* graph for your determination of the impulse. Assume the final velocity of the cart is zero.

3. As you did with the elastic collisions, determine the % difference between the impulse, $F\Delta t$, and the change in momentum, Δp, for each of the inelastic collisions. Compare your findings to those of others in your class. What can you conclude about these quantities?

4. From Newton's second law, derive the equation you have determined from the analysis of your data. Compare the fundamental units for both impulse and change in momentum.

EXTENSIONS

1. When you catch a fast-moving baseball, it hurts less when your hand "gives" a little than if you hold your hand stiff. Explain why this is so in terms of impulse and change in momentum.

2. Now cars are made to crumple during a collision. Explain how this works in terms of impulse and change in momentum.

3. Suppose you had used a stiffer spring in the experiment. Describe how the shape of the force *vs.* time graph would differ from that which you observed.

INSTRUCTOR INFORMATION

Impulse and Momentum

This lab is written with the assumption that the instructor will engage the students in discussions at critical junctures. These discussions can take place with the entire class or with individual lab groups. The icon at left indicates where these discussions should occur.

The Microsoft Word files for the student pages can be found on the CD that accompanies this book. See *Appendix A* for more information.

OBJECTIVES

In this experiment, the student objectives include

- Collect force, velocity, and time data for a cart experiencing different types of collisions.
- Determine an expression for the change in momentum, Δp, in terms of the force and duration of a collision.

During this experiment, you will help the students

- Manipulate graphs including selecting variables for axes on graphs, grouping graphs, and using the Integral and Statistics tools in Logger *Pro* or LabQuest App.
- Recognize that the change in momentum of an object is a function of both the force acting on it and the duration of the interaction.
- Define the impulse as the area under a force *vs.* time graph.
- Develop the impulse-change in momentum expression $F\Delta t = m\Delta v$.
- Recognize that the change in momentum for an object that undergoes an inelastic collision is half of that for an elastic collision.

REQUIRED SKILLS

In order for student success in this experiment, it is expected that students know how to

- Zero a Motion Detector and a Dual-Range Force Sensor. Zeroing a motion detector is addressed in Activity 2.
- Reverse the direction of a Motion Detector or Dual-Range Force sensor in the software.

EQUIPMENT TIPS

While there are a number of ways one could collect suitable data for this experiment, this version uses a Dynamics Track, standard cart, Dual Range Force Sensor (DFS) and a Motion Detector. Experiment 10B uses a photogate and cart picket fence to obtain timing data. The accessories in the Bumper and Launcher Kit (hoop springs, clay and holder) greatly facilitate the study of both elastic and inelastic collisions. The bumper that accompanies the DFS creates a collision that is too brief to allow a satisfactory analysis of the relationship between impulse and change in momentum.

Mount the Dual-Range Force Sensor upside down to the Dynamics Track Bracket. Then attach the assembly to the track as shown in Figures 1 and 2.

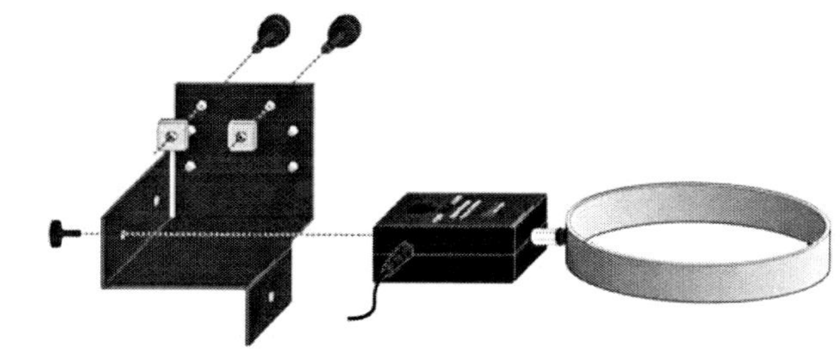

Figure 1

***Note:** Shown inverted for assembly.*

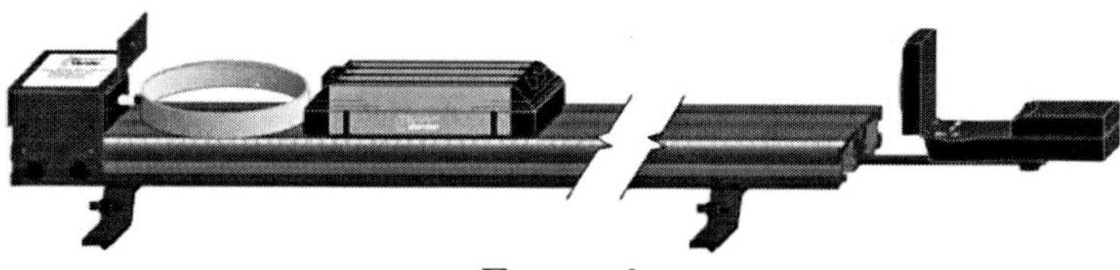

Figure 2

If this kit is not available, connect a hoop spring to the DFS and affix the sensor to the Dynamics Track using the aluminum rod that comes with the DFS and a 1/4" hex nut, as shown in Figure 3.

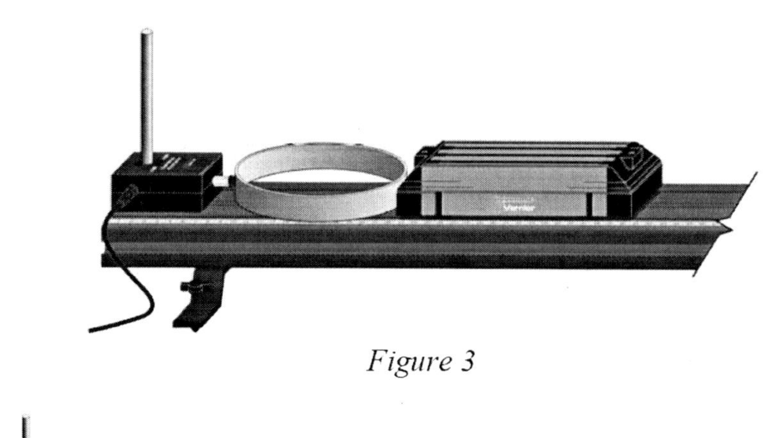

Figure 3

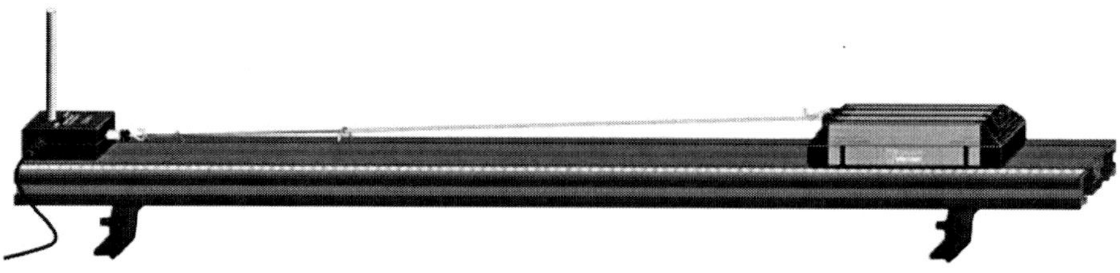

Figure 4

An alternative to the spring is to use a length of elastic cord or a long rubber band with a length of string. Attach the elastic cord to the cart and the string to the hook on a DFS, as shown in Figure 4. The string should be long enough to allow the cart to travel nearly the length of the track before the elastic cord brings the cart to a stop and reverses its direction. Make sure the cart reverses direction before it gets too close to the motion detector.

PRE-LAB DISCUSSION

This experiment should be performed after students have a thorough grasp of Newton's laws of motion. Explain to students that this experiment will give them the opportunity to study momentum, p, which is defined as the product of the mass and velocity of an object $p = mv$.

Demonstrate the motion of a cart along a level track, pointing out that, so long as the sum of the forces acting on the cart is zero, its velocity and hence its momentum remains constant. They should recognize this as a case of Newton's first law. From Newton's second law, students should predict that a constant force acting on the cart will produce uniform acceleration. However, they may have difficulty relating a change in velocity to the *changing* force that occurs during a collision.

Demonstrate a collision between a cart and the hoop spring (see Figure 2). Ask the students to describe how the force applied by the spring changes over the period of time the cart is in contact with the spring. Inform them that this experiment affords them the opportunity to study how the force and duration of a collision is related to the change in momentum of the cart.

LAB PERFORMANCE NOTES

When the motion detector and force sensor are connected to the data-collection interface and Logger *Pro* or LabQuest App is started, the default graph screen shows both force *vs.* time and position *vs.* time graphs. Advise students to change the vertical axis label of the position-time graph to *velocity*. Very high sampling rates (500 Hz) are recommended so that sufficient data points to produce a smooth *F-t* graph are obtained during the short duration of the collision.

Note: In Logger *Pro*, this poses no difficulties. LabQuest App warns that this sampling rate is too high. During data collection in this experiment, the LabQuest App will take extra time to process the data. It may appear that the application is not running. You'll see an indicator showing that it is processing the data. It may take up to 30 seconds for the *F-t* and *v-t* graphs to appear. Remind students that changing the setting on the force sensor in LabQuest App requires resetting the data-collection parameters.

Students can check to see if they have leveled the track by examining the velocity-time graph prior to the collision. Encourage them to try collisions at different initial velocities, but make sure that they realize that the hoop spring should slow and then reverse the velocity of the cart smoothly. Keep in mind that the default setting of the force sensor registers a push as negative.

SAMPLE RESULTS AND POST-LAB DISCUSSION – ELASTIC

Steps 1–2

You may have to show students using Logger *Pro* how to group the two graphs. To do this, shift-click to select both graphs, then choose Group Graphs (X-Axis) from the Page menu. Then, when you use the Examine tool, it becomes easier to select the same time interval on each graph. The graph in Figure 5 shows how the velocity of the cart changes during a collision with the lighter (~25 N/m) hoop spring. Because of frictional forces, some slowing of the cart is inevitable. To minimize the effect of friction on the analysis of the change in velocity resulting from the force exerted by the spring, students should select the range of velocities corresponding to the interval during which the spring is acting on the cart. Students can then use the Statistics tool to find the velocity of the cart just before and just after the collision. From the mass and the change in velocity, students can calculate the change in momentum of the cart.

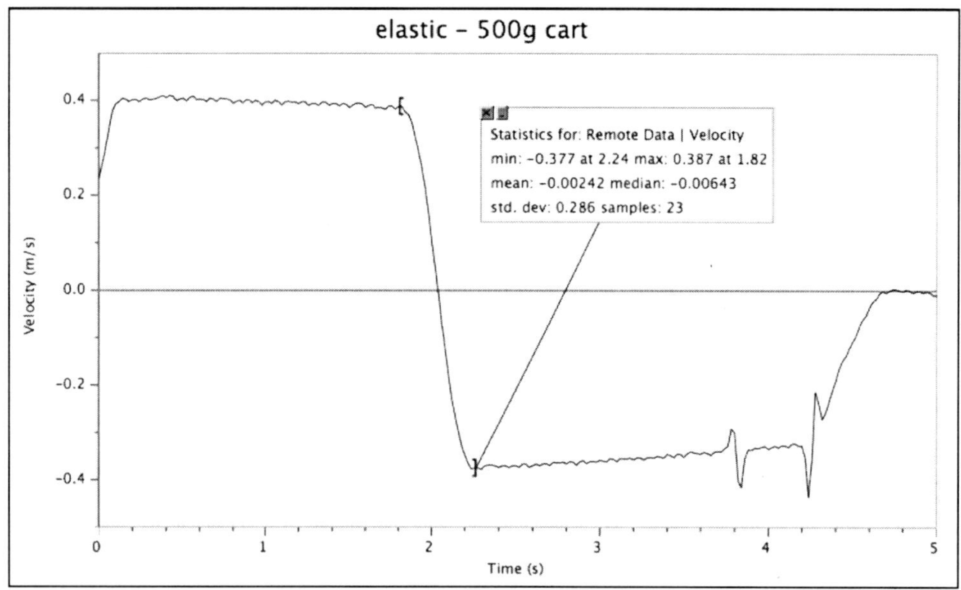

Figure 5 v-t graph for collision with hoop spring

Step 3

In previous examples in the study of graphs in kinematics or energy, the shape of the region under curve was regular, so determining the area was straightforward. In the case of the *F-t* graph (Figure 6), the area under the curve is roughly triangular, but use of the Integral tool yields a much better value for the area. Once students understand that this value represents the accumulation of the areas of many tiny rectangles, each the product of the force and the duration, you can give this quantity its physics name: *impulse.*

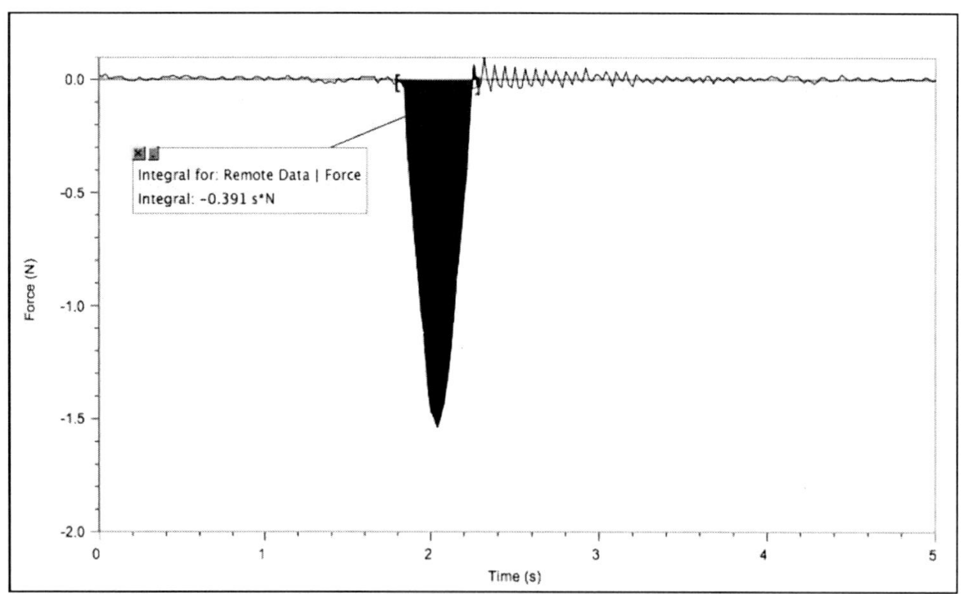

Figure 6 F-t graph for collision with hoop spring

Steps 4–5

Values for elastic collisions are listed below.

Mass (kg)	Δv (m/s)	Δp (kg-m/s)	$F\Delta t$ (N-s)	% difference
0.500	–0.764	–0.382	–0.391	2.4
0.700	–0.644	–0.453	–0.465	2.9
1.00	–0.738	–0.738	–0.753	2.0

Students should conclude that the impulse and the change in momentum are the same. The minor differences are due to frictional losses and limitations of the sensors.

SAMPLE RESULTS AND POST-LAB DISCUSSION – INELASTIC

Steps 1–2

Students should note that there is a small set of "bumps" in the F-t graph representing the bounce that occurred during the inelastic collision. This portion of the graph should be included in the analysis of the net impulse delivered to the cart (see Figure 7).

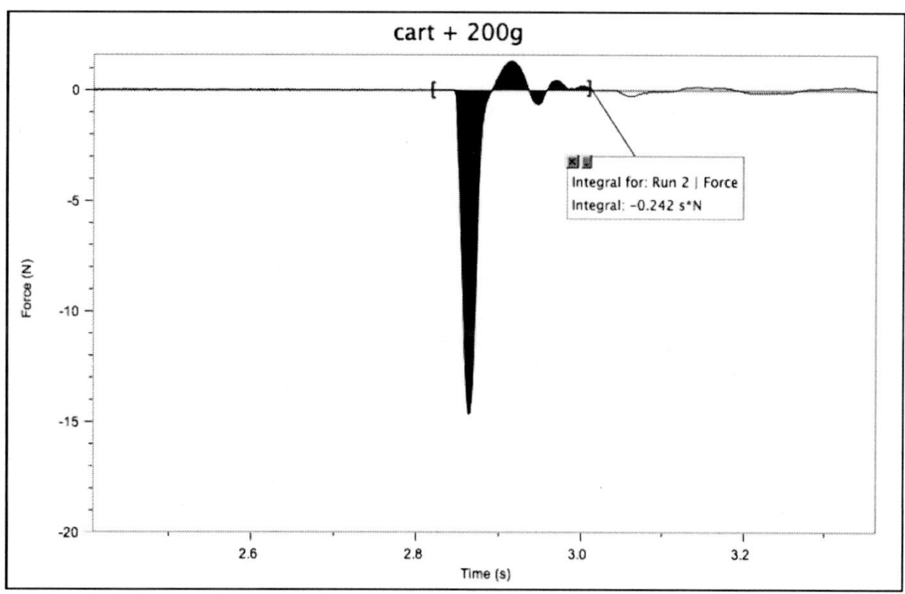

Figure 7 F-t graph for inelastic collision

Step 3

Values for inelastic collisions are listed below.

Mass (kg)	Δv (m/s)	Δp (kg-m/s)	$F\Delta t$ (N-s)	% difference
0.510	–0.322	–0.164	–0.170	3.7
0.710	–0.334	–0.237	–0.242	2.1
1.01	–0.312	–0.315	–0.312	1.0

Students should conclude that the impulse and the change in momentum are equal for inelastic collisions as well. The mathematical expression for this relationship is $F\,\Delta t = m\,\Delta v$.

Step 4

The most common expression of Newton's second law is $F_{net} = m\,a$. Substituting $\Delta v/\Delta t$ for a yields $F_{net} = m\,\Delta v/\Delta t$. Rearranging yields $F_{net}\Delta t = m\,\Delta v$.

We have made the assumption that the only force acting on the cart was provided by the spring or the clay. The basic units of the variables in this equation agree.

$$\mathrm{N\cdot s = kg\cdot \frac{m}{s}}$$

$$\mathrm{\frac{kg\cdot m}{s^2}\cdot s = kg\cdot \frac{m}{s}}$$

Having performed the experiment, students should be able to answer the pre-lab questions.

1. Airbags are used to prolong the collision. Since the impulse provided to the driver is the same whether the steering wheel or airbag stops the driver, increasing the duration (Δt) reduces the force needed to bring the driver's velocity to zero.

2. An airbag that did not deflate when the driver struck it would cause the driver to bounce backwards, thus increasing both the change in momentum and the force applied to the driver.

EXTENSIONS

1. By "giving" during a catch, the impact time is increased, thus reducing the force exerted on your hand.

2. An automobile body designed to crumple in a collision both reduces the change in momentum (because bouncing is reduced) and reduces the force acting on the car (because the time of impact is increased). Both are beneficial to the well-being of passengers.

3. The duration of the collision would have been shorter had a stiffer spring been used. So, the peak force applied would have been larger. This would have the effect of making the impulse a shorter duration and higher magnitude.

Experiment
10B

Impulse and Momentum

INTRODUCTION

You are no doubt familiar with everyday uses of the term *momentum*; e.g., a sports team that has begun to exert superiority over an opponent is said to have gained "momentum." However, in physics, this term has a precise definition: momentum, p, is the product of the mass and velocity of an object $p = mv$.

You have learned that a net force is required to change the *velocity* of an object. In this experiment you will examine how the *momentum* of a cart changes as a force acts on it. This will enable you to determine the relationship between force, the length of time the force is applied and the change in the momentum of the cart.

OBJECTIVES

In this experiment, you will

- Collect force, velocity, and time data as a cart experiences different types of collisions.
- Determine an expression for the change in momentum, Δp, in terms of the force and duration of a collision.

MATERIALS

Vernier data-collection interface	standard cart
Logger *Pro* or LabQuest App	Cart Picket Fence
Vernier Photogate and bracket	string
Vernier Dual-Range Force Sensor	(optional) Vernier Bumper and Launcher Kit
Vernier Dynamics Track	(optional) elastic cord

PRE-LAB QUESTIONS

1. In a car collision, the driver's body must change speed from a high value to zero. This is true whether or not an airbag is used, so why use an airbag? How does it reduce injuries?
2. Suppose airbags were not vented to allow the gas inside to escape, but remained inflated (like a balloon). Would they be as effective in protecting a passenger in a collision?

PROCEDURE

1. Attach a Photogate to the Dynamics Track using the bracket. Position the photogate about 30 cm from one end of the track[1]. Place the cart picket fence on the cart, as shown in Figure 2.

[1] Locate the photogate so that the cart picket fence has passed completely through the gate before the cart makes contact with the spring.

2. Adjust the leveling screws on the feet as needed to level the track. To make sure the track is level, give the cart a gentle push. It should reach the opposite end of the track without a noticeable change in velocity.

3. Connect the photogate and the Dual-Range Force Sensor (DFS) to the interface and start the data-collection program.

4. Set up data collection.

   Using Logger *Pro*

   a. Use time-based data-collection mode. Change the data-collection rate to 500 samples/second (250 samples/second if using a LabPro) and the data-collection length to 5 seconds.
   b. Choose Set Up Sensors ▶ Show All Interfaces from the Experiment menu.
   c. Click the image of the Photogate and choose Gate Timing.

   Using LabQuest as a standalone device

   a. Tap Mode. Select Time Based data collection. Set the data-collection rate to 500 samples/second and the duration to 5 seconds.
   b. In the same mode dialog, scroll down to the Photogate Mode section and expand the section by tapping on the arrow.
   c. Choose Gate Timing. The default flag size of 5 cm is correct.
   d. Select OK to accept these settings.

   With either program, make the necessary adjustments so that two graphs, force *vs.* time and velocity *vs.* time, appear in the graph window.

Part 1 Elastic collisions

5. Replace the hook end of the force sensor with the hoop spring bumper.[2] Attach the force sensor to the bumper launcher assembly as shown in Figure 1. Then attach the bumper launcher assembly to the end of the track as shown in Figure 2.

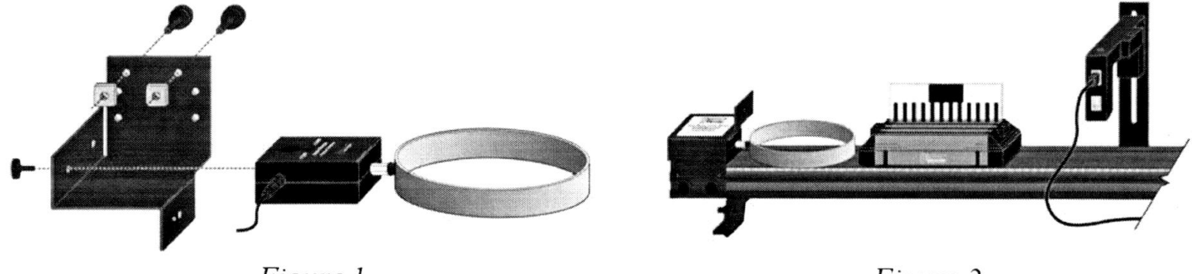

Figure 1 *Figure 2*

Note: Shown inverted for assembly.

6. Practice launching the cart with your finger so that when it collides with the hoop spring bumper, it slows to a stop and reverses direction smoothly. An abrupt collision will not yield satisfactory data.

[2] If this type spring is not available, your instructor will show you how to use an alternate arrangement to collect the data for this experiment.

7. Zero the force sensor. Adjust the position of the photogate so that the 5 cm flag on the picket fence clears the photogate before the cart touches the hoop spring bumper.

8. Start data collection, then launch the cart toward the hoop spring bumper. In LabQuest App, you must stop data collection to see the graphs.

9. Collect data for at least three elastic collisions, varying the mass of the cart. Be sure to store the data for each run.

Part 2 Inelastic collisions

1. Replace the hoop spring bumper with one of the clay holders from the Bumper and Launcher Kit. Attach cone-shaped pieces of clay to both the clay holder and to the front of the cart as shown in Figure 3.

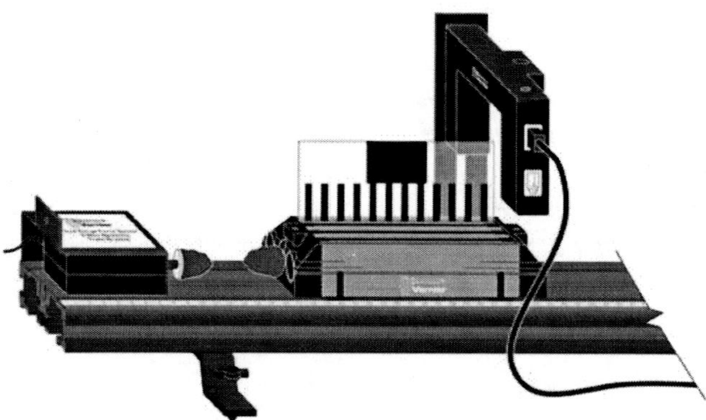

Figure 3

2. Practice launching the cart with your finger so that when the clay "nose" on the front of the cart collides with the clay on the force sensor, the cart comes to a stop without bouncing. A collision that is too jarring will not yield satisfactory data.

3. Set the switch on the force sensor to the 50 N position, reset the data-collection parameters as you did in Step 4, then re-zero the force sensor. Begin collecting data, then gently launch the cart through the photogate.

4. Collect data as before for at least three inelastic collisions, varying the mass of the cart. Be sure to store the data for each run. Re-shape the clay on both the force sensor and the front of the cart after each collision.

EVALUATION OF DATA

Part 1 Elastic collisions

1. In Logger *Pro*, use the Examine tool to determine the initial and final velocities of the cart as it passed through the gate. In LabQuest App, you will see two points on the velocity *vs.* time graph. Select the region of the graph containing these points and choose Statistics from the Analyze menu. The maximum and minimum values are the initial and final velocities.

2. Determine the change in velocity, $v_f - v_i$, of the cart. Keep in mind that one of these values must be negative. From Δv and the mass of the cart, determine its change in momentum, Δp.

3. As you learned in kinematics, the area under a curve often has physical significance. In the case of the *F-t* graph, the area of the region corresponding to when the cart was in contact with the spring is the product of the average force and the time during which the spring was interacting with the cart. You can determine this area by choosing Integral from the Analyze menu. In your class discussion you will give a name to this quantity.

4. Compare the value (both magnitude and sign) of the quantity you determined in Step 3 with the change in momentum of the cart.

5. Perform similar analyses for your remaining elastic collisions. Determine the % difference between the impulse, $F\,\Delta t$, and the change in momentum, Δp for each of the collisions. Compare your findings to those of others in your class. What can you conclude about these quantities?

Part 2 Inelastic collisions

1. In Logger *Pro*, use the Examine tool to determine the initial velocity of the cart as it passed through the gate. The final velocity of the cart is zero. In LabQuest App, you will see a single point on the velocity *vs.* time graph. Place the cursor on this point and read the value of the velocity.

2. Assuming the final velocity is zero, determine the change in velocity, $v_f - v_i$, of the cart. From this value and the mass of the cart, determine the change in momentum, Δp, of the cart.

3. Because some "bouncing" is unavoidable, you should discuss how to select an appropriate interval of the *F-t* graph for your determination of the impulse.

4. As you did with the elastic collisions, determine the % difference between the impulse, $F\,\Delta t$, and the change in momentum, Δp, for each of the inelastic collisions. Compare your findings to those of others in your class. What can you conclude about these quantities?

5. From Newton's second law, derive the equation you have determined from the analysis of your data. Compare the fundamental units for both impulse and change in momentum.

EXTENSIONS

1. When you catch a fast-moving baseball, it hurts less when your hand "gives" a little than if you hold your hand stiff. Explain in terms of impulse and change in momentum.

2. Now cars are made to crumple during a collision. Explain in terms of impulse and change in momentum.

3. Suppose you had used a stiffer spring in the experiment. Describe how the shape of the force *vs.* time graph would differ from that which you observed.

INSTRUCTOR INFORMATION

Experiment
10B

Impulse and Momentum

This lab is written with the assumption that the instructor will engage the students in discussions at critical junctures. These discussions can take place with the entire class or with individual lab groups. The icon at left indicates where these discussions should occur.

The Microsoft Word files for the student pages can be found on the CD that accompanies this book. See *Appendix A* for more information.

OBJECTIVES

In this experiment, the student objectives include

- Collect force, velocity, and time data for a cart experiencing different types of collisions.
- Determine an expression for the change in momentum, Δp, in terms of the force and duration of a collision.

During this experiment, you will help the students

- Manipulate graphs including selecting variables for axes on graphs, grouping graphs, and using the Integral tools in Logger *Pro* or LabQuest App.
- Recognize that the change in momentum of an object is a function of both the force acting on it and the duration of the interaction.
- Define the impulse as the area under a force *vs.* time graph.
- Develop the impulse-change in momentum expression
- Recognize that the change in momentum for an object that undergoes an inelastic collision is half of that for an elastic collision.

REQUIRED SKILLS

In order for student success in this experiment, it is expected that students know how to

- Set up a photogate in Gate mode. This is addressed in Activity 4.
- Zero a Dual-Range Force Sensor.

EQUIPMENT TIPS

While there are a number of ways one could collect suitable data for this experiment, this version uses a Dynamics Track, standard cart, Dual Range Force Sensor (DFS), and a Photogate and cart picket fence. Experiment 10A uses a Motion Detector to obtain timing data. The accessories in the Bumper and Launcher Kit (hoop springs, clay and holder) greatly facilitate the study of both elastic and inelastic collisions. The bumper that accompanies the DFS provides a collision that is too brief to allow a satisfactory analysis of the relationship between impulse and change in momentum.

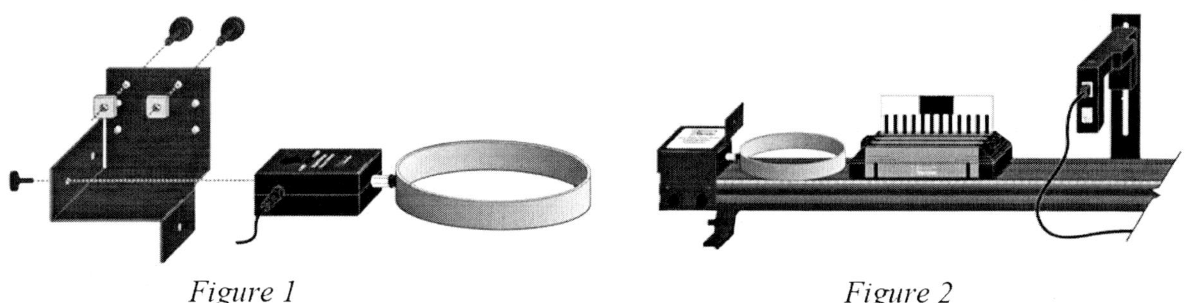

Figure 1 *Figure 2*

Note: Shown inverted for assembly.

Mount the Dual-Range Force Sensor upside down to the Dynamics Track Bracket. Then attach the assembly to the track as shown in Figures 1 and 2. Position the photogate so that the flag on the cart passes through it completely before the cart contacts the spring.

If this kit is not available, one can connect a hoop spring to the DFS and affix the sensor to the Dynamics Track using the aluminum rod that comes with the DFS and a 1/4" hex nut, as shown in Figure 3.

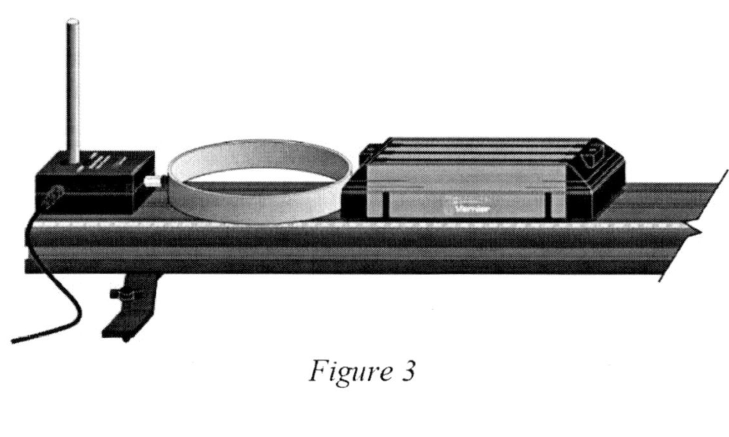

Figure 3

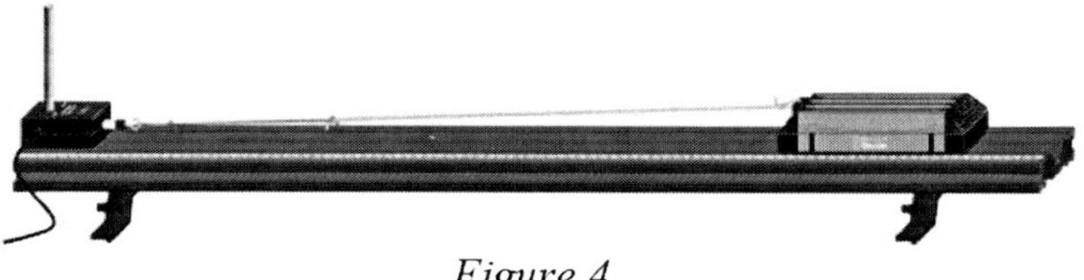

Figure 4

An alternative to the spring is to use a length of elastic cord or a long rubber band with a length of string. Attach the elastic cord to the cart and the string to the hook on a DFS, as shown in Figure 4. The string should be long enough to allow the cart to travel nearly the length of the track before the elastic cord brings the cart to a stop and reverses its direction. Make sure the flag on the picket fence passes completely through the photogate before the cart reverses direction.

PRE-LAB DISCUSSION

This experiment should be performed after students have a thorough grasp of Newton's laws of motion. Explain to students that this experiment will give them the opportunity to study momentum, *p*, which is defined as the product of the mass and velocity of an object $p = mv$.

Demonstrate the motion of a cart along a level track, pointing out that, so long as the sum of the forces acting on the cart is zero, its velocity and hence its momentum remains constant. They should recognize this as a case of Newton's first law. From Newton's second law, students should predict that a constant force acting on the cart will produce uniform acceleration. However, they may have difficulty relating a change in velocity to the *changing* force that occurs during a collision.

Demonstrate a collision between a cart and the hoop spring, as shown in Figure 2. Ask the students to describe how the force applied by the spring changes over the period of time the cart is in contact with the spring. Inform them that this experiment affords them the opportunity to study how the force and duration of a collision is related to the change in momentum of the cart.

LAB PERFORMANCE NOTES

When the photogate and force sensor are connected to the data-collection interface and Logger *Pro* is started, the default graph screen shows a number of graphs. Advise students to view only the force *vs.* time and velocity *vs.* time graphs. Because they are using two sensors in different modes, they must first choose Time Based and set the sampling rate to 500 Hz (250 Hz if using a LabPro) and reduce the duration to 5 seconds. This high sampling rate insures that sufficient data points to produce a smooth *F-t* graph are obtained during the short duration of the collision. Then, they must choose Set Up Sensors ▶ Show All Interfaces from the Experiment menu, click on the image of the photogate and select Gate Timing in order to obtain velocity data.

A similar process is used in LabQuest App. First, selecting Gate as the Photogate Mode in Photogate timing sets the application to determine the velocity of the cart as it passes through the gate. Next, returning to the Mode window and selecting Time Based, then choosing a sampling rate of 500 Hz for 5 seconds allows LabQuest App to produce a smooth *F-t* graph. In the graph window, change the vertical axis label of the gate-state *vs.* time graph to velocity.

Note: In LabQuest App, you must tap the Stop button to end data collection and see the graphs. Remind students that changing the setting on the force sensor in LabQuest App requires resetting the data collection parameters.

Make sure that students realize that they must launch the cart so that it passes through the photogate *before* it undergoes a collision. Encourage students to try collisions at different initial velocities, but make sure that they realize that the hoop spring should slow and then reverse the velocity of the cart smoothly.

SAMPLE RESULTS AND POST-LAB DISCUSSION – ELASTIC

Steps 1–2

The velocities reported by either Logger *Pro* or LabQuest App are absolute values. Students will have to decide which direction is positive and keep track of signs. This convention must be consistent with the values reported by the force sensor. Keep in mind that the default setting of the force sensor registers a push as negative.

Step 3

In previous examples in the study of graphs in kinematics or energy, the shape of the region under curve was regular, so determining the area was straightforward. In the case of the *F-t* graph (Figure 5), the area under the curve is roughly triangular, but use of the Integral tool yields a much better value for the area. Once students understand that this value represents the accumulation of the areas of many tiny rectangles, each the product of the force and the duration, you can give this quantity its physics name: *impulse.*

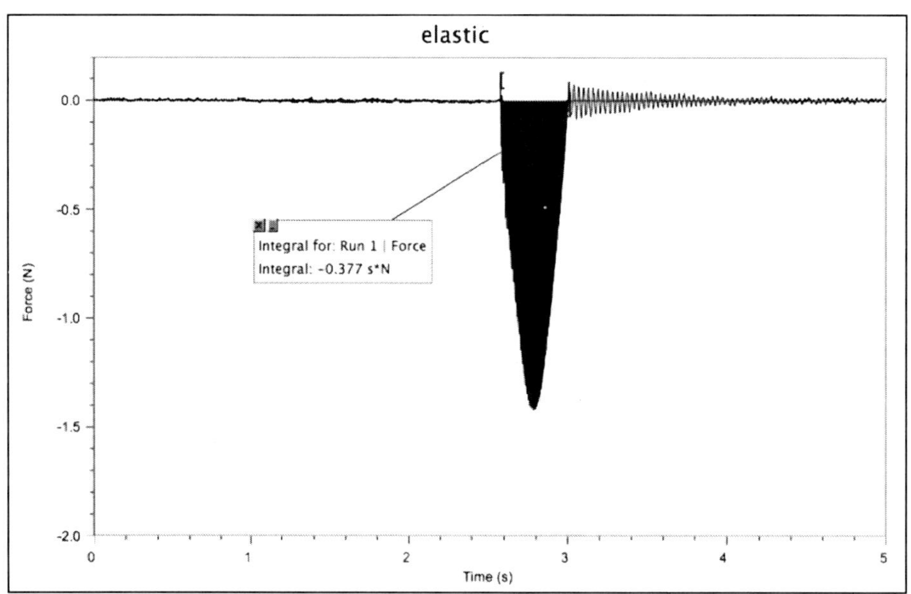

Figure 5 F-t graph–collision with hoop spring

Steps 4–5

Values for elastic collisions are listed below.

Mass (kg)	v_i (m/s)	v_f (m/s)	Δv (m/s)	Δp (kg-m/s)	$F\Delta t$ (N-)s	% difference
0.500	0.370	–0.352	–0.722	–0.361	–0.377	4.43
0.700	0.302	–0.287	–0.589	–0.412	–0.423	2.67
1.00	0.350	–0.327	–0.677	–0.677	–0.683	0.89

Students should conclude that the impulse and the change in momentum are equal. The minor differences are due to frictional losses and limitations of the sensors.

SAMPLE RESULTS AND POST-LAB DISCUSSION – INELASTIC

Steps 1–2

Students can assume that the final velocity in these collisions is zero. They should take care that the value they determine for Δv is consistent with the value reported by the force sensor.

Step 3

Students should note that there is a small set of "bumps" in the *F-t* graph representing the bounce that occurred during the "inelastic" collision. This portion of the graph should be included in the analysis of the net impulse delivered to the cart (see Figure 6).

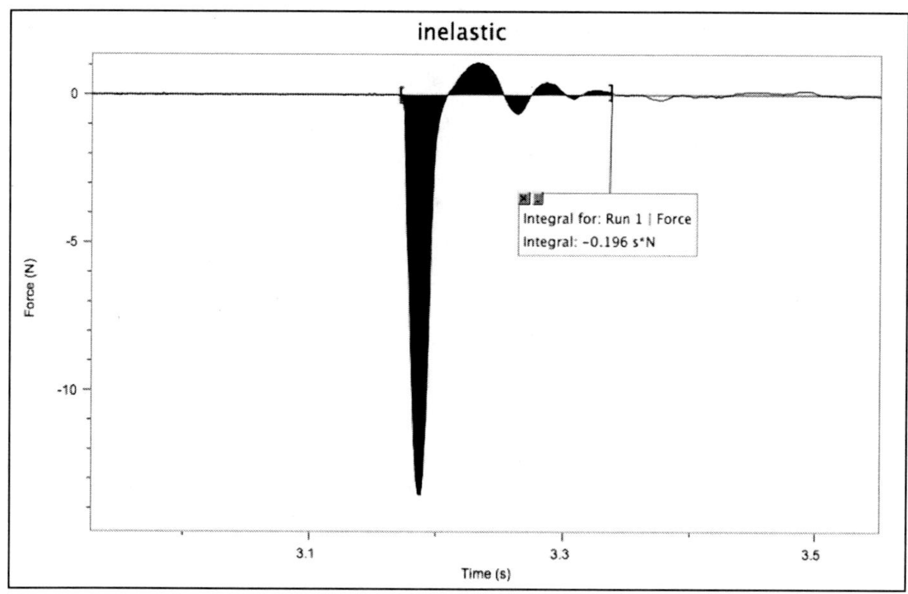

Figure 6 F-t graph for inelastic collision

Step 4

Values for inelastic collisions are listed below.

Mass (kg)	Δv (m/s)	Δp (kg-m/s)	$F\Delta t$ (N-s)	% difference
0.510	–0.391	–0.199	–0.196	1.51
0.710	–0.259	–0.184	–0.177	3.80
1.01	–0.286	–0.289	–0.282	2.42

Students should conclude that the impulse and the change in momentum are equal for inelastic collisions as well. The mathematical expression for this relationship is $F\Delta t = m\Delta v$.

Step 5

The most common expression of Newton's second law is $F_{net} = m\,a$. Substituting $\frac{\Delta v}{\Delta t}$ for a yields $F_{net} = m\frac{\Delta v}{\Delta t}$. Rearranging yields $F_{net}\Delta t = m\,\Delta v$.

We have made the assumption that the only force acting on the cart was provided by the spring or the clay. The basic units of the variables in this equation agree.

$$N \cdot s = kg \cdot \frac{m}{s}$$

$$\frac{kg \cdot m}{s^2} \cdot s = kg \cdot \frac{m}{s}$$

Having performed the experiment, students should be able to answer the pre-lab questions.

1. Airbags are used to prolong the collision. Since the impulse provided to the driver is the same whether the steering wheel or airbag stops the driver, increasing the duration (Δt) reduces the force needed to bring the driver's velocity to zero.

2. An airbag that did not deflate when the driver struck it would cause the driver to bounce backwards, thus increasing both the change in momentum and the force applied to the driver.

EXTENSION

1. By "giving" during a catch, the impact time is increased, thus reducing the force exerted on your hand.

2. An automobile body designed to crumple in a collision both reduces the change in momentum (because bouncing is reduced) and reduces the force acting on the car (because the time of impact is increased). Both are beneficial to the well-being of passengers.

3. The duration of the collision would have been shorter had a stiffer spring been used. So, the peak force applied would have been larger. This would have the effect of making the impulse a shorter duration and higher magnitude.

Experiment

11A

Momentum and Collisions

INTRODUCTION

You may have learned that a moving object possesses kinetic energy. Momentum is another property of an object, related to its mass and velocity, that is useful to describe its behavior. Momentum, *p*, is the product of the mass and velocity of an object, $p = mv$.

You may have learned an external force produces a change in the momentum of an object. If we consider as our system two carts that undergo a collision, then any forces they exert on one another are *internal* to the system. In this experiment you will examine the momentum of both carts before and after collisions to see what effect, if any, these forces have on the momentum of a *system*.

OBJECTIVES

In this experiment, you will

- Collect velocity-time data for two carts experiencing different types of collisions.
- Compare the system momentum before and after collisions.
- Compare the kinetic energy of the system before and after collisions.

MATERIALS

Vernier data-collection interface
Logger *Pro* or LabQuest App
two Vernier Motion Detectors
two Motion Detector brackets
neodymium magnets and Velcro®
  patches for carts
Vernier Dynamics Track
standard cart
plunger cart
500 g standard lab mass

PRE-LAB QUESTION

Consider a head-on collision between a cue ball and a billiard ball initially at rest. Sketch a velocity-time graph for each ball for the interval shortly before until shortly after the collision. Justify your predictions for the final velocity of each billiard ball.

PROCEDURE

1. Attach the Motion Detectors to the brackets and position them at opposite ends of the Dynamics Track.
2. If your motion detectors have a switch, set each of them to Track.

3. Adjust the leveling screws on the feet as needed to level the track. To make sure the track is level, place a cart on the track and give it a gentle push. It should not slow more in one direction than in the other.
4. Connect both motion detectors to the interface and start the data-collection program. Make the necessary adjustments so that a velocity *vs.* time graph for each detector is shown in the graph window.
5. Make sure that each of your carts has the neodymium magnets at one end and the Velcro patches at the other. Place both carts, linked with their Velcro patches, in the center of the track. Zero both motion detectors and reverse the direction of one of them.
6. Begin collecting data, then gently push the linked carts towards one of the motion detectors (see Figure 1). Be sure to keep your hands out of the way of the motion detectors. Catch the carts before they run off the track. The velocity-time graphs from each detector should be nearly mirror images of one another; they will also show a slight decrease in velocity due to friction. Adjust the level of the track until this decrease appears to be nearly the same in both directions.

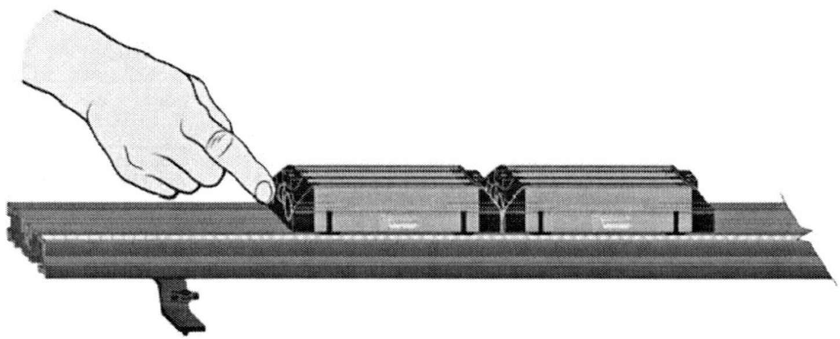

Figure 1

Part 1 Elastic collisions

7. Reverse the carts so that their magnet ends face one another. Separate them by about 40 cm. Practice launching one cart toward the other so that at closest approach they exert forces on each other without touching. A jarring collision will not yield satisfactory data.
8. Place the target cart in the middle of the track. Position the other cart at least 20 cm in front of one of the motion detectors.
9. Start data collection. Then, when you hear the motion detectors clicking, launch one of the carts toward the other. Because momentum, like velocity, is a vector quantity, check to see if the signs of the velocities match your experimental setup. If necessary, reverse the direction of one or both sensors.
10. In this experiment you are concerned with changes in momentum due to the collisions of the carts. Some slowing due to friction is inevitable. To minimize the effect of frictional losses in

your analysis, you should select short intervals of the *velocity-time* graphs just before and just after a collision. Then, choose Statistics from the Analyze menu and record the mean velocity of each cart for these intervals.

A data table has been provided for you. You may wish to use Logger *Pro* to help you record and analyze your data.

11. Collect data for up to six elastic collisions, varying the initial velocity and the mass of either cart. Try a collision in which both carts have an initial velocity, but different masses.

Part 2 Inelastic collisions

1. Reverse the carts so that the ends with the Velcro patches face one another. Practice launching one cart toward the other so that when they collide, the carts link smoothly and continue moving without a noticeable bounce. A jarring collision will not yield satisfactory data.

2. Collect data as before for at least three inelastic collisions, varying the initial velocity and the mass of either cart. Determine the velocity of the carts before and after the collision as you did in Part 1. Since both motion detectors provide velocity data after the collision, you will have to decide how to record the velocity of the linked carts.

Part 3 Explosions

1. Place the carts in the center of the track with the plunger end of one cart facing the other. Depress and lock the mechanism on the plunger cart. Position the carts so that they are touching.

2. Begin data collection, then give a quick tap to the release pin with something hard, such as the support rod for a force sensor, as shown in Figure 2. Catch the carts before they run off the track.

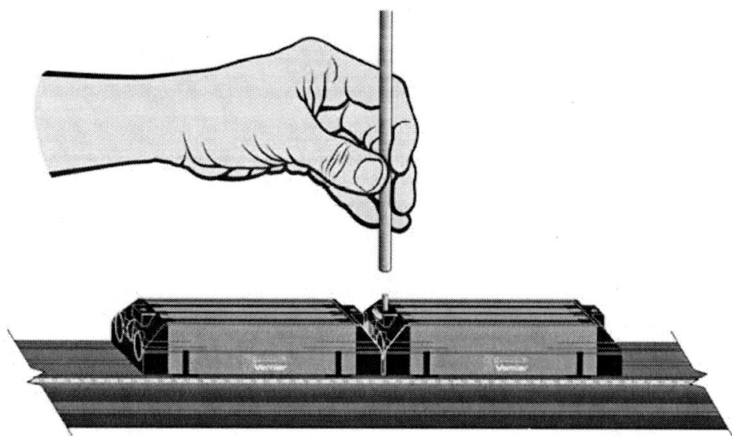

Figure 2

3. Repeat, varying the mass of either cart. Determine the velocity of the carts after the explosion as you did in Part 1.

EVALUATION OF DATA

Part 1 Elastic collisions

1. You can use the tables below to help with your evaluation of the momentum before and after the collisions of the carts.

	Cart 1			Cart 2		
Run	Mass (kg)	Initial velocity (m/s)	Final velocity (m/s)	Mass (kg)	Initial velocity (m/s)	Final velocity (m/s)
1						
2						
3						
4						
5						
6						

	Before			After			Ratio
Run	*p* of cart 1 (kg-m/s)	*p* of cart 2 (kg-m/s)	*p* of system (kg-m/s)	*p* of cart 1 (kg-m/s)	*p* of cart 2 (kg-m/s)	*p* of system (kg-m/s)	$\frac{p \text{ after}}{p \text{ before}}$
1							
2							
3							
4							
5							
6							

Another approach is to use Logger *Pro* to help you organize your calculations.

2. How does the total momentum of the system after the collision compare with that before the collision? Do your results agree with your expectations? Explain.

3. Calculate the total kinetic energy, $E_k = \frac{1}{2}mv^2$, of the system both before and after each of the collisions. How do these quantities compare?

Part 2 Inelastic collisions

1. You can use the tables below to help with your analysis of the momentum before and after the collision.

	Cart 1			Cart 2		
Run	Mass (kg)	Initial velocity (m/s)	Final velocity (m/s)	Mass (kg)	Initial velocity (m/s)	Final velocity (m/s)
1						
2						
3						

	Before			After			Ratio
Run	*p* of cart 1 (kg-m/s)	*p* of cart 2 (kg-m/s)	*p* of system (kg-m/s)	*p* of cart 1 (kg-m/s)	*p* of cart 2 (kg-m/s)	*p* of system (kg-m/s)	$\frac{p \text{ after}}{p \text{ before}}$
1							
2							
3							

How does the total momentum of the system after the collision compare to that before the collision? Is the agreement in these inelastic collisions as good as that in the elastic collisions? Try to account for any differences.

2. Calculate the total kinetic energy of the system both before and after each of the collisions. How do these quantities compare? Compare your findings with those of others in your class.

3. We have used "elastic" to describe collisions in which the objects bounce, and "inelastic" to describe collisions in which the objects stick. Based on your comparison of the kinetic energy before and after collisions, provide a more conceptual definition of these descriptors.

Part 3 Explosions

1. You can use the tables below to help with your analysis of the momentum before and after the collision.

	Cart 1			Cart 2		
Run	Mass (kg)	Initial velocity (m/s)	Final velocity (m/s)	Mass (kg)	Initial velocity (m/s)	Final velocity (m/s)
1						
2						
3						

	Before			After			Ratio
Run	*p* of cart 1 (kg-m/s)	*p* of cart 2 (kg-m/s)	*p* of system (kg-m/s)	*p* of cart 1 (kg-m/s)	*p* of cart 2 (kg-m/s)	*p* of system (kg-m/s)	% diff
1							
2							
3							

How does the total momentum of the system after the explosion compare to that when the carts were stationary? Report any discrepancy as a percentage of the momentum of one of the carts.

2. Calculate the total kinetic energy of the system both before and after each of the explosions. How do you account for the increase in kinetic energy?

Experiment
11A

INSTRUCTOR INFORMATION

Momentum and Collisions

This lab is written with the assumption that the instructor will engage the students in discussions at critical junctures. These discussions can take place with the entire class or with individual lab groups. The icon at left indicates where these discussions should occur.

The Microsoft Word files for the student pages and a sample Logger *Pro* data-analysis file can be found on the CD that accompanies this book. See *Appendix A* for more information.

OBJECTIVES

In this experiment, the student objectives include

- Collect velocity- time data for two carts experiencing different types of collisions.
- Compare the system momentum before and after collisions.
- Compare the kinetic energy of the system before and after collisions.

During this experiment, you will help the students

- Recognize that the momentum of a system is a conserved quantity; i.e., when no external agent acts on a system, its momentum remains unchanged during collisions.
- Recognize that elastic collisions are ones in which both kinetic energy and momentum are conserved.

REQUIRED SKILLS

In order for student success in this experiment, it is expected that students know how to

- Zero a Motion Detector. This is addressed in Activity 2.
- Manipulate graphs, including selecting variables for axes on graphs, grouping graphs, and using the Statistics tools in Logger *Pro*. This is addressed in Activity 3.

EQUIPMENT TIPS

While there are several ways one could collect suitable data for this experiment, this version uses a Dynamics Track, standard and plunger carts (including the magnets and Velcro patches that come with the carts), and two Motion Detectors. Experiment 11B uses Photogates and cart picket fences to obtain timing data.

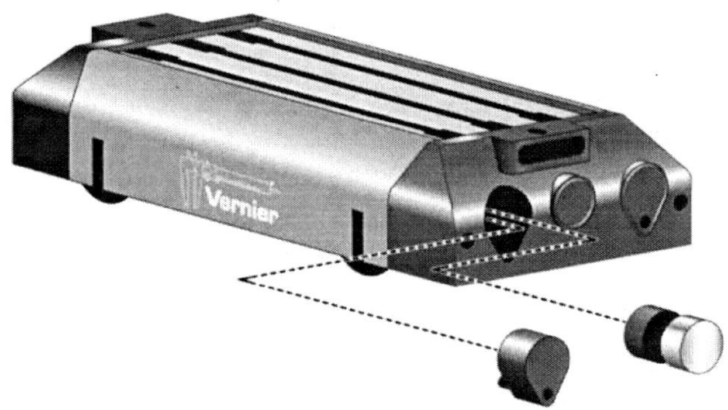

Figure 1

Prepare the carts in advance with the neodymium magnets placed in the cart end caps (see Figure 1) to allow elastic collisions without the carts physically touching one another. Placing the Velcro patches on the other ends of the carts allows for inelastic collisions.

Some slowing of the carts due to friction is unavoidable; its effect can be assessed by placing two carts in the center of the track, zeroing the motion detectors, then collecting data for a gentle push in either direction. An examination of the velocity-time graphs should help students make adjustments to level the track (see Figure 2).

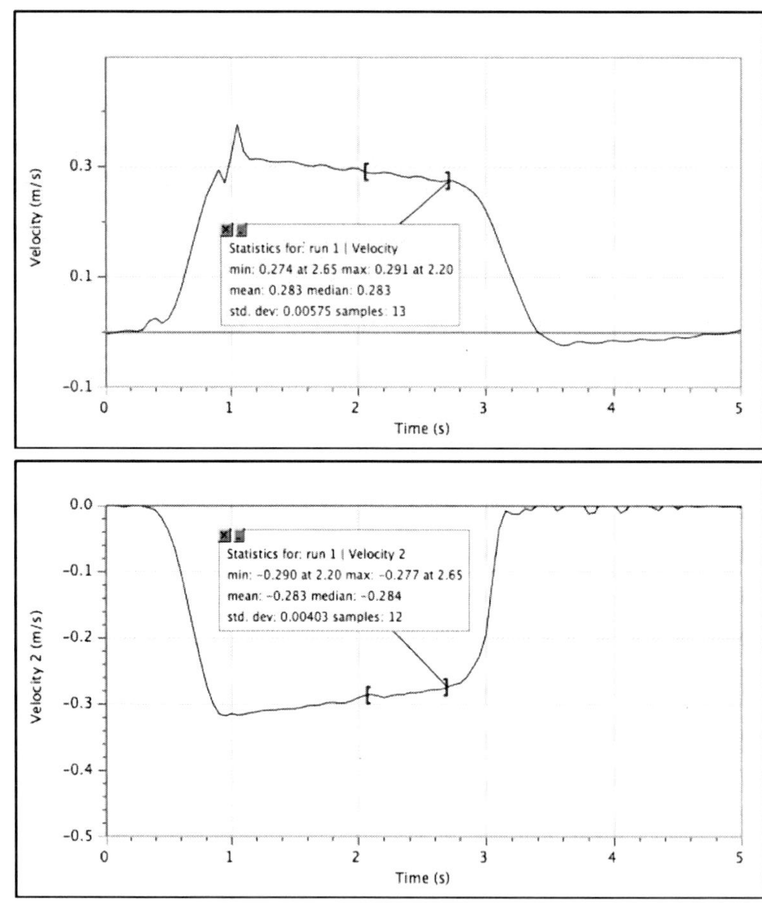

Figure 2

PRE-LAB DISCUSSION

This experiment should be performed after students have a thorough grasp of Newton's laws of motion. Be sure that students know that momentum, p, is defined as the product of the mass and velocity of an object $p = mv$. Explain to students that this experiment will give them the opportunity to study the effect of collisions on the momentum of a system of objects. Do not tell them that momentum is a conserved quantity. They should be able to determine this from the evaluation of the data they obtain in the experiment.

Demonstrate the motion of a cart along a level track, pointing out that, so long as the sum of the forces acting on the cart is zero, its velocity and hence its momentum remains constant. They should recognize this as a case of Newton's first law. Next, demonstrate an elastic collision of two carts by pushing one cart towards a stationary one. Students should note that the momentum of each changes as a result of the collision. Ask them to consider whether the momentum of the system has changed. This would be a good time for students to reflect on the pre-lab question.

LAB PERFORMANCE NOTES

When the motion detectors are connected to the data-collection interface and Logger *Pro* or LabQuest App is started, students should make the necessary adjustments so they can view the velocity-time graph from each detector.

It is essential that students define the positive direction and make adjustments to the motion detectors so that the velocity readings reflect their convention. Students need to be careful to include the sign as well as the magnitude of the velocity readings when they record data. Encourage students to determine the initial and final velocity of both carts after each run. They can record the data in the table in the lab. A sample Logger *Pro* file, 11A Momentum calculations.cmbl, provided on the CD that accompanies this book, may help students with the analysis of the data. Even if the students do not use this file, it would be instructive for them to make a plot of system momentum-final *vs.* system momentum-initial and perform a linear fit on their data.

It takes some practice to launch a cart so that a smooth collision occurs. Students need to exercise greater care when they launch both carts at one another. Encourage students to try to add the 500 g mass to either cart and to launch the colliding cart from both directions. Students also need to take care to keep their hands out of the way of the motion detectors after they push the cart(s).

Because of frictional forces, some slowing of the carts is inevitable. To minimize the effect of friction on their evaluation of data, students should select a short interval just before and after the collision. They can then use the Statistics tool to find the mean velocity of the carts during these intervals (see Figure 3). In this example, both the initial velocity of cart 2 and the final velocity of cart 1 were zero. Students can use the mass and velocity data to calculate the momentum of the carts before and after the collision.

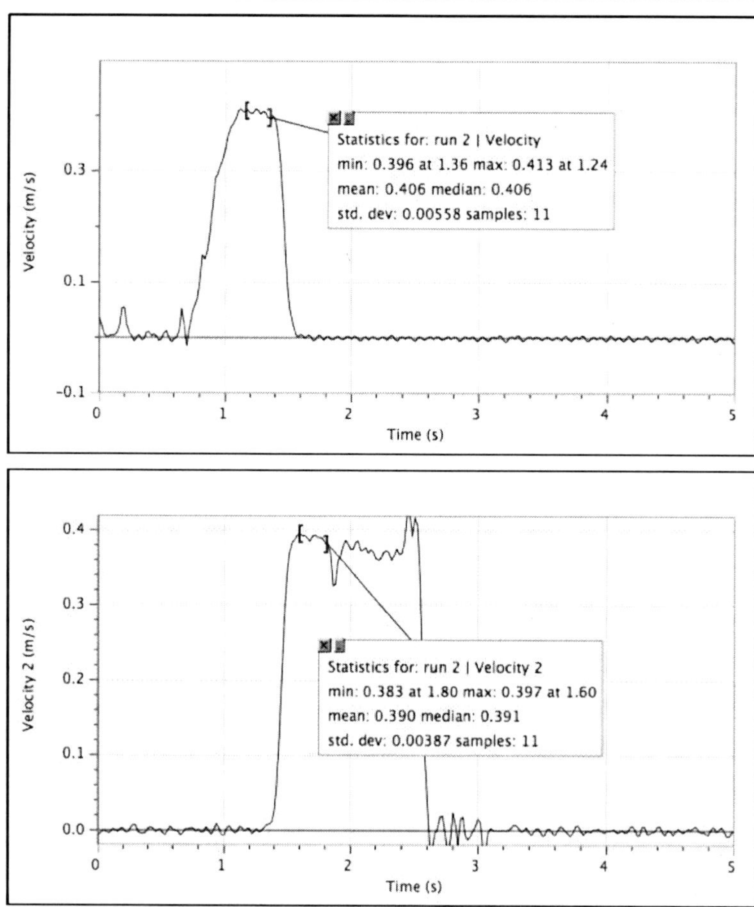

Figure 3 Determining velocity of carts

SAMPLE RESULTS AND POST-LAB DISCUSSION – ELASTIC

Step 1

Sample data for Part 1

	Cart 1			Cart 2		
Run	Mass (kg)	Initial velocity (m/s)	Final velocity (m/s)	Mass (kg)	Initial velocity (m/s)	Final velocity (m/s)
1	0.510	0.405	0	0.510	0	0.391
2	1.01	0.332	0.101	0.510	0	0.425
3	0.510	0.416	–0.0833	1.01	0	0.246
4	0.510	0	–0.356	1.01	–0.289	–0.070
5	0.510	0.184	–0.226	0.510	–0.252	0.166
6	1.01	0.205	–0.038	0.510	–0.199	0.290

Sample momenta for Part 1

	Before			After			Ratio
Run	*p* of cart 1 (kg-m/s)	*p* of cart 2 (kg-m/s)	*p* of system (kg-m/s)	*p* of cart 1 (kg-m/s)	*p* of cart 2 (kg-m/s)	*p* of system (kg-m/s)	*p* after / *p* before
1	0.207	0	0.207	0	0.199	0.199	96.5%
2	0.335	0	0.335	0.102	0.217	0.319	95.1%
3	0.212	0	0.212	–0.042	0.248	0.206	97.1%
4	0	–0.292	–0.292	–0.182	–0.071	–0.252	86.4%
5	0.094	–0.129	–0.035	–0.115	0.085	–0.031	88.2%
6	0.207	–0.199	0.106	–0.039	0.148	0.109	103.5%

Step 2

Most students will see that, neglecting frictional losses, the system momentum after the collision agrees well with the system momentum before the collision. Aside from friction, no other forces act on the carts, so the momentum ought to remain constant.

Step 3

Kinetic energy calculations

	Before			After			Ratio
Run	E_k of cart 1 (J)	E_k of cart 2 (J)	E_k of system (J)	E_k of cart 1 (J)	E_k of cart 2 (J)	E_k of system (J)	E_k after / E_k before
1	0.0418	0	0.0418	0	0.0390	0.0390	93.2%
2	0.0557	0	0.0557	0.0052	0.0461	0.0512	92.0%
3	0.0441	0	0.0441	0.0018	0.0306	0.0323	73.3%
4	0	0.0422	0.0422	0.0323	0.0025	0.0348	82.5%
5	0.0086	0.0162	0.0248	0.0130	0.0070	0.0201	80.5%
6	0.0212	0.0101	0.0313	0.0007	0.0214	0.0222	70.8%

For the simple collisions (runs 1 and 2) in which the frictional slowing did not affect the final velocity of either cart, the kinetic energy of the system is nearly conserved. Note that the greatest discrepancies occurred in runs 3 and 6, in which the final velocity of one or the other cart was very low. To be able to really answer this question, students will have to wait until they examine the sticky collisions.

SAMPLE RESULTS AND POST-LAB DISCUSSION – INELASTIC

Step 1

Sample data for Part 2

	Cart 1			Cart 2		
Run	Mass (kg)	Initial velocity (m/s)	Final velocity (m/s)	Mass (kg)	Initial velocity (m/s)	Final velocity (m/s)
1	0.510	0.463	0.226	0.510	0	0.226
2	1.01	0.354	0.228	0.510	0	0.228
3	0.510	0.464	0.151	1.01	0	0.151

Sample momenta for Part 2

	Before			After			Ratio
Run	p of cart 1 (kg-m/s)	p of cart 2 (kg-m/s)	p of system (kg-m/s)	p of cart 1 (kg-m/s)	p of cart 2 (kg-m/s)	p of system (kg-m/s)	$\frac{p \text{ after}}{p \text{ before}}$
1	0.236	0	0.236	0.115	0.115	0.231	97.6%
2	0.358	0	0.358	0.230	0.116	0.347	96.9%
3	0.237	0	0.237	0.077	0.153	0.230	97.0%

Having considered this question in Part 1, students should conclude that, neglecting frictional losses, the system momentum after the collision agrees well with the system momentum before the collision. Aside from friction, no other forces act on the carts, so the momentum ought to remain constant.

Step 2

Kinetic energy calculations

	Before			After			Ratio
Run	E_k of cart 1 (J)	E_k of cart 2 (J)	E_k of system (J)	E_k of cart 1 (J)	E_k of cart 2 (J)	E_k of system (J)	$\frac{E_k \text{ after}}{E_k \text{ before}}$
1	0.0547	0	0.0547	0.0130	0.0130	0.0260	47.7%
2	0.0633	0	0.0633	0.0263	0.0133	0.0395	62.4%
3	0.0549	0	0.0549	0.0058	0.0115	0.0173	31.6%

Step 3

Students should conclude that, even though momentum is conserved in inelastic collisions, there is a significant loss in kinetic energy. Some of this energy is transferred to other accounts (thermal, noise, etc). This is a good time to revisit question 4 from Part 1. If we use "elastic" to describe collisions in which the kinetic energy of the system is conserved, then we should view collisions on a continuum, with "elastic" and "inelastic" at the extremes. Even collisions that appear to be "bouncy" are not completely elastic.

SAMPLE RESULTS AND POST-LAB DISCUSSION – EXPLOSIONS

Step 1

	Before			After			Ratio
Run	*m* cart 1 (kg)	*m* of cart 2 (kg)	*p* of system (kg-m/s)	*p* of cart 1 (kg-m/s)	*p* of cart 2 (kg-m/s)	*p* of system (kg-m/s)	% diff
1	0.51	0.51	0	–0.292	0.286	–0.006	2.1%
2	1.01	0.51	0	–0.290	0.294	0.004	–1.3%
3	0.51	1.01	0	–0.305	0.308	0.003	–0.8%

Students should conclude that the system momentum after the explosion is very nearly the same as it was when the carts were stationary – zero.

Step 2

If students have studied energy storage and transfer prior to performing this lab, they should conclude that the elastic energy stored in the compressed spring in the plunger cart was transferred to the kinetic energy of the carts after the explosion.

GRAPH OF SYSTEM MOMENTUM

Whether or not students used Logger *Pro* to help them analyze the data, you should encourage them to plot system momentum-final *vs.* system momentum-initial and perform a linear fit on their data. The slope of this graph is very close to one, indicating that system momentum is conserved (see Figure 4).

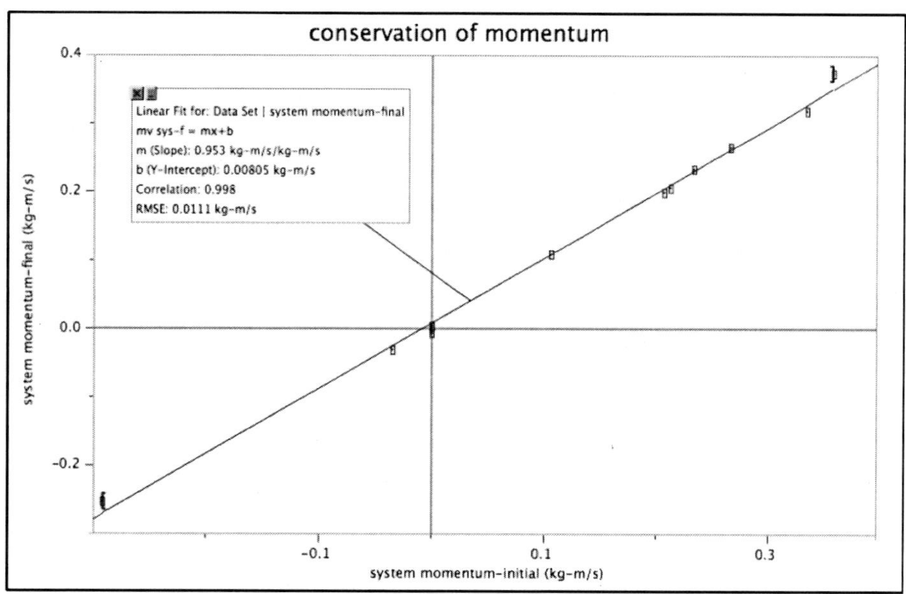

Figure 4

Experiment
11B
Momentum and Collisions

INTRODUCTION

You may have learned that a moving object possesses kinetic energy. Momentum is another property of an object, related to its mass and velocity, which is useful to describe its behavior. Momentum, p, is the product of the mass and velocity of an object, $p = mv$.

You may have learned that an external force produces a change in the momentum of an object. If we consider as our system two carts that undergo a collision, then any forces they exert on one another are *internal* to the system. In this experiment you will examine the momentum of both carts before and after collisions to see what effect, if any, these forces have on the momentum of a *system*.

OBJECTIVES

In this experiment, you will

- Collect velocity-time data for two carts experiencing different types of collisions.
- Compare the system momentum before and after collisions.
- Compare the kinetic energy of the system before and after collisions.

MATERIALS

Vernier data-collection interface
Logger *Pro* or LabQuest App
two Vernier Photogates
two track brackets
two Cart Picket Fences
500 g standard lab mass
Vernier Dynamics Track
standard cart
plunger cart
neodymium magnets and Velcro® patches for carts

PRE-LAB QUESTION

Consider a head-on collision between a cue ball and a billiard ball initially at rest. Sketch a velocity-time graph for each ball for the interval shortly before until shortly after the collision. Justify your predictions for the final velocity of each billiard ball.

PROCEDURE

1. Attach the Photogates to the brackets and position them approximately 30 cm apart in the middle of the track. Connect both photogates to the interface and start the data-collection program.

2. Set up data collection.

   Using Logger *Pro*

   a. Choose Set Up Sensors ▶ Show All Interfaces from the Experiment menu.
   b. Click on the image of each Photogate and choose Gate Timing.
   c. Make the data table window wider so that you can more easily read the velocity data. You will not need the graphs.
   d. After each run, stop data collection before you collect data for the next run.

   Using LabQuest as a standalone device

   a. Tap Mode, then select Gate as the Photogate Mode. The default distance of 0.05 m is appropriate for this experiment.
   b. If one or both carts pass through a gate more than once, you can find the velocities from the data table.

3. Adjust the leveling screws on the feet as needed to level the track. To make sure the track is level, place a Cart Picket Fence on a cart so that the 5 cm bar will block the photogate, as shown in Figure 1. Begin collecting data, then give the cart a gentle push. The data-collection program will report the speed of the cart as it passes though each gate. Note the fraction: v_f / v_i for launches in each direction and adjust the leveling feet until this fraction is nearly the same in each direction.

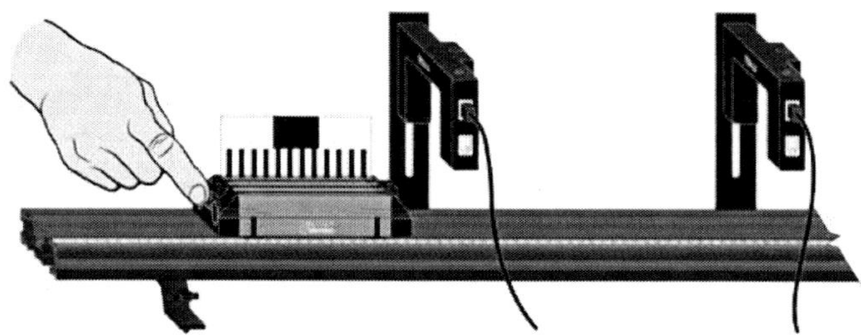

Figure 1

4. Make sure that each of your carts has the neodymium magnets at one end and the Velcro patches at the other.

Part 1 Elastic collisions

5. Place both carts on the track so that their magnet ends face one another. Place another picket fence on the second cart. Position this cart between the photogates, but nearer the second gate. Position the first cart outside of the photogates. Practice launching this cart toward the second so that, at closest approach, the carts repel one another without touching. A jarring collision will not yield satisfactory data.

6. Start data collection, then launch the cart outside of the photogates towards the other. Be sure to catch the target cart before it runs off the track.

7. Read the cart velocities as they pass through each of the photogates from the data table in either Logger *Pro* or LabQuest App. Collect data for up to six elastic collisions, varying the initial velocity and the mass of either cart. Be sure to record the mass and the initial and final velocities of each cart for each collision. Because momentum, like velocity, is a vector quantity, you need to be mindful of the signs of the velocity.

   A data table has been provided for you. You may wish to use Logger *Pro* to help you record and analyze your data.

8. In some collisions, one of the carts should pass through both gates. If the cart stops before its second pass through the gate, you will have to increase the initial velocity of the launched cart.

Part 2 Inelastic collisions

1. Reverse the carts so that the ends with the Velcro patches face one another. Practice launching one cart toward the other so that when they collide, the carts link smoothly and continue moving without a noticeable bounce. A jarring collision will not yield satisfactory data.

2. Collect data as before for at least three inelastic collisions, varying the initial velocity and the mass of either cart. Be sure to record the mass and the initial and final velocities of each cart for each collision. Since both carts pass through the second photogate, you will have to decide how to record the velocity of the linked carts.

Part 3 Explosions

1. Place the carts in the center of the track with the plunger end of one cart facing the other. Depress and lock the mechanism on the plunger cart. Position the carts so that they are touching, and are midway between the photogates.

2. Start data collection, then give a quick tap to the release pin with something hard like the support rod for a force sensor (see Figure 2). Catch the carts before they run off the track.

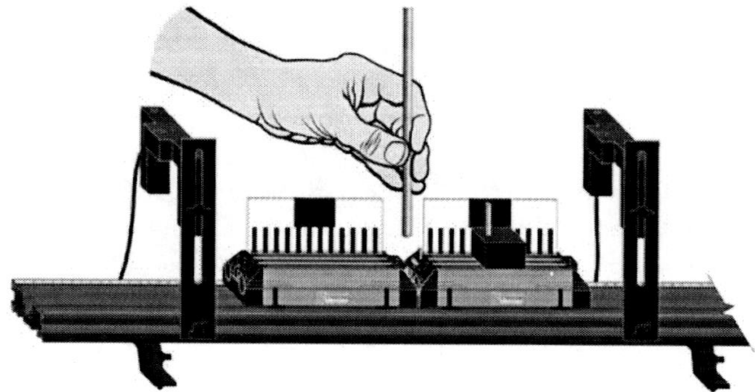

Figure 2

3. Repeat, varying the mass of either cart.

EVALUATION OF DATA

Part 1 Elastic collisions

1. In this experiment you are concerned with changes in momentum due to the collisions of the carts. While you took steps to minimize the effect of frictional losses, keep in mind that some slowing due to friction is inevitable.

2. You can use the tables below to help with your evaluation of the momentum before and after the collision.

	Cart 1			Cart 2		
Run	Mass (kg)	Initial velocity (m/s)	Final velocity (m/s)	Mass (kg)	Initial velocity (m/s)	Final velocity (m/s)
1						
2						
3						
4						
5						
6						

	Before			After			Ratio
Run	*p* of cart 1 (kg-m/s)	*p* of cart 2 (kg-m/s)	*p* of system (kg-m/s)	*p* of cart 1 (kg-m/s)	*p* of cart 2 (kg-m/s)	*p* of system (kg-m/s)	*p* after / *p* before
1							
2							
3							
4							
5							
6							

Another approach would be for you to use Logger *Pro* to help you organize your calculations.

3. How does the total momentum of the system after the collision compare with that before the collision? Do your results agree with your expectations? Explain.

4. Calculate the total kinetic energy, $E_k = \frac{1}{2}mv^2$, of the system both before and after each of the collisions. How do these quantities compare?

Part 2 Inelastic collisions

1. Note that the carts slow somewhat due to frictional losses between the time the target and the colliding cart pass through the second photogate. Decide how to report a reasonable value for the final velocity of the linked carts.

	Cart 1			Cart 2		
Run	Mass (kg)	Initial velocity (m/s)	Final velocity (m/s)	Mass (kg)	Initial velocity (m/s)	Final velocity (m/s)
1						
2						
3						

	Before			After			Ratio
Run	*p* of cart 1 (kg-m/s)	*p* of cart 2 (kg-m/s)	*p* of system (kg-m/s)	*p* of cart 1 (kg-m/s)	*p* of cart 2 (kg-m/s)	*p* of system (kg-m/s)	*p* after / *p* before
1							
2							
3							

2. How does the total momentum of the system after the collision compare to that before the collision? Is the agreement in these inelastic collisions as good as that in the elastic collisions? Try to account for any differences.

3. Calculate the total kinetic energy of the system both before and after each of the collisions. How do these quantities compare? Compare your findings with those of others in your class.

4. We have used "elastic" to describe collisions in which the objects bounce, and "inelastic" to describe collisions in which the objects stick. Based on your comparison of the kinetic energy before and after collisions, provide a more conceptual definition of these descriptors.

Part 3 Explosions

1. You can use the tables below to help with your analysis of the momentum before and after the collision.

	Cart 1			Cart 2		
Run	Mass (kg)	Initial velocity (m/s)	Final velocity (m/s)	Mass (kg)	Initial velocity (m/s)	Final velocity (m/s)
1						
2						
3						

	Before			After			Ratio
Run	*p* of cart 1 (kg-m/s)	*p* of cart 2 (kg-m/s)	*p* of system (kg-m/s)	*p* of cart 1 (kg-m/s)	*p* of cart 2 (kg-m/s)	*p* of system (kg-m/s)	% diff
1							
2							
3							

How does the total momentum of the system after the explosion compare to that when the carts were stationary? Report any discrepancy as a percentage of the momentum of one of the carts.

2. Calculate the total kinetic energy of the system both before and after each of the explosions. How do you account for the increase in kinetic energy?

INSTRUCTOR INFORMATION

Experiment **11B**

Momentum and Collisions

This lab is written with the assumption that the instructor will engage the students in discussions at critical junctures. These discussions can take place with the entire class or with individual lab groups. The icon at left indicates where these discussions should occur.

The Microsoft Word files for the student pages and a sample Logger *Pro* data-analysis file can be found on the CD that accompanies this book. See *Appendix A* for more information.

OBJECTIVES

In this experiment, the student objectives include

- Collect velocity-time data for two carts experiencing different types of collisions.
- Compare the system momentum before and after collisions.
- Compare the kinetic energy of the system before and after collisions.

During this experiment, you will help the students

- Recognize that the momentum of a system is a conserved quantity; i.e., when no external agent acts on a system, its momentum remains unchanged during collisions.
- Recognize that elastic collisions are ones in which both kinetic energy and momentum are conserved.

REQUIRED SKILLS

In order for student success in this experiment, it is expected that students know how to set up a photogate in Gate mode. This is addressed in Activity 4.

EQUIPMENT TIPS

While there are several ways one could collect suitable data for this experiment, this version uses a Dynamics Track, standard and plunger carts (including the magnets and Velcro patches that come with the carts), photogates and cart picket fences. Experiment 11A uses two Motion Detectors to obtain timing data.

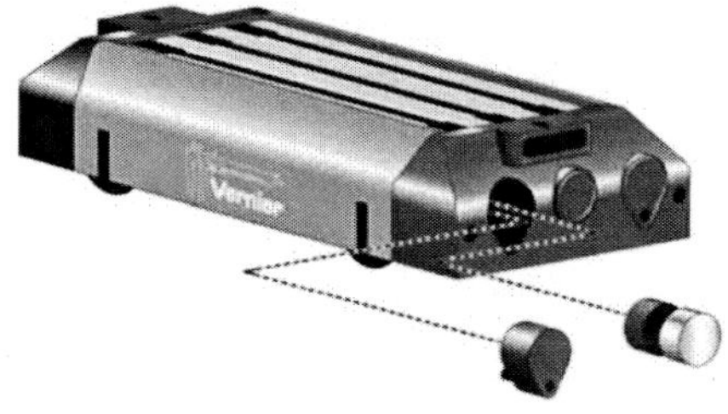

Figure 1

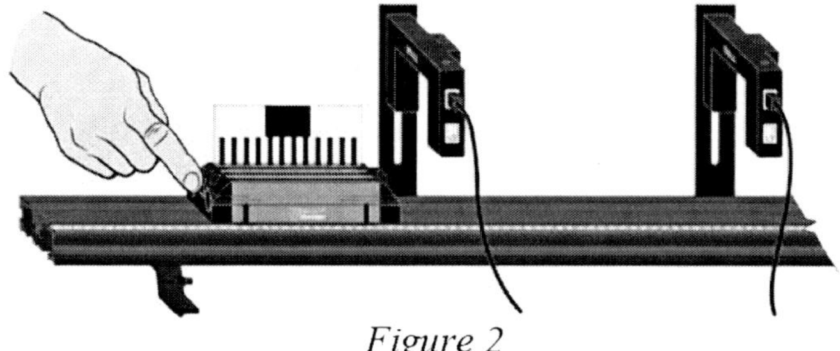

Figure 2

The carts should be prepared in advance with the neodymium magnets placed in the cart end caps (see Figure 1) to allow elastic collisions without the carts physically touching one another. Placing the Velcro patches on the other ends of the carts allows for inelastic collisions. Some slowing due to friction is unavoidable; its effect can be assessed by placing a cart outside of a pair of photogates, then collecting data for a gentle push in either direction (see Figure 2). The decrease in velocity from gate 1 to gate 2 should be approximately the same in both directions.

PRE-LAB DISCUSSION

This experiment should be performed after students have a thorough grasp of Newton's laws of motion. Be sure that students know that momentum, p, is defined as the product of the mass and velocity of an object, $p = mv$. Explain to students that this experiment will give them the opportunity to study the effect of collisions on the momentum of a system of objects. Do not tell them that momentum is a conserved quantity. Students should be able to determine this from the evaluation of the data they obtain in the experiment.

Demonstrate the motion of a cart along a level track, pointing out that, so long as the sum of the forces acting on the cart is zero, its velocity and hence its momentum remains constant. They should recognize this as a case of Newton's first law. Next, demonstrate an elastic collision of two carts by pushing one cart towards a stationary one. Students should note that the momentum of each changes as a result of the collision. Ask them to consider whether the momentum of the *system* has changed. This would be a good time for students to reflect on the pre-lab question.

LAB PERFORMANCE NOTES

When the photogates are connected to the data-collection interface and Logger *Pro* or LabQuest App is launched, students need to use the Gate Timing mode of data-collection. Explicit instructions are provided in the student version of the lab. They will not need the graphs for this experiment. Instead, they should expand the data table in Logger *Pro* so that they can see the relevant columns, as shown in Figure 3 below. In LabQuest App, they should tap on the Table tab to read the velocity values from the data table.

	Latest						
	Time (s)	State 1	GT 1 (s)	v 1 (m/s)	State 2	GT 2 (s)	v 2 (m/s)
1	2.87519	1					
2	2.99041	0	0.11521	0.434			
3	3.23940				1		
4	3.41787				0	0.17847	0.280
5	3.61823	1					
6	4.01713	0	0.39889	0.125			
7							

Figure 3

It is essential that students define the positive direction and make sure that the signs of the velocities of each cart reflect this convention when they record data. A sample Logger *Pro* file, 11B Momentum calculations.cmbl, provided on the CD that accompanies this manual, may help students with the analysis of the data. Even if the students do not use this file, it would be instructive for them to make a plot of system momentum-final *vs.* system momentum-initial and perform a linear fit on their data.

It takes some practice to launch a cart so that a smooth collision occurs. Students need to exercise greater care when they launch both carts at one another; initial velocities that are too great result in derailing the carts. Encourage students to try to add the 500 g mass to either cart and to launch the colliding cart from both directions.

SAMPLE RESULTS AND POST-LAB DISCUSSION – ELASTIC

Step 1

Students can use the mass and initial and final velocity of each cart values to calculate the momentum of the carts before and after the collision. They should keep in mind that, due to frictional forces, some slowing of the carts is inevitable. If students were careful in Step 3, this problem should be minimized.

Step 2

Sample data for Part 1

	Cart 1			Cart 2		
Run	Mass (kg)	Initial velocity (m/s)	Final velocity (m/s)	Mass (kg)	Initial velocity (m/s)	Final velocity (m/s)
1	0.51	0.398	0	0.51	0	0.382
2	0.51	0	–0.338	0.51	–0.356	0
3	0.51	0	–0.527	1.01	–0.424	–0.091
4	0.51	–0.459	0.105	1.01	0	–0.292
5	1.01	0.474	0.119	0.51	0	0.584
6	1.01	0.235	–0.121	0.51	–0.319	0.381

Sample momenta for Part 1

	Before			After			Ratio
Run	*p* of cart 1 (kg-m/s)	*p* of cart 2 (kg-m/s)	*p* of system (kg-m/s)	*p* of cart 1 (kg-m/s)	*p* of cart 2 (kg-m/s)	*p* of system (kg-m/s)	$\frac{p \text{ after}}{p \text{ before}}$
1	0.203	0.000	0.203	0	0.195	0.195	96.0%
2	0.000	–0.182	–0.182	–0.172	0.000	–0.172	94.9%
3	0.000	–0.428	–0.428	–0.269	–0.092	–0.361	84.2%
4	–0.234	0.000	–0.234	0.054	–0.295	–0.241	103.1%
5	0.479	0.000	0.479	0.120	0.298	0.418	87.3%
6	0.237	–0.163	0.074	–0.122	0.194	0.072	96.6%

Step 3

Students should see that, neglecting frictional losses, the system momentum after the collision agrees well with the system momentum before the collision. Aside from friction, no other forces act on the carts, so the momentum ought to remain constant.

Step 4

Kinetic energy calculations

	Before			After			Ratio
Run	E_k of cart 1 (J)	E_k of cart 2 (J)	E_k of system (J)	E_k of cart 1 (J)	E_k of cart 2 (J)	E_k of system (J)	$\frac{E_k \text{ after}}{E_k \text{ before}}$
1	0.0404	0.0000	0.0404	0.0291	0.0372	0.0372	92.1%
2	0.0000	0.0323	0.0323	0.0708	0.0000	0.0291	90.1%
3	0.0000	0.0908	0.0908	0.0028	0.0042	0.0750	82.6%
4	0.0537	0.0000	0.0537	0.0072	0.0431	0.0459	85.4%
5	0.1135	0.0000	0.1135	0.0074	0.0870	0.0941	83.0%
6	0.0279	0.0259	0.0538	0.0291	0.0370	0.0444	82.5%

For the simple collisions (Runs 1 and 2) in which the frictional slowing did not affect the final velocity of either cart, the kinetic energy of the system is nearly conserved. Note that greater discrepancies occurred in Runs 3–6, in which the final velocity of one or the other cart was relatively low. To be able to really answer this question, students will have to wait until they examine the sticky collisions.

SAMPLE RESULTS AND POST-LAB DISCUSSION – INELASTIC

Steps 1–2

The slowing due to friction results in a relatively large difference between the velocities reported for each cart as they pass through the second photogate. To minimize this effect, suggest to students that they should report the velocity of the target cart as the final velocity of both carts.

Sample data for Part 2

	Cart 1			Cart 2		
Run	Mass (kg)	Initial velocity (m/s)	Final velocity (m/s)	Mass (kg)	Initial velocity (m/s)	Final velocity (m/s)
1	0.51	0.364	0.171	0.51	0	0.171
2	1.01	0.362	0.229	0.51	0	0.229
3	0.51	0.435	0.133	1.01	0	0.133

Sample momenta for Part 2

	Before			After			Ratio
Run	*p* of cart 1 (kg-m/s)	*p* of cart 2 (kg-m/s)	*p* of system (kg-m/s)	*p* of cart 1 (kg-m/s)	*p* of cart 2 (kg-m/s)	*p* of system (kg-m/s)	$\frac{p \text{ after}}{p \text{ before}}$
1	0.186	0	0.186	0.087	0.087	0.174	94.0%
2	0.366	0	0.366	0.231	0.117	0.348	95.2%
3	0.222	0	0.222	0.068	0.134	0.202	91.1%

Having considered this question in Part 1, students should conclude that, neglecting frictional losses, the system momentum after the collision agrees well with the system momentum before the collision. Aside from friction, no other forces act on the carts, so the momentum ought to remain constant.

Step 3

Kinetic energy calculations

	Before			After			Ratio
Run	E_k of cart 1 (J)	E_k of cart 2 (J)	E_k of system (J)	E_k of cart 1 (J)	E_k of cart 2 (J)	E_k of system (J)	$\frac{E_k \text{ after}}{E_k \text{ before}}$
1	0.0338	0	0.0338	0.0075	0.0075	0.0149	44.1%
2	0.0662	0	0.0662	0.0265	0.0134	0.0399	60.2%
3	0.0483	0	0.0483	0.0045	0.0089	0.0134	27.9%

Step 4

Students should conclude that, even though momentum is conserved in inelastic collisions, there is a significant loss in kinetic energy compared with the "bouncy" collisions. Some of this energy is transferred to other accounts (thermal, noise, etc). This is a good time to revisit Question 4 from Part 1. If we use "elastic" to describe collisions in which the kinetic energy of the system is conserved, then we should view collisions on a continuum, with "elastic" and "inelastic" at the extremes. Even collisions that appear very "bouncy" are not completely elastic.

SAMPLE RESULTS AND POST-LAB DISCUSSION – EXPLOSIONS

Step 1

	Before			After			
Run	*m* cart 1 (kg)	*m* of cart 2 (kg)	*p* of system (kg-m/s)	*p* of cart 1 (kg-m/s)	*p* of cart 2 (kg-m/s)	*p* of system (kg-m/s)	% diff
1	0.51	0.51	0	–0.290	0.294	0.005	–1.6%
2	1.01	0.51	0	–0.319	0.316	–0.003	0.9%
3	0.51	1.01	0	–0.324	0.334	0.010	–3.2%

Students should conclude that the system momentum after the explosion is very nearly the same as it was when the carts were stationary – zero.

Step 2

If students have studied energy storage and transfer prior to performing this lab, they should conclude that the elastic energy stored in the compressed spring in the plunger cart was transferred to the kinetic energy of the carts after the explosion.

GRAPH OF SYSTEM MOMENTUM

Whether students used Logger *Pro* to help them analyze the data, you should encourage them to plot system momentum-final *vs.* system momentum-initial and perform a linear fit on their data. The slope of this graph is very close to one, indicating that system momentum is conserved.

Computer Experiment
12A
Centripetal Acceleration

INTRODUCTION

The typical response when one hears the word acceleration is to think of an object changing its speed. You have also learned that velocity has both magnitude and direction. So, an object traveling at constant speed in a circular path is undergoing an acceleration. In this experiment you will develop an expression for this type of acceleration.

OBJECTIVES

In this experiment, you will

- Analyze velocity vectors of an object undergoing uniform circular motion to determine the direction of the acceleration vector at any given moment.
- Collect force, velocity, and radius data for a mass undergoing uniform circular motion.
- Analyze the force *vs.* velocity, force *vs.* mass, and force *vs.* radius graphs.
- Determine the relationship between force, mass, velocity, and radius for an object undergoing uniform circular motion.
- Use this relationship and Newton's second law to determine an expression for centripetal acceleration.

MATERIALS

Vernier data-collection interface
Logger *Pro*
Vernier Photogate
Dual-Range Force Sensor
Vernier Centripetal Force Apparatus
masses

PRE-LAB INVESTIGATION

Tie something soft (such as a stopper) to a one-meter length of string. Taking care not to hit anyone nearby, swing the stopper so that it travels in a horizontal circular path over your head. Feel the tension force you must apply in order to keep the stopper moving at a nearly constant speed. Now, see what effect varying the speed of the stopper or the length of the string has on the force you apply to keep the stopper moving in a circular path. Record your observations.

Your instructor will lead a discussion that will enable you to determine the direction of the acceleration vector for an object moving at constant speed in a circular path. For this experiment, you will use an apparatus that will allow you to measure the force acting on an object undergoing circular motion that is more uniform than you could achieve by swinging it.

PART 1 – FORCE *VS.* VELOCITY

PROCEDURE

1. Attach a Dual-Range Force Sensor and a Vernier Photogate to the Vernier Centripetal Force Apparatus (CFA), as shown in Figure 1.

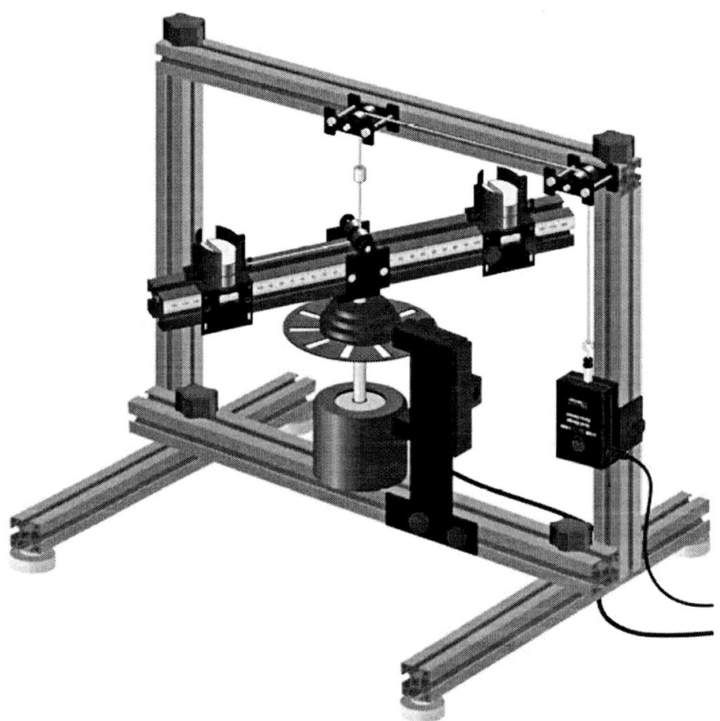

Figure 1

2. Connect the force sensor and the photogate to the interface.

3. Set up data collection.
   a. Open the experiment file 12A Centripetal Acceleration.cmbl. Data collection has been set up so that Logger *Pro* calculates the distance the carriage on the beam has travelled during its circular motion. You can examine the formula by double-clicking on the column header for Distance.
   b. Because Distance column calculation depends on the radius of the circular path, you *must* change the value of the radius parameter in Logger *Pro* whenever you move the carriage.
   c. Press the spacebar to stop collecting data when you think it is appropriate to do so.

4. Determine the mass of the sliding mass carriage. Add mass to both the sliding and fixed mass carriages as directed by your instructor. The mass of the sliding and fixed carriages should be the same so that the beam is balanced. Record the total mass of the sliding carriage and extra mass.

5. Position the fixed carriage so that its center is 10 cm from the axis of rotation. Adjust the position of the force sensor on the rail so that, when the line is taut, the center of the sliding mass carriage is also at 10 cm. Make sure that the parameter, radius, in Logger *Pro* is set to 0.10 m.

6. Zero the force sensor.

7. Spin the beam by twisting the knurled spindle of the CFA with your fingers. When the force reaches 5–8 N, begin collecting data. When you stop data-collection, use the friction between your hand and the knurled spindle to stop the beam.

EVALUATION OF DATA

1. Choose Next Page from the Page menu. Note that the vertical axis displays Force-interpolated; these are values that Logger *Pro* has interpolated from the values of force measured by the sensor.

2. Write a statement that describes the relationship between the force acting on the carriage and its tangential velocity.

3. If your graph of force *vs*. velocity is not linear, take steps to modify a column so as to obtain a linear relationship. Choose New Graph from the Insert menu, choose the new column variable for the horizontal axes, and then rearrange the graphs on the page.

4. Write the equation of the line that best fits your linearized graph. Simplify the units of your slope as much as possible. Save your Logger *Pro* file.

5. Compare the value of the slope of the linearized graph with that obtained by other groups in your class. Speculate about what might be responsible for any differences in the slopes.

PART 2 – INVESTIGATING THE EFFECT OF MASS AND RADIUS

When a quantity (in this case, force) is a function of more than one variable, it is usually the case that the slope of the graph is related to the parameters held constant during the experiment. Examine the units of the slope of your graph of F *vs*. v^2. Write an expression involving mass and radius that has the same units as that of your slope. Substitute the known values of these parameters; how closely does the value of this expression agree with that of your slope? Predict the effect of doubling the mass on the value of the slope. What effect would doubling the radius have on the slope? You can test your conclusions by varying first the mass and then the radius as follows.

PROCEDURE

1. Store your first run.

2. Change the mass on both the fixed and sliding carriages and record the value of the total mass of the sliding carriage and any extra masses. Return to Page 1 of your experiment file.

3. Re-zero the force sensor, then spin the beam as you did before. Once the force reaches 5–8 N, begin collecting data. When you stop data collection, stop the beam as you did in Part 1. Store this run.

4. Change the system mass again and record the value of the total mass of the sliding carriage and any extra masses, then repeat Step 3.

5. Return the mass on both the fixed and sliding carriages to the original value used in Part 1.

6. Decrease the radius of both the sliding and fixed mass carriages by 2–3 cm. Record the value of the radius.

7. Re-zero the force sensor, then spin the beam as you did before. Once the force reaches 5–8 N, begin collecting data. When you stop data collection, stop the beam. Store this run.

8. Now, change the radius so that it is 2–3 cm greater than your initial value. Record the value of the radius, then repeat Step 7.

EVALUATION OF DATA

1. Return to Page 2 of your Logger *Pro* file.

2. Select More on the vertical axis on the Force-interpolated *vs.* velocity graph and select the interpolated force for the three runs in which mass was varied. On the graph you should see a family of curves.

3. Now do the same for the Force-interpolated *vs.* velocity^2 (F-i *vs.* v^2) graph. Perform linear fits on all three sets of data. Record the value of the slope of each of the equations of the lines. What relationship appears to exist between the value of the slope and the total mass of the sliding carriage?

4. To study the effect of changing the radius, select More on the vertical axis of the F-i *vs.* v^2 graph. De-select Force-interpolated for the runs you examined in Step 3. Now, select it for one of the runs in which you changed the radius.

5. Because the velocity was calculated using the value of the radius, you must set this parameter to the radius used for each run you wish to examine. Perform a linear fit on the data for the desired run. Compare the value of the slope for this run to that for your first run ($r = 0.10$ m).

6. Repeat Steps 4 and 5 for another run in which you changed the radius. Does the change in the radius have the expected effect on the value of the slope? Compare your findings with those of other groups in class.

7. Write an equation relating the net force, mass, radius and velocity of a system undergoing circular motion.

8. Use what you have learned in Steps 3–5 of this section and Newton's second law to write an equation for the acceleration of the object undergoing circular motion. Use your text or a web resource to determine the meaning of the term "centripetal."

EXTENSION

Wikipedia warns that the centripetal force is not to be confused with centrifugal force. It describes the latter as a fictitious or inertial force. Describe an example of such a force that you have experienced and how this interaction might better be explained in terms of centripetal force.

Centripetal Acceleration

INTRODUCTION

The typical response when one hears the word acceleration is to think of an object changing its speed. You have also learned that velocity has both magnitude and direction. So, an object traveling at constant speed in a circular path is undergoing an acceleration. In this experiment you will develop an expression for this type of acceleration.

OBJECTIVES

In this experiment, you will

- Analyze velocity vectors of an object undergoing uniform circular motion to determine the direction of the acceleration vector at any given moment.
- Collect force, velocity, and radius data for a mass undergoing uniform circular motion.
- Analyze the force *vs.* velocity, force *vs.* mass, and force *vs.* radius graphs.
- Determine the relationship between force, mass, velocity, and radius for an object undergoing uniform circular motion.
- Use this relationship and Newton's second law to determine an expression for centripetal acceleration.

MATERIALS

LabQuest
Vernier Photogate
Vernier Dual-Range Force Sensor
Vernier Centripetal Force Apparatus
masses

PRE-LAB INVESTIGATION

Tie something soft (such as a stopper) to a one-meter length of string. Taking care not to hit anyone nearby, swing the stopper so that it travels in a horizontal circular path over your head. Feel the tension force you must apply in order to keep the stopper moving at a nearly constant speed. Now, see what effect varying the speed of the stopper or the length of the string has on the force you apply to keep the stopper moving in a circular path. Record your observations.

Your instructor will lead a discussion that will enable you to determine the direction of the acceleration vector for an object moving at constant speed in a circular path. For this experiment, you will use an apparatus that will allow you to measure the force acting on an object undergoing circular motion that is more uniform than you could achieve by swinging it.

PART 1 – FORCE *VS.* VELOCITY

PROCEDURE

1. Attach a Dual-Range Force Sensor and a Vernier Photogate to the Vernier Centripetal Force Apparatus (CFA), as shown in Figure 1.

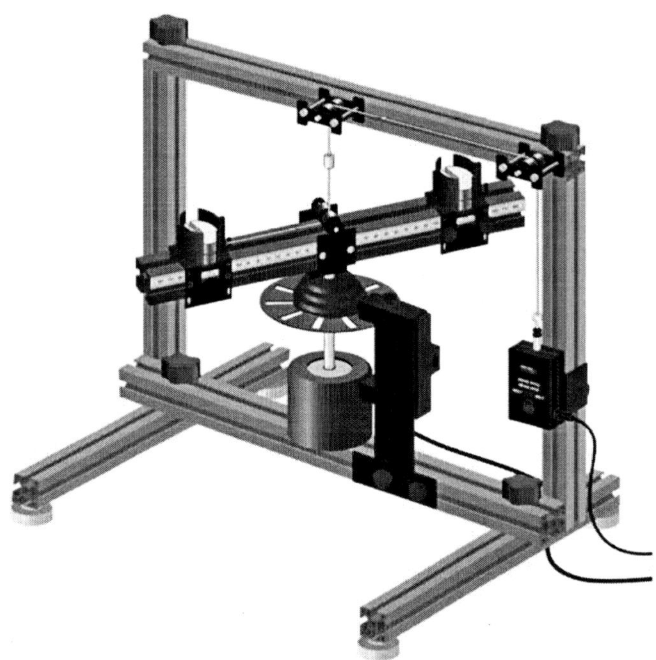

Figure 1

2. Connect the force sensor and the photogate to LabQuest.

3. Set up data collection.
   a. Tap Mode. The default Photogate Mode▶Motion works for this experiment. Select the User defined button. The distance to be entered is 1/10 of the circumference of the mass path. Because the radius in part 1 is 10 cm, this value should be 0.0628 m.
   b. To control when data collection stops, tap the option labeled "with the Stop button" in the field labeled "End data collection." Tap OK.
   c. From the Graph menu select Graph 1. Change the axes of the graph so that it shows force *vs.* velocity.
   d. Tap the Meter tab.

4. Determine the mass of the sliding mass carriage. Add mass to both the sliding and fixed mass carriages as directed by your instructor. The mass of the sliding and fixed carriages should be the same so that the beam is balanced. Record the total mass of the sliding carriage and extra mass.

5. Position the fixed carriage so that its center is 10 cm from the axis of rotation. Adjust the position of the force sensor on the rail so that, when the line is taut, the center of the sliding mass carriage is also at 10 cm.

6. Zero the force sensor.

7. Spin the beam by twisting the knurled spindle of the CFA with your fingers. When the force reaches 5–8 N, begin collecting data. Let data collection continue for 20-40 s. When you stop data collection, use the friction between your hand and the knurled spindle to stop the beam.

EVALUATION OF DATA

1. Examine your graph. Write a statement that describes the relationship between the force acting on the carriage and its tangential velocity.
2. If your graph of force *vs.* velocity is not linear, create a calculated column that will enable you to obtain a linear relationship.
3. From the Graph menu select Graph 1. Set the graph up to show the linearized plot.
4. Write the equation of the line that best fits your linearized graph. Determine the units of the slope; simplify these as much as possible. Save your LabQuest App file.
5. Compare the value of the slope of the linearized graph with that obtained by other groups in your class. Speculate about what might be responsible for any differences in the slopes.

PART 2 – INVESTIGATING THE EFFECT OF MASS AND RADIUS

When a quantity (in this case, force) is a function of more than one variable, it is usually the case that the slope of the graph is related to the parameters held constant during the experiment. Examine the units of the slope of your graph of F *vs.* v^2. Write an expression involving mass and radius that has the same units as that of your slope. Substitute the known values of these parameters; how closely does the value of this expression agree with that of your slope? Predict the effect of doubling the mass on the value of the slope. What effect would doubling the radius have on the slope? You can test your conclusions by varying first the mass and then the radius as follows.

PROCEDURE

1. Change the mass on both the fixed and sliding carriages and record the value of the total mass of the sliding carriage and any extra masses.
2. Re-zero the force sensor, then spin the beam as you did before. Once the force reaches 5–8 N, begin collecting data. When you stop data collection, stop the beam as you did in Part 1.
3. Perform a linear fit to the data on the F *vs.* v^2. Record the value of the slope for the fit.
4. Change the system mass again and record the value of the total mass of the sliding carriage and any extra masses, then repeat Step 2 and 3.
5. Return the mass on both the fixed and sliding carriages to the original value used in Part 1. Decrease the radius of both the sliding and fixed mass carriage by 2-3 cm. Record the value of the radius. Adjust the value of the User defined distance in the Mode window so that it is 1/10 of the circular path of the mass carriage. A warning will appear. Choose to discard the data.
6. Re-zero the force sensor, then spin the beam as you did before. Once the force reaches 5–8 N, begin collecting data. When you stop data collection, stop the beam.
7. Set up Graph 1 to be a plot of F *vs.* v^2. Perform a linear fit to the data. Record the value of the slope for the fit.

8. Now, change the radius so that it is 2–3 cm greater than your initial value. Be sure to adjust the User defined distance as you did in Step 5. Record the value of the radius, then repeat Step 6 and 7.

EVALUATION OF DATA

1. Examine the value of the slope of each of the first three runs. What relationship appears to exist between the value of the slope and the total mass of the sliding carriage?

2. To study the effect of changing the radius, select one of the runs in which you changed the radius from its original value of 0.10 m.

3. Compare the value of the slope for this run to that for your first run (r = 0.10 m).

4. Repeat Steps 2 and 3 for another run in which you changed the radius. Does the change in the radius have the expected effect on the value of the slope? Compare your findings with those of other groups in class.

6. Write an equation relating the net force, mass, radius and velocity of a system undergoing circular motion.

7. Use what you have learned in Steps 1–5 and Newton's second law to write an equation for the acceleration of the object undergoing circular motion. Use your text or a web resource to determine the meaning of the term "centripetal."

EXTENSION

Wikipedia warns that the centripetal force is not to be confused with centrifugal force. It describes the latter as a fictitious or inertial force. Describe an example of such a force that you have experienced and how this interaction might better be explained in terms of centripetal force.

Experiment
12A

INSTRUCTOR INFORMATION

Centripetal Acceleration

In this version of the experiment, students use the Centripetal Force Apparatus (CFA) to develop an expression for centripetal acceleration. While the object gradually slows during data collection, for any short interval, the object undergoes nearly uniform circular motion in a horizontal plane. This simplifies the treatment of centripetal acceleration in the pre-lab discussion.

The approach used in the alternate version (12B) employs readily available equipment, but it requires careful pre-lab discussion to help students understand that we are treating the arc of a pendulum bob as a portion of a circular path.

This lab is written with the assumption that the instructor will engage the students in discussions at critical junctures. These discussions can take place with the entire class or with individual lab groups. The icon at left indicates where these discussions should occur.

The Microsoft Word files for the student pages can be found on the CD that accompanies this book. See *Appendix A* for more information.

OBJECTIVES

In this experiment, the student objectives include

- Analyze velocity vectors of an object undergoing uniform circular motion to determine the direction of the acceleration vector at any given moment.
- Collect force, velocity, and radius data for a mass undergoing uniform circular motion.
- Analyze the force *vs.* velocity, force *vs.* mass, and force *vs.* radius graphs.
- Determine the relationship between force, mass, velocity, and radius for an object undergoing uniform circular motion.
- Use this relationship and Newton's second law to determine an expression for centripetal acceleration.

During this experiment, you will help the students

- Recognize that the instantaneous velocity of an object undergoing uniform circular motion is tangent to the circle at any point in its path.
- Recognize that the acceleration of such an object is perpendicular to the instantaneous velocity; thus it is directed toward the center of the circle.

REQUIRED SKILLS

In order for student success in this experiment, it is expected that students know how to

- Create calculated columns in the software. This is addressed in Activities 1 and 3.
- Use calculated columns to linearize graphs. This is addressed in Activity 1.

EQUIPMENT TIPS

Finding a way to collect suitable data for the analysis called for in this experiment has long been the bane of physics teachers. The Vernier Centripetal Force Apparatus allows students to measure the force acting on an object undergoing a gradually slowing circular motion while controlling the independent variables mass and radius.

Secure a Dual-Range Force Sensor to the vertical support of the CFA. A Vernier Photogate should be attached to the bottom rail, positioned so that the spokes of the disk will interrupt the beam as the spindle rotates (see Figure 1).

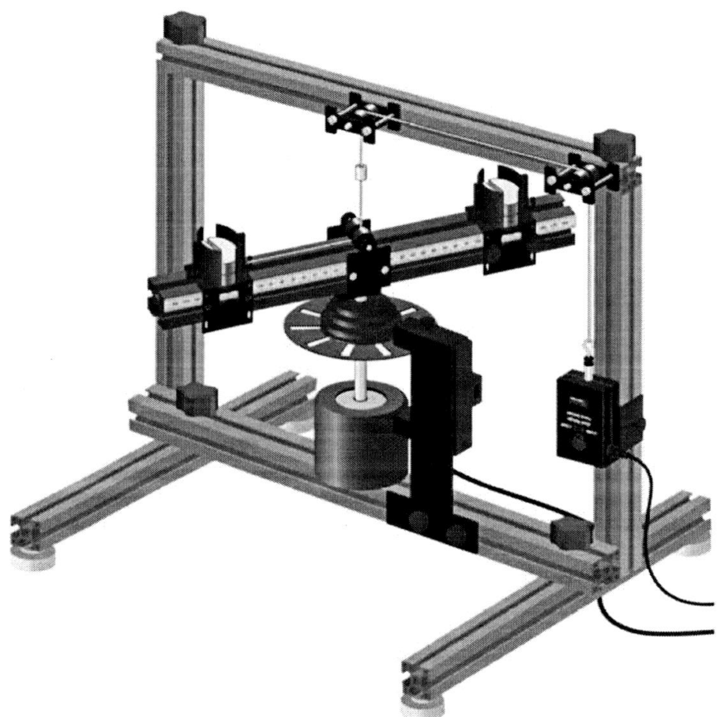

Figure 1

To minimize "noise" in the data due to wobbling of the beam, the following steps should be taken to make sure the apparatus is level.

- Remove the sliding mass carriage from the beam. Move the fixed carriage to the end of the beam and add a 100 g mass to it.
- If the apparatus is not level, the beam will swing so that it oscillates around the lowest point in its path. Adjust the leveling feet until the beam no longer has a preferred orientation.
- Replace the sliding carriage on the beam and proceed with the experiment.

PRE-LAB DISCUSSION

This experiment should be performed only after students are thoroughly versed in vector addition, the use of force diagrams and have explored Newton's second law. Start by demonstrating uniform circular motion, whirling a soft object on the end of a string in a horizontal plane overhead. Ask students if the velocity of the object is changing. If they at first note only that the speed is constant, remind them that velocity also depends on direction. Since the velocity is constantly changing, the bob is constantly accelerating; a force is required to produce this acceleration. If your lab space allows for it, it would be worthwhile for students to experience how the tension force they must provide to keep a stopper in uniform circular motion depends on the speed and the radius of the circle.

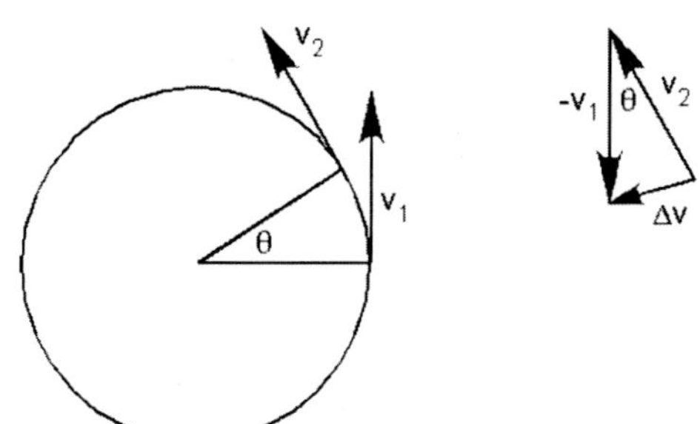

After their initial investigation, it is important to help students recognize that the acceleration at any given instant is perpendicular to the tangential velocity. Since $a = \Delta v/\Delta t$, the difference of vectors v_2 and v_1, taken at the mid-time (the point midway between the origin of the two vectors) is directed toward the center of the circle. As Δt goes to 0, the vector $\vec{a}$ is perpendicular to $\vec{v}$ at that instant.

LAB PERFORMANCE NOTES

Because rotational kinematics is not addressed until the next experiment (13 Rotational Dynamics) this experiment uses units of linear velocity (m/s) rather than those of angular velocity (rad/s). Whether students use Logger *Pro* or LabQuest App, they must provide the value of the radius for the software to correctly calculate the velocity. A Logger *Pro* file 12A Centripetal Acceleration.cmbl, is provided on the CD that accompanies this book for students to use in collecting data. A ten spoke wheel is mounted to the axle that supports the spinning horizontal beam. The spokes of the wheel interrupt the photogate beam. The time generated by these events is used to calculate the velocity of the carriage. It is instructive for your students to understand this calculation.

Double click the Distance column heading in the Data Table. The Calculated Column Options dialog box shows the following equation for the distance calculation.

```
stepColumnBased("Gate State", 0, 0.628*Radius, 1, 1)
```

This equation determines how far the carriage moves as successive spokes of the wheel break the beam. That distance is "0.628*Radius", where "Radius" is the radius of the motion of the carriage and "0.628" is the circumference of the motion (2π radians) divided by 10 spokes.

Logger *Pro* users must enter the value of the radius in the parameter field; LabQuest App users need to enter 1/10 of the circumference of the mass path (initially 0.0628 m) as the value of **User defined** in Photogate Timing mode. They must adjust this value each time they change the radius in Part 2. With either software, students must stop collecting data manually – by pressing the spacebar in Logger *Pro* or tapping the Stop button in LabQuest App. It is left to students to decide when to do so. This provides an interesting challenge to students. If they stop too quickly, the resulting force *vs.* velocity graph appears somewhat linear, but has a significant negative vertical intercept (See Figure 2). Should students make this mistake, ask them what was the value of the force when the beam was motionless. Encourage them to repeat the run, collecting data for a longer time. As a rule of thumb, students should obtain a recognizable upward-opening parabola for the force *vs.* velocity graph if they stop when the velocity is approximately one-third of its original value.

Assign different masses to student groups so there will be some variation in the values of the slope of their F *vs.* v^2 graphs. Instruct students to spin the beam up to speed by twisting the knurled spindle rather than by applying a force to the beam itself. The latter approach tends to introduce a wobble in the beam that gives rise to erratic force *vs.* velocity data.

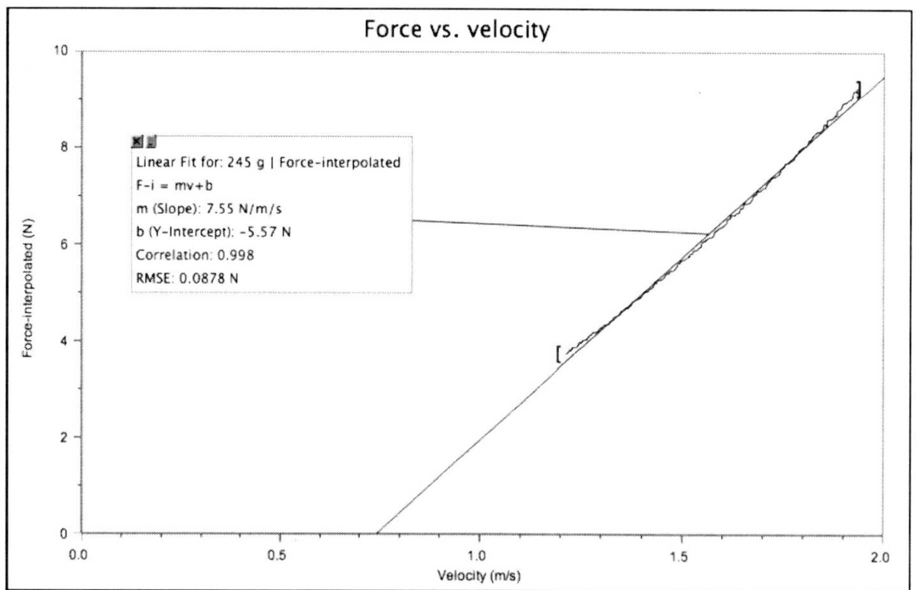

Figure 2 Stopping too soon

After students have performed the evaluation of data for Part 1, they need to perform additional runs in Part 2 to test their ideas about the relationships among the slope, mass and radius. Varying mass is no problem, but it is important that students remember to adjust the parameter **radius** when they analyze the runs in which they changed the radius from its initial value.

SAMPLE RESULTS AND POST-LAB DISCUSSION–PART 1

Steps 1–2

The plot of Force-interpolated *vs.* velocity appears to be a portion of an upward-opening parabola indicating that force is proportional to the square of velocity.

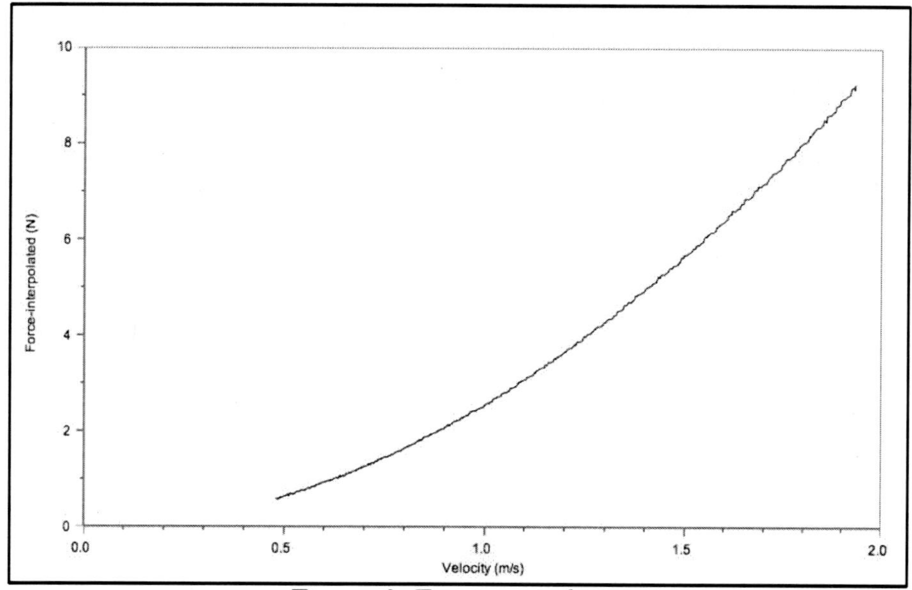

Figure 3 Force vs. velocity

Steps 3–4

By this point, students should be familiar with "linearizing" a graph. If not, or if they need a reminder, see the file, 10-1 Linearization.cmbl, in the Tutorial folder in Logger *Pro*.

Students should add a new calculated column for velocity squared and insert a new graph of force *vs.* velocity squared on the page (see Figure 4).

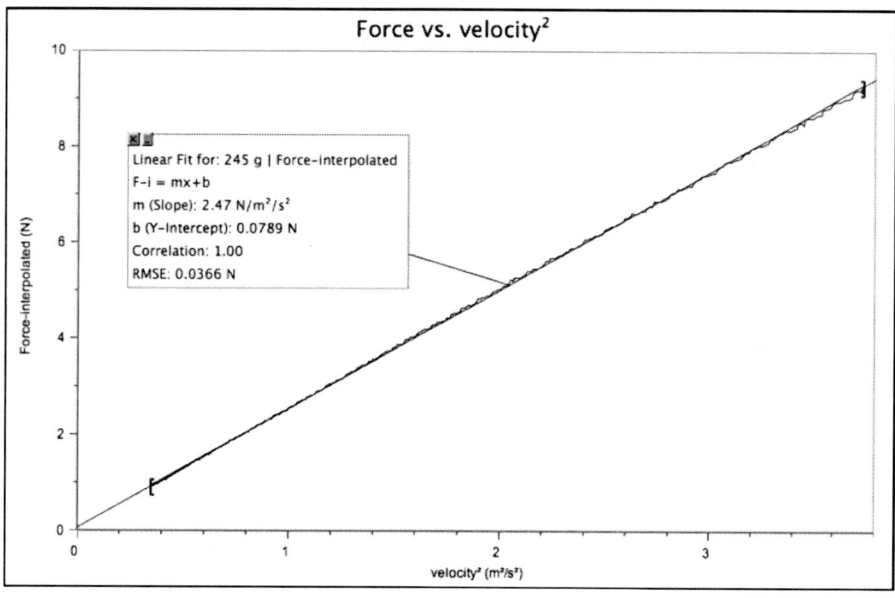

Figure 4 Force vs. velocity²

The equation representing the relationship shown in the graph is $F = \left(2.47\frac{\text{N}}{\text{m}^2/\text{s}^2}\right)v^2 - 0.0789\text{N}$.

This intercept is negligible. When the units of the slope are simplified, one obtains kg/m. Students might note that these are the units for mass and length and speculate that the slope is related to the variables (mass and radius) held constant in this part of the experiment. Whether they do or not at this time, they will have the opportunity to examine this relationship in Part 2.

Step 5

When students compare the slope of the lines of best fit, they should find that the greater the mass, the steeper the slope.

SAMPLE RESULTS AND POST-LAB DISCUSSION–PART 2

Steps 1–3

When students select Force-interpolated for the three data sets in which they have changed the mass and then perform linear fits, they should obtain a graph similar to that in Figure 5. Students should readily see that the slope of the lines of best fit are all nearly 10 times as large as the mass in kilograms. This not only suggests that the slope is proportional to mass, but also that the slope could be equal to *m/r*. They will confirm this relationship in the following steps.

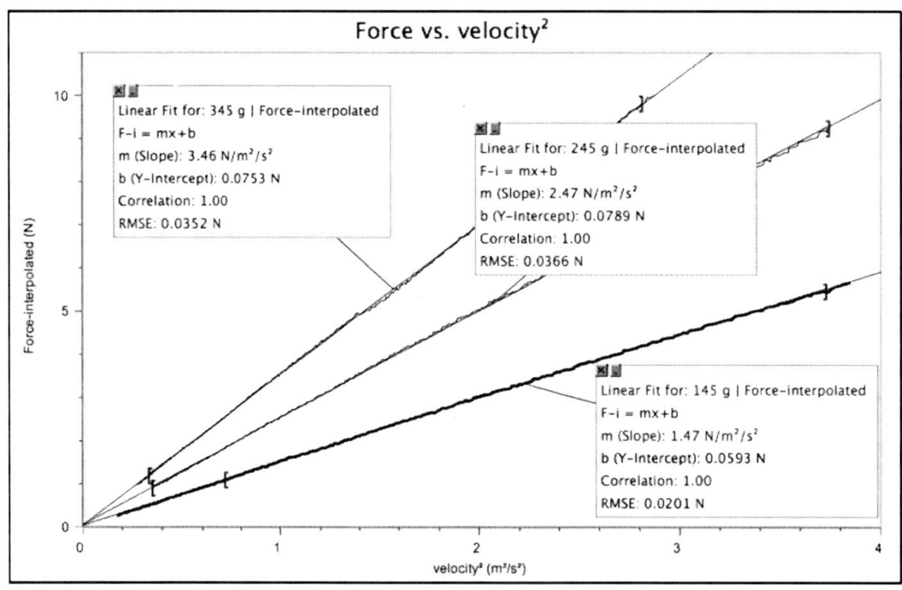

Figure 5 Varying mass with constant radius of 10 cm

Steps 4–6

After students have de-selected Force-interpolated for the runs in which they varied the mass, they should select a run in which they have changed the radius. It is important that they adjust the radius parameter before they attempt a linear fit on the new run. When they do so, they should obtain a graph like the one shown in Figure 6.

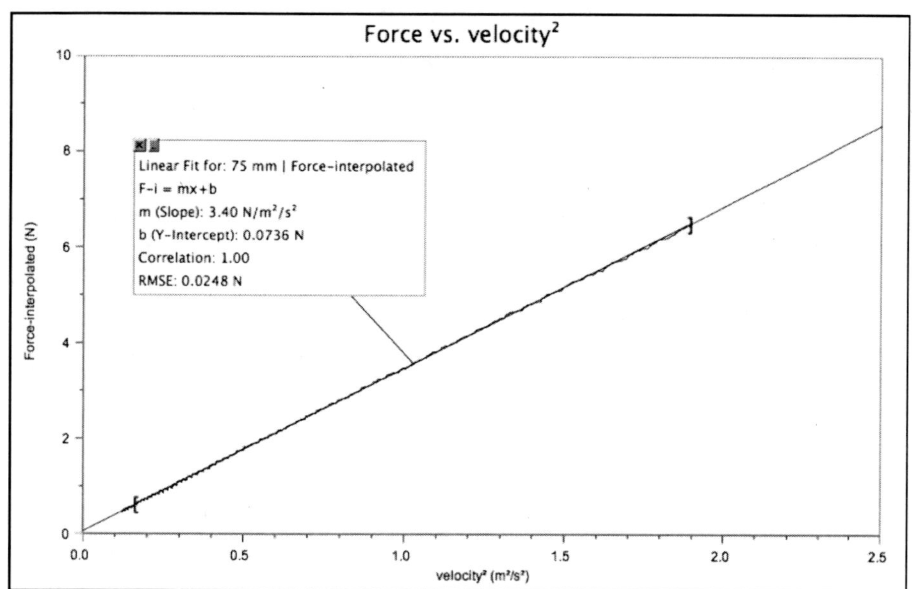

Figure 6 Varying the radius

Note that the value of the slope for this run (r = 0.075 m) is larger than that for the run with m = 0.245 kg shown in Figure 5. This further supports the hypothesis that the slope of the graph is inversely related to the radius.

It is important to note that students should examine only one run at a time because the value of the slope of each line changes as one changes the radius parameter. For the run in the graph shown in Figure 6, the value of m/r is $0.245\,\text{kg}/0.075\,\text{m} = 3.27\,\text{kg/m}$. The slope of the graph is within 4% of this value. Similar correspondence is found for the run when the radius is decreased.

For a more robust analysis, students could pool the slope and radius data from runs in which the mass was the same. They could then plot slope *vs.* radius and find that the slope of the F vs. v^2 graph is inversely proportional to the radius.

Step 7

This analysis suggests that the overall relationship is

$$F_{net} = \frac{m}{r} \cdot v^2$$

Rearranging terms gives the commonly written equation for centripetal force

$$F_{net} = m\frac{v^2}{r}.$$

Step 8

Students should recall from the pre-lab discussion that even though the *speed* of an object undergoing uniform circular motion is constant, its *velocity* is constantly changing. From Newton's second law, students should deduce that the acceleration of an object undergoing uniform circular motion is given by $a = v^2/r$. Physicists use the term "centripetal" to describe a force or acceleration directed toward the center of the circle.

EXTENSION

A centrifugal (or center-fleeing) force is the fictitious force one feels as a result of one's perspective from a non-inertial reference frame. If you were on a merry-go-round, you would need some force to keep you moving in a circular path; otherwise, you would fly off in a line tangent to the circle at that instant. Your tendency to continue moving in a straight-line path makes you feel as if some force is pushing off the merry-go-round.

Experiment
12B

Centripetal Acceleration

INTRODUCTION

The typical response when one hears the word acceleration is to think of an object changing its speed. You have also learned that velocity has both magnitude and direction. So, an object traveling at constant speed in a circular path is undergoing an acceleration. In this experiment you will develop an expression for this type of acceleration.

OBJECTIVES

In this experiment, you will

- Analyze velocity vectors of an object undergoing uniform circular motion to determine the direction of the acceleration vector at any given moment.
- Collect force, velocity, and radius data for a mass swinging as a pendulum.
- Analyze the force *vs.* velocity and force *vs.* radius graphs.
- Determine the relationship between force, mass, velocity, and radius when the force is perpendicular to the velocity.
- Use this relationship and Newton's second law to determine an expression for centripetal acceleration.

MATERIALS

Vernier data-collection interface
Logger *Pro* or LabQuest App
Vernier Photogate
Vernier Dual-Range Force Sensor
right-angle clamp
mass hanger
string
ring stand or support rod
rod to support Force Sensor
slotted lab masses
metric tape

PRE-LAB INVESTIGATION

Tie something soft (such as a stopper) to a one-meter length of string. Taking care not to hit anyone nearby, swing the stopper so that it travels in a horizontal circular path over your head. Feel the tension force you must apply in order to keep the stopper moving at a nearly constant speed. Now, see what effect varying the speed of the stopper or the length of the string has on the force you apply to keep the stopper moving in a circular path. Record your observations.

Your instructor will lead a discussion that will enable you to determine the direction of the acceleration vector for an object moving at constant speed in a circular path. For this experiment, you will treat the arc of a pendulum bob as a portion of a circular path. In your pre-lab discussion, you will decide at what point in the motion of the pendulum you should record its velocity and the tension force acting on it.

PROCEDURE

1. Secure a Dual-Range Force Sensor to a support rod connected with a right angle clamp to a longer rod or ring stand, as shown in Figure 1. The sensor should be in the vertical position.

2. Tie a mass hanger to one end of a length of string (0.7 to 1.0 m) and make a loop at the other end that will allow the hanger to swing freely from the hook on the force sensor.

3. Place a Photogate on the floor beneath the mass hanger so that the hanger (plus the additional mass you place on it) can swing freely through the arms of the Photogate, as shown in Figure 2.

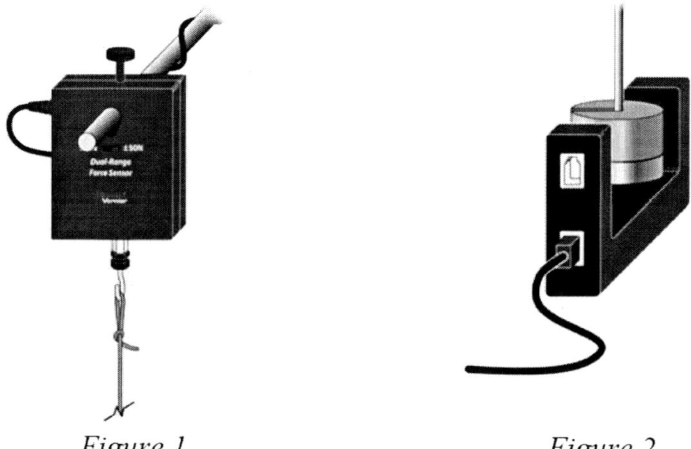

Figure 1 *Figure 2*

4. Connect the force sensor and the photogate to the interface and start the data-collection program.

5. Set up data collection.

   Using Logger *Pro*

   a. Use time-based data-collection mode. Change the data-collection rate to 50 samples/second and the data-collection length to 10 seconds.
   b. Choose Set Up Sensors ▶ Show All Interfaces from the Experiment menu.
   c. Click the image of the Photogate and choose Gate Timing.
   d. Click again on the image of the Photogate to set the Distance or Length to the measured diameter of the object that will block the Photogate as it swings through its arms.

   Using LabQuest as a standalone device

   a. Tap Mode. Select Time Based data collection. Set the data-collection rate to 50 samples/second, and the duration to 10 seconds.
   b. In the same mode dialog, scroll down to the Photogate Mode section, and expand the section by tapping on the arrow.
   c. Choose Gate Timing. Change the length to the measured diameter of the object that will block the Photogate as it swings through its arms.
   d. Select OK to accept these settings.

With either program, make the necessary adjustments so that two graphs, force *vs.* time and velocity *vs.* time, appear in the graph window.

Part 1 Force *vs.* velocity

1. Record the mass of the object serving as your pendulum bob. For this experiment, we are treating the hanger and weight as an idealized bob, that is, as a point mass. Its center of mass (CM) is the position where you could theoretically balance it on your finger if it were lying sideways. Measure and record the distance from the hook on the force sensor to the center of mass of the bob that will swing through the photogate.

2. Pull the bob back 15 to 20° from its resting position and release it. If it appears to be swinging smoothly through the photogate, begin collecting data; otherwise, repeat.
3. Select a region of the force *vs.* time graph where the maximum force appears to be nearly constant for at least three cycles. Choose Statistics and record the maximum value of the force.
4. Choose Statistics for this same portion of the velocity *vs.* time graph and record the maximum velocity.
5. Repeat Steps 2–4, gradually increasing the distance you pull the object back from the vertical, until you have at least seven data points. Your final velocity reading should be at least four times as large as your initial value.

Part 2 Force *vs.* radius

6. Measure and record the distance from the hook on the force sensor to the center of mass of the bob that swings through the photogate. This initial value of the radius should be at least 0.9 m.
7. Pull the bob back about 15–20° from its resting position and release it. Observe the trace on the velocity *vs.* time graph. In order to keep the velocity relatively constant as you vary the radius, choose some value of the velocity that appears to be easily reproducible and record this value.
8. Pull the bob back as you did before and release it. When its velocity nears your target value, select a region of the force *vs.* time graph and record the maximum value of the force as you did in Step 3.
9. Shorten the length of string connecting your mass hanger to the force sensor, measure and record the distance from the hook on the force sensor to the center of mass of the bob. Repeat Step 8.
10. Continue until you have at least six data points. Your minimum radius should be less than half of your starting value.

EVALUATION OF DATA

While it is possible to evaluate the data for this experiment in LabQuest App, it is easier to use Logger *Pro*. With LabQuest App, you should disconnect the sensors and start a new file for each part. In Logger *Pro*, you should disconnect the interface from the computer before starting a new file.

Part 1 Force *vs.* velocity

1. Start a new file and manually enter the data you have recorded.
2. Write a statement that describes the relationship between the peak tension force acting on the object and its maximum velocity.
3. If your graph of F *vs.* v_{max} is not linear, take steps to modify a column so as to produce a linear relationship. When you have done so, *save your file* and sketch your original and then linearized graph.
4. Write the equation of the line that best fits your linearized graph. Simplify the units of your slope as much as possible.

Part 2 Force *vs.* radius

5. Start a new file and manually enter the data you have recorded.
6. Write a statement that describes the relationship between the peak tension force acting on the object and the radius of the arc described by the swinging object.
7. If your graph of F *vs.* r is not linear, take steps to modify a column so as to produce a linear relationship. When you have done so, *save your file* and sketch your original and then linearized graph.
8. Write the equation of the line that best fits your linearized graph. Simplify the units of your slope as much as possible.

Part 3 Consolidation of your findings

9. Prepare a summary of your analysis of data (whiteboard or chart paper) for either or both parts of the experiment, as directed by your instructor. Include the original and linearized graph and the equation of the line of best fit. Also report the system mass and the value of the parameter you held constant. In your class discussion you will compare your findings with those of other groups.
10. If your linearized graphs have a relatively large vertical intercept, the relationship between the variables is not proportional. Discuss what adjustment you need to make and why this adjustment is appropriate. Open your saved files and make the necessary adjustments in order to obtain graphs depicting proportional relationships. If possible, print these graphs.
11. When a quantity (in this case, net force) is a function of more than one variable, it is usually the case that the slope of the graph is related to the parameters held constant during the experiment. Examine the units of the slope of your graph of F_{net} *vs.* v^2; simplify these to find an expression involving these parameters. How closely does the value of this expression agree with that of the slope? From what you have learned, write an equation for the net force in terms of all three variables. Confirm this by performing a similar analysis of your F_{net} *vs.* $1/r$ graph.

12. Use what you have learned in Step 11 and Newton's second law to write an equation for the acceleration of the object at the point in its path when the net force acts perpendicular to the object's velocity. Use your text or a web resource to determine the meaning of the term "centripetal."

EXTENSION

Wikipedia warns that the centripetal force is not to be confused with centrifugal force. It describes the latter as a fictitious or inertial force. Describe an example of such a force that you have experienced and how this interaction might better be explained in terms of centripetal force.

Experiment
12B

INSTRUCTOR INFORMATION

Centripetal Acceleration

This version of the experiment provides students the opportunity to develop an expression for centripetal acceleration without the need for special apparatus (like the Vernier Centripetal Force Apparatus). The approach used in this experiment employs readily available equipment, but it requires careful pre-lab discussion to help students understand that we are treating the arc of a pendulum bob as a portion of a circular path.

In the other version of the experiment (12A), students use the Centripetal Force Apparatus to collect the data. While the object gradually slows during data collection, for any short interval, the object undergoes nearly uniform circular motion in a horizontal plane. This simplifies the treatment of centripetal acceleration in the pre-lab discussion.

This lab is written with the assumption that the instructor will engage the students in discussions at critical junctures. These discussions can take place with the entire class or with individual lab groups. The icon at left indicates where these discussions should occur.

The Microsoft Word files for the student pages can be found on the CD that accompanies this book. See *Appendix A* for more information.

OBJECTIVES

In this experiment, the student objectives include

- Analyze velocity vectors of an object undergoing uniform circular motion to determine the direction of the acceleration vector at any given moment.
- Collect force, velocity, and radius data for a mass swinging as a pendulum.
- Analyze the force *vs.* velocity and force *vs.* radius graphs.
- Determine the relationship between force, mass, velocity and radius when the net force is perpendicular to the velocity.
- Use this relationship and Newton's second law to determine an expression for centripetal acceleration.

During this experiment, you will help the students

- Recognize that the instantaneous velocity of an object undergoing uniform circular motion is tangent to the circle at any point in its path.
- Recognize that the acceleration of such an object is perpendicular to the instantaneous velocity; thus it is directed toward the center of the circle.
- Recognize that the arc described by a pendulum can be treated as a portion of a circular path.
- Recognize that the tension force acting on a pendulum bob at the bottom its arc is perpendicular to the velocity at that instant.
- Recognize the differences between the motion of a pendulum bob and uniform circular motion.

REQUIRED SKILLS

In order for student success in this experiment, it is expected that students know how to

- Zero a force sensor.
- Create calculated columns in the software. This is addressed in Activities 1 and 3.
- Use calculated columns to linearize graphs. This is addressed in Activity 1.
- Set up a photogate in Gate mode. This is addressed in Activity 4.

EQUIPMENT TIPS

Secure a Dual-Range Force Sensor to a support rod connected with a right angle clamp to a longer rod or ring stand as shown in Figure 1. The sensor should be in the vertical position. There should be sufficient clearance between the string and the edge of the table so that the mass hanger and weight can freely swing between the arms of a photogate placed on the floor beneath the force sensor, as shown in Figure 2.

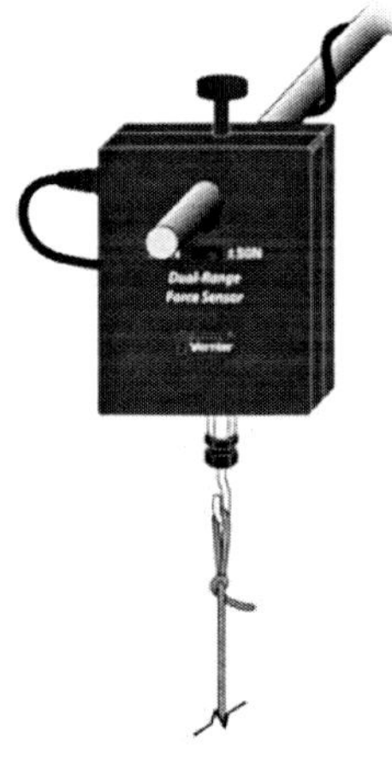

Figure 1

Figure 2

PRE-LAB DISCUSSION

This experiment should be performed only after students are thoroughly versed in vector addition, the use of force diagrams and have explored Newton's second law. Start by demonstrating uniform circular motion, whirling a soft object on the end of a string in a horizontal plane overhead. Ask students if the velocity of the object is changing. If they at first note only that the speed is constant, remind them that velocity also depends on direction. Since the velocity is constantly changing, the bob is constantly accelerating; a force is required to produce this acceleration. If your lab space allows for it, it would be worthwhile for students to experience how the tension force they must provide to keep a stopper in uniform circular motion depends on the speed and the radius of the circle.

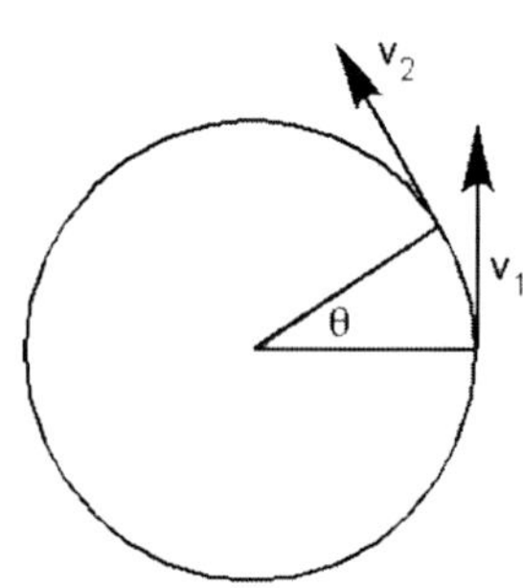

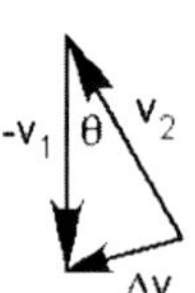

After their initial investigation, it is important to help students recognize that the acceleration at any given instant is perpendicular to the tangential velocity. Since $a = \Delta v/\Delta t$, the difference of vectors v_2 and v_1, taken at the mid-time (the point midway between the origin of the two vectors) is directed toward the center of the circle. As Δt goes to 0, the vector $\vec{a}$ is perpendicular to $\vec{v}$ at that instant.

While it is difficult to reliably measure the force acting of an object traveling in a horizontal circular path, we *can* measure the force acting on a pendulum bob as it moves in a vertical plane. You will help the students to see that the motion of a pendulum bob along a circular path. It is true that the velocity is not constant, but, at the bottom of the arc, the tension force acting on the bob is perpendicular to the tangential velocity at that point. At this point in the motion of the pendulum, the tension force will have its maximum value as the tangential velocity reaches its maximum. It is certainly true that the tension force is *not* the net force acting on the bob, but it would be instructive not to point this out at this time. It is fine if students recognize this fact at the start, but if they do not, they will learn that they must subtract the weight of the bob from the tension force to obtain the *net* force when they analyze their data.

Students must recognize that since the force depends on the mass, velocity, and the radius of the path, they must keep the other variables constant while they determine the relationship between force and velocity (or radius). In this experimental design, students use the same mass throughout. However, it would be useful for different groups to use different masses.

LAB PERFORMANCE NOTES

Students should have little difficulty with Part 1. For each value of the radius, they need to select a portion of both the force *vs.* time and velocity *vs.* time graphs, choose Statistics, and record the *maximum* value of each during this interval, as shown in Figures 3 and 4 below.

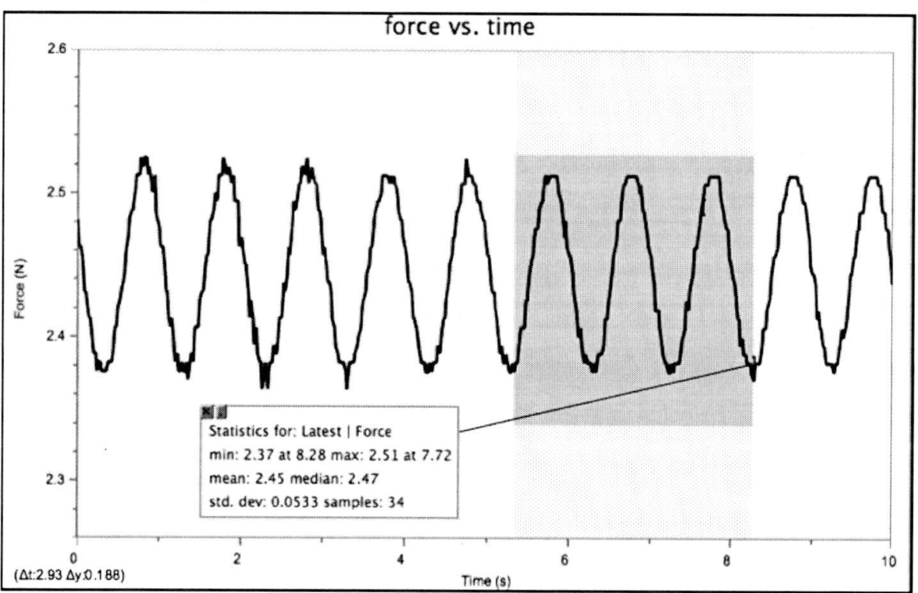

Figure 3

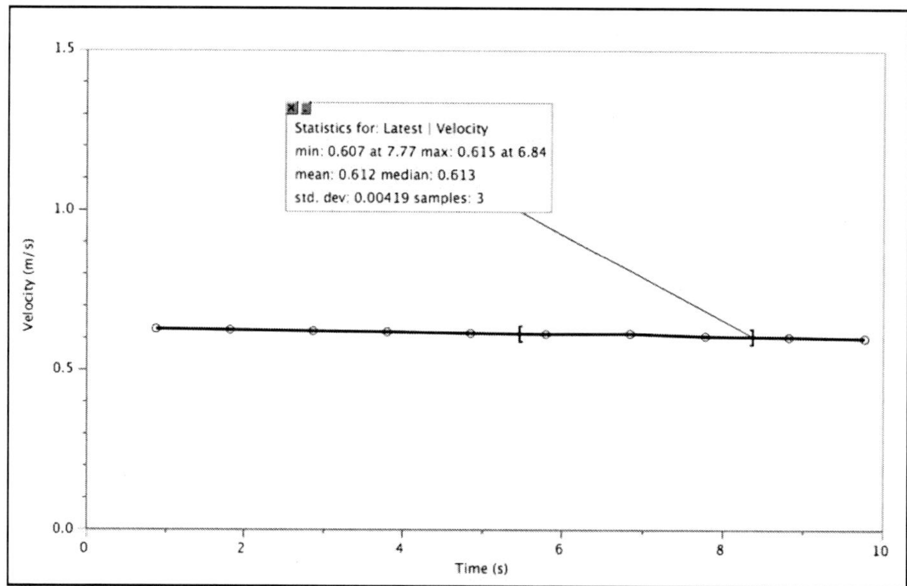

Figure 4

Part 2 is a bit more problematic, as students must try to keep velocity as constant as possible as they vary the radius. You may need to point out that the radius is not simply the length of the string, but the distance between the hook on the force sensor and the center of mass of the hanging weight acting as the bob. Advise the students that they should try to raise the bob the same amount as they pull it back and release it. They should choose some target value of velocity and, when a trial yields an acceptable velocity, they can record the force for that trial.

Once students have collected data for Parts 1 and 2, they should choose File ▶ New in either Logger *Pro* or LabQuest App to perform the Evaluation of Data.

SAMPLE RESULTS AND POST-LAB DISCUSSION

One way to expedite the post-lab discussion is to assign the treatment of Part 1 to some groups and Part 2 to the others.

Part 1 Force *vs.* velocity

A plot of force *vs.* velocity yields an upward-opening parabola as shown below.

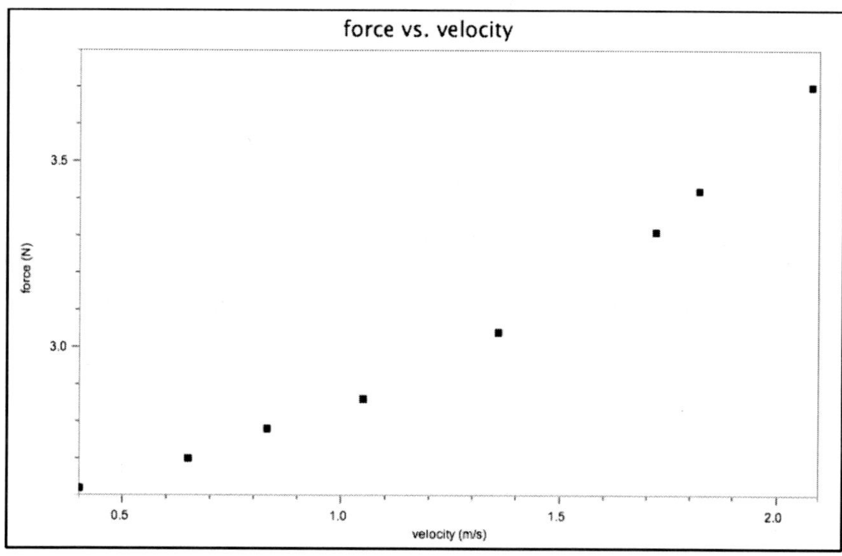

Figure 5

By this point, students should be familiar with "linearizing" a graph. If not, or if they need a reminder, see the file, 10-1 Linearization.cmbl, in the Tutorial folder in Logger *Pro*.

Students should add a new calculated column for velocity squared and display the graph of force *vs.* velocity squared. The linearized graph is displayed below.

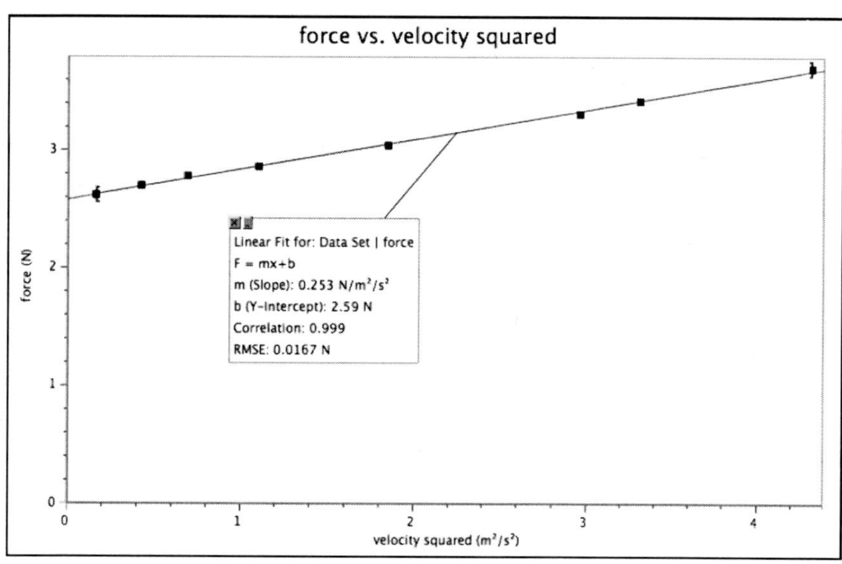

Figure 6

The equation representing the relationship shown in the graph (Figure 6) is $F = \left(0.253 \frac{\text{N}}{\text{m}^2/\text{s}^2}\right) v^2 + 2.59\ \text{N}$. When the units of the slope are simplified, one obtains kg/m. If students are concerned about the large value of the vertical intercept, tell them that this will be addressed in the discussion in Part 3.

Part 2 Force *vs.* radius

Students should see enough of a curve in the plot of the data that they will recognize that force and radius are inversely related (see Figure 7).

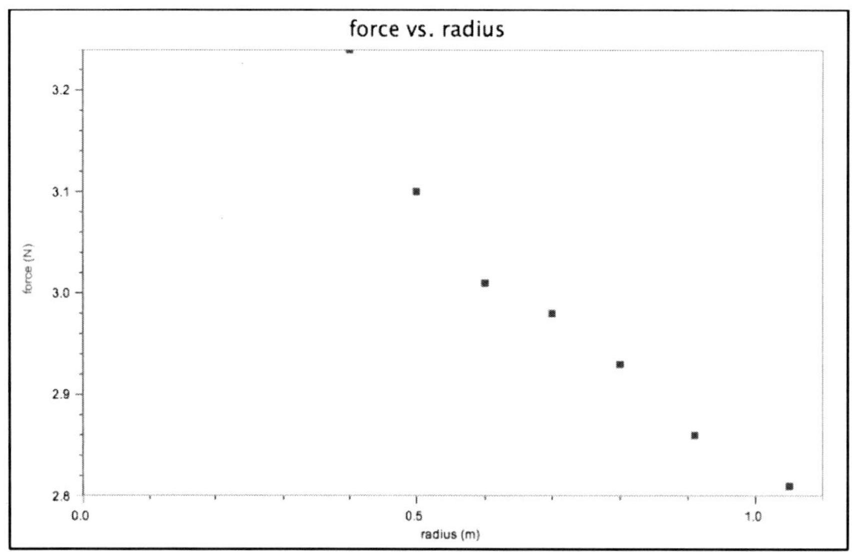

Figure 7

Students should add a new calculated column for 1/radius and display a graph of F vs. $1/r$ as shown in Figure 8.

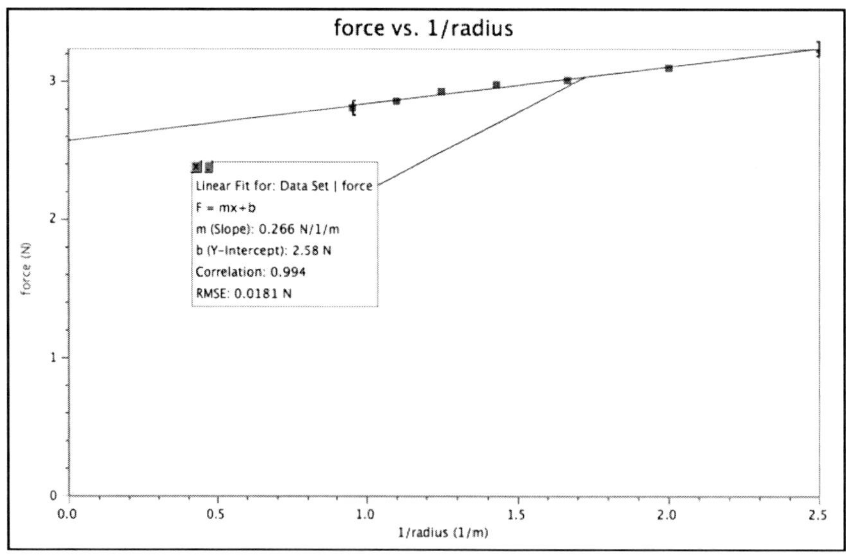

Figure 8

The equation representing the relationship shown in the graph (see Figure 8) is $F = \left(0.266 \frac{\text{N}}{1/\text{m}}\right)\frac{1}{r} + 2.58\ \text{N}$. When the units of the slope are simplified, one obtains kg·m^2·s^{-2}.

Part 3 Consolidation of your findings

Steps 9–10

If students have not already recognized that the explanation for the large vertical intercept is that the force measured by the force sensor is *not* the net force directed upward, suggest that they sketch a force diagram for the bob when it is at the bottom of its arc. When they realize that they must subtract the weight of the bob from the tension force, have them return to their saved files. They should add a new calculated column, net force, and use this as the vertical axis label for their linearized graph. Of course, if they have already realized that they should be using net force, they will have obtained a graph like that shown in Figure 9. This graph shows that that the net force is proportional to the inverse of the radius.

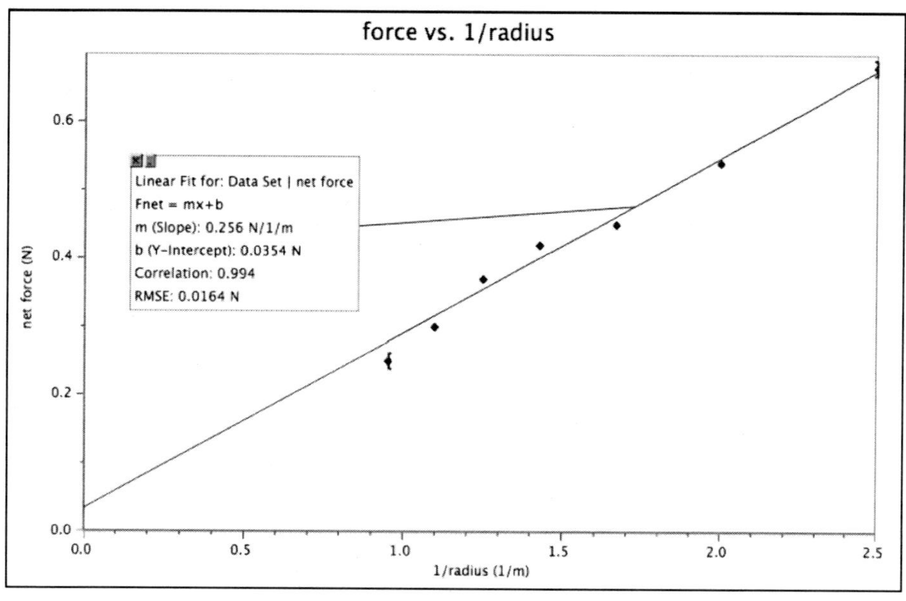

Figure 9

Step 11

Direct students to examine their net force vs. velocity squared graph. The variables held constant during data collection were mass and the radius. When the units of the slope are simplified, they reduce to kg/m, suggesting that the constant of proportionality might be m/r.

Comparison of the value for this fraction should show that there is close agreement between this value and the numerical value of the slope. This suggests that the overall relationship might be

$$F_{net} = \frac{m}{r} \cdot v^2 .$$

The equation for centripetal force is typically written as

$$F_{net} = m\frac{v^2}{r}.$$

Students could confirm this relationship by performing a similar analysis of the slope of the net force *vs.* 1/radius graph. There is close agreement between the value of mv^2 and the numerical value of the slope of this graph. Students should see that the relationship $F_{net} = mv^2 \cdot \frac{1}{r}$ is the same as the one they obtained earlier.

Step 12

From Newton's second law, students should deduce that the acceleration of the bob at its lowest point in the path is given by $a = v^2/r$. Physicists use the term "centripetal" to describe a force or acceleration directed toward the center of the circle.

EXTENSION

A centrifugal (or center-fleeing) force is the fictitious force one feels as a result of one's perspective from a non-inertial reference frame. If you were on a merry-go-round, you would need some force to keep you moving in a circular path; otherwise, you would fly off in a line tangent to the circle at that instant. Your tendency to continue moving in a straight-line path makes you feel as if some force is pushing off the merry-go-round.

Experiment

13

Rotational Dynamics

INTRODUCTION

When you studied Newtonian dynamics you learned that when an object underwent some form of translational motion (whether in a straight line, parabolic, or circular path), the net force applied to the object is proportional to the acceleration. The constant of proportionality is the mass of the accelerating object. When a torque (the rotational analogue to force), is applied to an object that is free to rotate, the object will undergo rotational acceleration. In this experiment, you will investigate the relationship between torque and angular acceleration.

OBJECTIVES

In this experiment, you will

- Collect angular acceleration data for objects subjected to a torque.
- Determine an expression for the torque applied to a rotating system.
- Determine the relationship between torque and angular acceleration.
- Relate the slope of a linearized graph to system parameters.
- Make and test predictions of the effect of changes in system parameters on the constant of proportionality.

MATERIALS

Vernier data-collection interface
Logger *Pro* or LabQuest App
ring stand
lightweight mass hanger
string
Vernier Rotary Motion Sensor
Vernier Rotary Motion Accessory kit
balance
drilled or slotted masses

PRE-LAB INVESTIGATION

1. Connect the sensor to the interface and launch the data-collection program. The default settings work fine for this investigation.

2. Tie a length of string to the edge of the largest pulley on the Rotary Motion Sensor. Pull the string taut and mark the string one meter from the point of attachment. Wind the string on the pulley until the one-meter mark is at the edge of the pulley.

3. Holding the sensor securely, start data collection, then pull the string so that it unwinds from the pulley.

4. Read the maximum angle on the graph of angle *vs.* time and record this value.

5. Repeat Steps 2–4, except that this time, wind the string in the opposite direction.

6. Now tie the string to the edge of the middle pulley on the sensor. Repeat Steps 2–4.

7. Determine the relationship between the angle through which the pulley turned, the linear distance the edge of the pulley traveled as the string was unwound, and the radius of the pulley. Compare your findings with others in class.

PROCEDURE

1. Connect the Ultra Pulley to the Swivel Mount and then to the Rotary Motion Sensor, as shown in Figure 1. Attach the sensor to a ring stand and position it so that a weight tied to the edge of the large pulley on the sensor and hanging over the Ultra Pulley can hang freely without touching the floor.

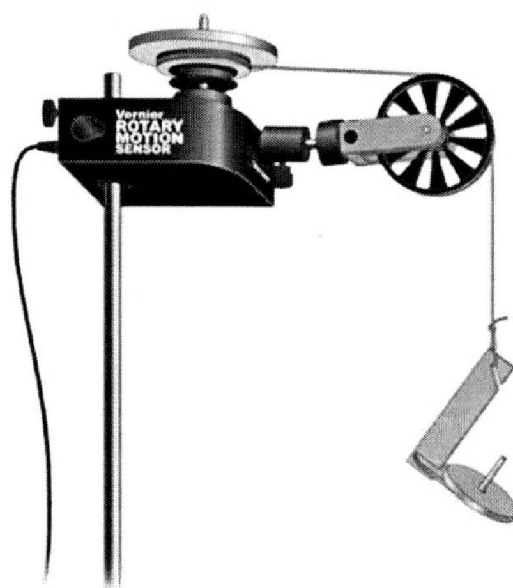

Figure 1

Part 1

2. Find the mass of one of the solid aluminum disks, then attach it to the 3-step pulley on the sensor. Record the radius of the pulley.

3. Open a new file in the data-collection program. The default data-collection rate is fine, but reduce the length of the experiment to five seconds.

4. Place the lightest mass available on the hanger, then wind the string onto the largest pulley on the Rotary Motion Sensor.

5. Start data collection, then release the hanging weight. Catch the hanger when the string has completely unwound.

6. To determine the angular acceleration of the disk, perform a linear fit on the appropriate portion of the angular velocity *vs.* time graph. Record this value in your lab notebook, along with the mass of the hanger and weight for each value you use.

7. Repeat Steps 4–6, increasing the mass of the hanging weight, until you have at least five different values of angular acceleration.

Part 2

8. Find the mass of the second solid aluminum disk. Using the longer machine screw and sleeve, attach both disks to the 3-step pulley on the sensor. Repeat Steps 3–7.

Part 3

9. Remove the disks from the sensor. Find the mass of each of the weights and the rod in the accessory kit. Attach each of the weights to opposite sides of the rod at a distance recommended by your instructor. Record this distance.

10. Attach the rod and weights to the sensor as shown in Figure 2. Repeat Steps 3–7 as you did in Parts 1 and 2.

Figure 2

EVALUATION OF DATA

In order to find a relationship between torque and angular acceleration, you need to know the value of the net torque acting on the system in each of the trials you performed. Since you were not able to measure the torque directly, you must derive an expression you can use to determine the torque from quantities that you *could* measure.

Consider, for a moment, an experiment in which you determined the relationship between net force and the acceleration of an object undergoing translational motion.

$$F_{net} = ma$$

If you did this experiment with a modified Atwood's apparatus, the force that the earth exerted on a hanging mass accelerated a cart on a track. However, as you may have found, the net force acting on the cart *while it was accelerating*, was less than the weight, mg, of the hanging mass. Consider why this was the case, sketch free body diagrams for both the cart and the hanging weight, then write an expression for the net force acting on the cart as it was accelerating.

From what you have learned about torque, τ, and the relationship between linear and angular acceleration, α, make the appropriate substitutions so you can derive a parallel expression for the net torque acting on the object on the rotating pulley. Your instructor may assist you in this derivation.

Use of Logger *Pro* will help you to calculate the value of the net torque in Parts 1–3 for your analysis. If you choose to perform the evaluation in LabQuest App, you will need to perform these calculations by hand.

Part 1

1. To evaluate the relationship between acceleration and force, disconnect the sensors from the interface and choose New from the File menu in Logger *Pro*.

2. Enter your values for the hanging mass and angular acceleration. Create a new calculated column finding the values of net torque. Your instructor may guide you in the design of this file.

3. Even though you investigated how angular acceleration responded to changes in the torque, in order to facilitate your analysis, plot a graph of net torque, τ, *vs.* angular acceleration, α.

4. If the relationship between net torque and angular acceleration appears to be linear, fit a straight line to your data.

5. Write a statement that describes the relationship between the net torque acting on the disk and its angular acceleration.

6. Write the equation that represents the relationship between the net torque, τ acting on the disk and its angular acceleration, α. Be sure to label this data set with the value of the mass of the disk.

Part 2

7. Choose New Data Set from the Data menu. Enter your values for the hanging mass and angular acceleration for the stacked disks. As you did in Step 2, create a new calculated column finding the values of the net torque the hanging mass applied to the disks.

8. Plot a graph of net torque, τ, *vs.* angular acceleration, α. If the relationship between net torque and angular acceleration appears to be linear, fit a straight line to your data. If possible, print a copy of the graph showing both data sets.

9. Write the equation that represents the relationship between the net torque, τ acting on the pair of disks and their angular acceleration, α. Be sure to label this data set with the value of the mass of the disk.

10. How does the slope of this equation compare to the one you obtained in Part 1?

As you are likely to have found before, the slope of a graph is usually some function of physical parameters of the system. For example, in the Newton's Second Law experiment, the slope of the graph of net force *vs.* acceleration is the *mass* of the object accelerated by the force. The greater the mass of the object, the larger was the force required to produce a given acceleration. In effect, the mass is a measure of the resistance to the change in the motion of the object. Physicists call this resistance to change in motion *inertia*. In this experiment, the slope is also a measure of the resistance of the object to undergo an acceleration; in the case of rotational motion it is known as the *moment of inertia*.

11. From the previous paragraph, you might suspect that the slope is a function of the mass. What evidence do you have that supports this hypothesis?

Part 3

12. Choose New Data Set from the Data menu. Enter your values for the hanging mass and angular acceleration for the rod and weights. As you did in Step 2, determine the values of the net torque the hanging mass applied to the rod and weights.

13. Plot a graph of net torque, τ, *vs*. angular acceleration, α. If the relationship between net torque and angular acceleration appears to be linear, fit a straight line to your data.

14. Write the equation that represents the relationship between the net torque, τ acting on the rod and weights and their angular acceleration, α.

Part 4 Further examination of moment of inertia

15. You might also expect that the slope of your graph (moment of inertia) is somehow related to the distance, r, of the object(s) from the axis of rotation. To test this hypothesis, you will need the data from several groups that positioned the weights at different distances along the rod. If that is not feasible, you will have to repeat Part 3 at least five more times, changing the distance the weights are from the center of rotation for each set of net torque *vs*. angular acceleration.

16. Open a new Logger *Pro* file and plot moment of inertia, I, *vs*. radius for the data collected in Part 3. If the relationship appears to be linear, fit a straight line to the data. If not, then take the necessary steps to modify a column so as to produce a linear relationship.

17. How does the slope of *this new* graph compare to the mass of the weights used? Suggest a possible expression for the moment of inertia, I, for the two point masses.

18. Try to account for the fact that your graph has a non-negligible intercept. Your instructor may guide you in making the necessary adjustment to your calculation of I.

19. Now, revisit your equations for Parts 1 and 2. How do the slopes of the lines compare to the expression you obtained for I in Step 17? Suggest a reason for any difference.

Experiment
13

INSTRUCTOR INFORMATION

Rotational Dynamics

This lab is written with the assumption that the instructor will engage the students in discussions at critical junctures. These discussions can take place with the entire class or with individual lab groups. The icon at left indicates where these discussions should occur. The calculations required in the Evaluation of Data are sufficiently complicated that the use of Logger *Pro* is strongly recommended. To perform the evaluation in LabQuest, students will have to calculate the torque in Parts 1–3 by hand in order to examine graphs of torque *vs.* angular acceleration.

The Microsoft Word files for the student pages can be found on the CD that accompanies this book. See *Appendix A* for more information.

OBJECTIVES

In this experiment, the student objectives include

- Collect angular acceleration data for objects subjected to a torque.
- Determine an expression for the torque applied to a rotating system.
- Determine the relationship between torque and angular acceleration.
- Relate the slope of a linearized graph to system parameters.
- Make and test predictions of the effect of changes in system parameters on the constant of proportionality.

During this experiment, you will help the students

- Recognize the relationships between θ, ω, and α and their translational counterparts.
- Relate moment of inertia of a disk to its mass and the radius of rotation.
- Relate moment of inertia of a collection of point masses to their mass and the radius of rotation.

REQUIRED SKILLS

In order for student success in this experiment, it is expected that students know how to

- Create new calculated columns to determine values to use in graphing. This is addressed in Activity 1.
- Perform linear fits to graphs. This is addressed in Activity 1.

EQUIPMENT TIPS

In order for students to investigate the relationship between net torque and angular acceleration, they need to use a Rotary Motion Sensor and the Rotational Motion Accessory Kit. Connect the Ultra Pulley to the Swivel Mount and then to the Rotary Motion Sensor, as shown in Figure 1. Attach the sensor to a ring stand and position it so that a weight tied to the edge of the large pulley on the sensor and hanging over the Ultra Pulley can hang freely without touching the floor. For the data sets in this experiment, the hanging weights ranged from 15 to 65 g.

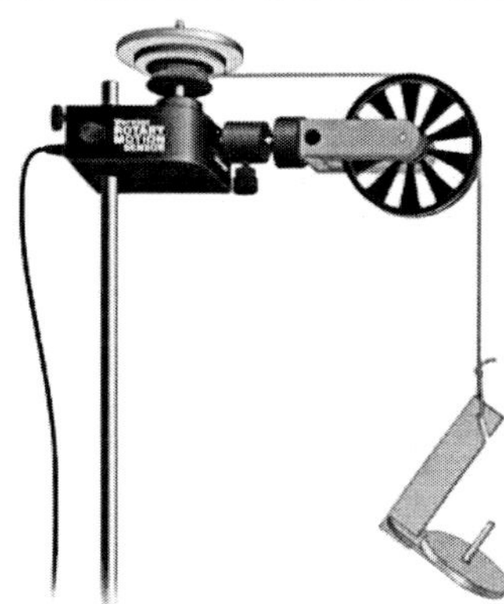

Figure 1

PRE-LAB DISCUSSION

This experiment should be performed only after students have been introduced to θ, ω, and α as the rotational analogs to displacement, linear velocity, and linear acceleration. They should have also had the opportunity to explore the relationship between net force, mass, and acceleration, ideally with an approach similar to that employed in Experiment 4 in this manual.

The pre-lab investigation is designed to familiarize students with the rotary motion sensor and to review the concept that the linear displacement of a point on a rotating body is the product of the angular displacement and the distance between that point and the axis of rotation. From here, it should not be a stretch of the imagination for students to see that one could determine *angular* acceleration in a manner similar to that used to find *linear* acceleration.

LAB PERFORMANCE NOTES

When the Rotary Motion Sensor is connected to the data-collection interface and Logger *Pro* or LabQuest App is started, the default graph screen shows both angle *vs.* time (θ-t) and velocity *vs.* time (ω-t) graphs. Students can determine the angular acceleration of the rotating body by finding the slope of the appropriate portion of the ω-t graph, as shown in Figure 2.

Depending on the availability of equipment and the depth of the investigation of rotational dynamics you wish your students to perform, you have two options. Evaluation of the data for Parts 1 and 2 of the experiment will enable students to conclude that net torque is directly proportional to angular acceleration and that the constant of proportionality is somehow related to the mass. You can then simply define the moment of inertia for various objects.

Performance of Part 3 and the subsequent evaluation of data should help students determine the relationship between the moment of inertia, system mass, and radius for both point masses and for disks. You will need to assign different values of radii for the point masses to the lab groups so that they can use class data in their analysis of the relationship between moment of inertia and radius. As a time-saving alternative, you could collect the data yourself for Part 3 for various radii and provide these data to your students to perform the analysis called for in Part 4.

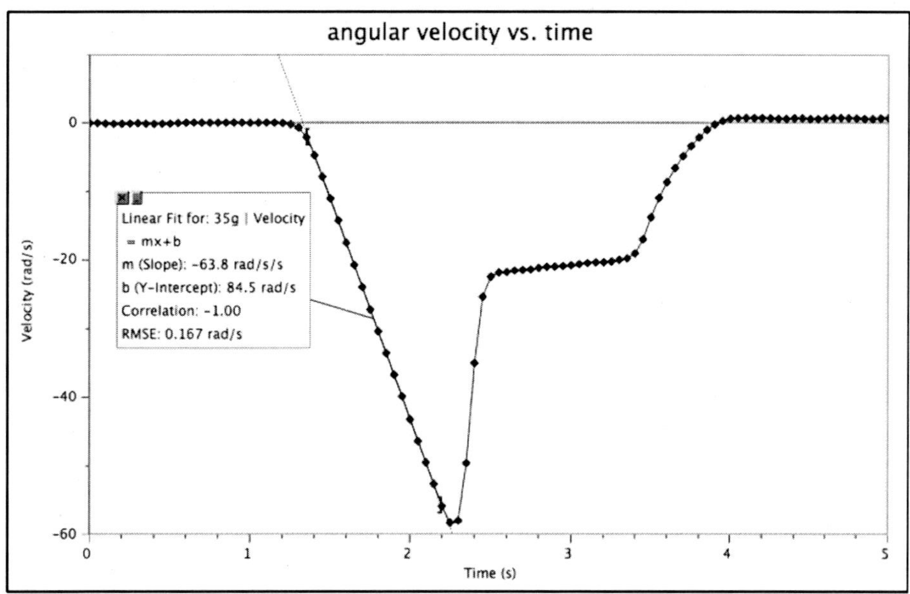

Figure 2

Postpone, for now, discussion of how students will determine the net torque applied to the object on the sensor pulley. Students can simply record in their lab notebooks the mass of the towing weight for each trial they perform.

SAMPLE RESULTS AND POST-LAB DISCUSSION

Determination of net torque

In order to find a relationship between torque and angular acceleration, students must know the value of the torque acting on the system in each of the trials they have performed. Hints are provided in the student version, but you may need to help students with the derivation of the expression for net torque. Since students cannot measure the torque directly, they will determine the torque from quantities that they *can* measure.

Consider, for a moment, Experiment 4, in which students determined the relationship between net force and the acceleration of an object undergoing translational motion.

$$F_{net} = ma$$

The force that the earth exerted on a hanging mass accelerated a cart on a track. However, as they noticed, the net force acting on the cart *while it was accelerating*, was less than the weight, *mg*, of the hanging mass. This occurred because *both* the hanging mass, *m*, and the much larger cart mass, *M*, were being accelerated by this weight.

$$mg = (M + m)a$$

The *net force* acting on the cart alone (*M*) is given by $mg - ma$.

Since torque, τ, is the cross product of the force and the radius, and linear acceleration, a, is the product of the radius and the angular acceleration, α, an expression for net torque can be determined by making the appropriate substitutions in the expression for net force.

$$\text{net force } = mg - ma$$
$$\text{net torque } = r(mg - m\alpha r) \text{ or } rm(g - \alpha r)$$

PARTS 1 AND 2 – DISKS

Steps 1–4, 7–8

The plots of net torque *vs.* angular acceleration for both the single disk and stacked disks are shown in Figure 3.

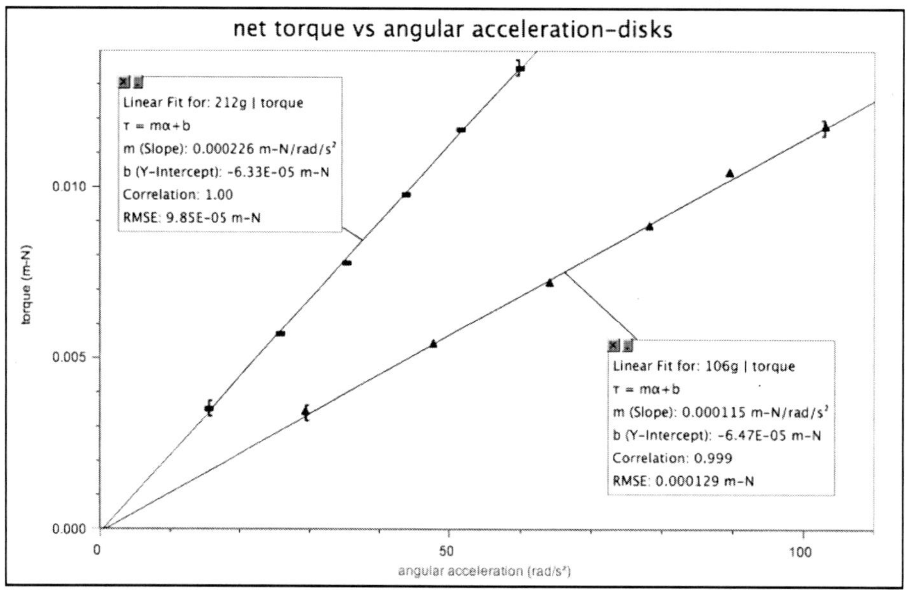

Figure 3 $\tau - \alpha$ *graph for disks*

Step 5

Students should conclude that the net torque is directly proportional to the angular acceleration.

Steps 6 and 9

The equation that describes the relationship between net torque and angular acceleration for the single disk shown in Figure 3 is

$$\tau = (1.15 \times 10^{-4} \text{ m} \cdot \text{N} \cdot \text{s}^2)\alpha$$

For the two stacked disks, the equation is

$$\tau = (2.26 \times 10^{-4} \text{ m} \cdot \text{N} \cdot \text{s}^2)\alpha$$

In both cases the vertical intercept is negligible.

Steps 10 and 11

Because students might suspect that the slope of the graph is related to the mass of the object undergoing acceleration, they should not be surprised that the slope of the second equation is nearly double that of the first.

If you choose to stop here, then you should introduce the term *moment of inertia, I,* as the slope of the graph of net torque *vs.* angular acceleration. For a disk, $I = \frac{1}{2}mr^2$. The slope of each line in the graph is slightly greater than the value of *I* calculated from mass and radius of the disk alone. This is due to the fact that the pulley (with a non-negligible mass) is also undergoing acceleration. If the pulley mass were included in the calculation of *I*, the slopes of the lines in the graph are much closer to the calculated value of *I*.

PART 3 – MASSES ON ROD

Steps 12–14

Students should find that the net torque is still proportional to the angular acceleration. However, as Figure 4 shows, the value of the slope is quite different from that obtained for the disks.

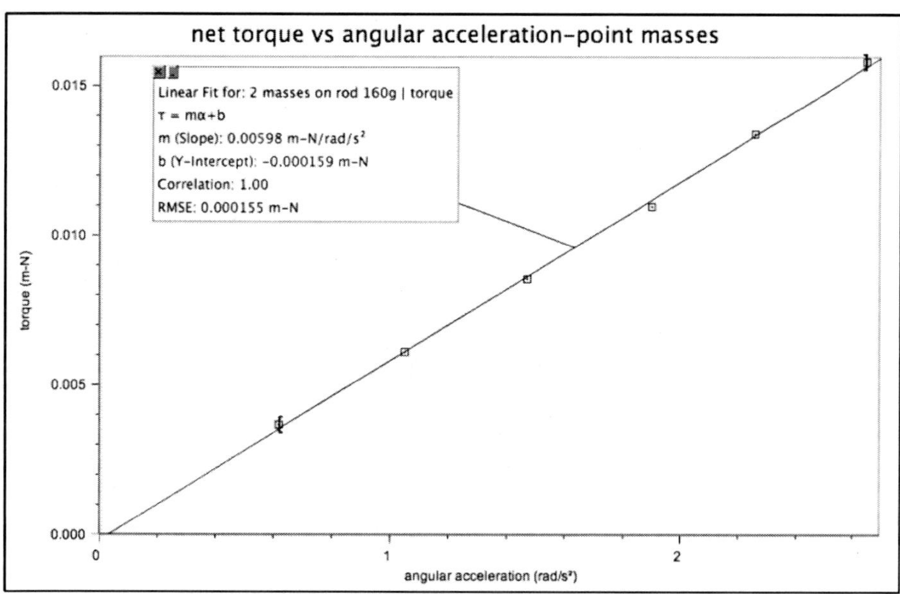

Figure 4 $\tau-\alpha$ graph for point masses

The equation of the line for this graph is

$$\tau = (5.98 \times 10^{-3}\,\text{m} \cdot \text{N} \cdot \text{s}^2)\alpha$$

The combined mass of the weights on the rod is less than the mass of the two disks, yet the slope is more than 20 times as large as that for the two disks. Clearly, some factor besides mass must figure into the moment of inertia. In the next part, students examine the relationship between moment of inertia and radius.

PART 4 – FURTHER EXAMINATION OF MOMENT OF INERTIA

Steps 15 & 16

A plot of moment of inertia *vs.* radius for the weights attached to the rod appears to be a top-opening parabola. By this point, students should be familiar with "linearizing" a graph. If not, or if they need a reminder, see 10-1 Linearization.cmbl in the Tutorial folder in Logger *Pro*.

Students should add a new calculated column for radius squared and display the graph of moment of inertia *vs.* radius squared on a new page. The linearized graph is displayed below.

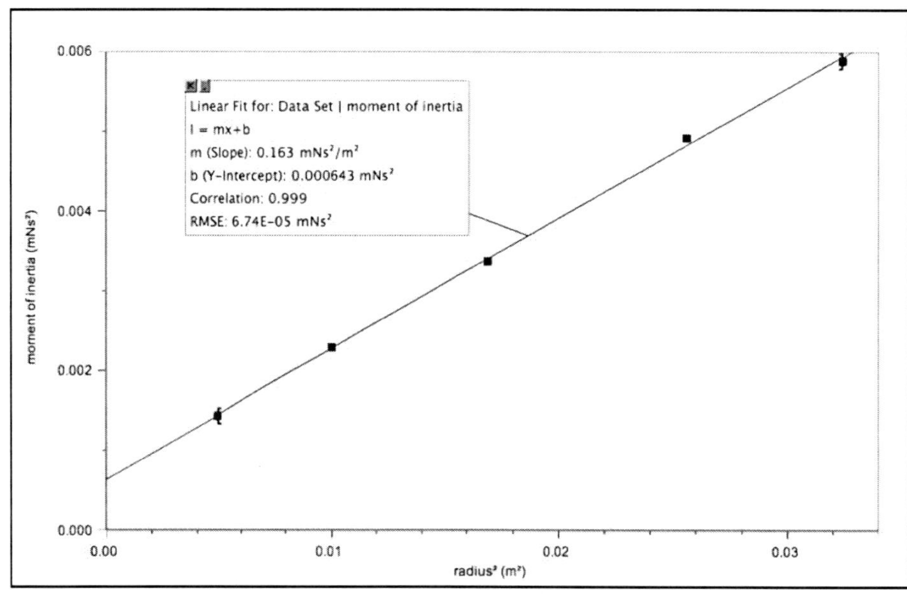

Figure 5

Students should conclude that there is a linear relationship between the moment of inertia and the square of the distance that the point masses were from the center of rotation.

Step 17

Students saw in Parts 1 and 2 that the moment of inertia was related to the mass, and have now found that it is also is related to the square of the radius of rotation. They should now examine the units of the slope of this graph; when they do, they will find that they reduce to units of mass.

$$\frac{\text{m} \cdot \text{N} \cdot \text{s}^2}{\text{m}^2} = \frac{\text{m} \dfrac{\text{kg} \cdot \text{m}}{\text{s}^2} \text{s}^2}{\text{m}^2} = \text{kg}$$

Now, if they compare the slope of the graph in Figure 5 to the combined mass of the weights attached to the rod, they will find that they agree to within 2%. This suggests that the expression for the moment of inertia, I, of point masses is mr^2.

Step 18

Students should be concerned about the non-negligible intercept in the graph in Figure 5. Reflection on the meaning of the intercept suggests that the moment of inertia of this system would not be zero even if the point masses were at the center or rotation. See if you can get the students to suggest a possible explanation. They ought to conclude that the rod to which the weights were attached had a non-negligible moment of inertia – this contributed to the overall value of I for the system. The value for the rod, calculated from $I = \frac{1}{12} m l^2$, is 5×10^{-4} kg m^2, which very nearly accounts for the intercept.

Step 19

Students should now revisit the graph of net torque *vs.* angular acceleration for the disks. They will find that the slope is approximately $\frac{1}{2} mr^2$. This expression is reasonable because the mass of a disk is distributed evenly across the entire radius.

Experiment

14

Conservation of Angular Momentum

INTRODUCTION

In your study of linear momentum, you learned that, in the absence of an unbalanced external force, the momentum of a system remains constant. In this experiment, you will examine how the *angular* momentum of a rotating system responds to changes in the moment of inertia, I.

OBJECTIVES

In this experiment, you will

- Collect angle *vs.* time and angular velocity *vs.* time data for rotating systems.
- Analyze the θ-t and ω-t graphs both before and after changes in the moment of inertia.
- Determine the effect of changes in the moment of inertia on the angular momentum of the system.

MATERIALS

Vernier data-collection interface
Logger *Pro* or LabQuest App
Vernier Rotary Motion Sensor
Vernier Rotary Motion Accessory Kit
ring stand or vertical support rod
balance
metric ruler

PROCEDURE

1. Mount the Rotary Motion Sensor to the vertical support rod. Place the 3-step Pulley on the rotating shaft of the sensor so that the largest pulley is on top. Measure the mass and diameter of the aluminum disk with the smaller hole. Mount this disk to the pulley using the longer machine screw sleeve (see Figure 1).

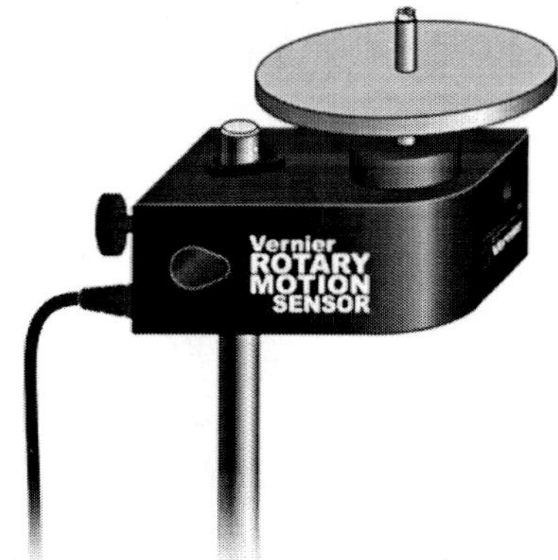

Figure 1

2. Connect the sensor to the data-collection interface and begin the data-collection program. The default data-collections settings are appropriate for this experiment.

3. Spin the aluminum disk so that it is rotating reasonably rapidly, then begin data collection. Note that the angular velocity gradually decreases during the interval in which you collected data. Consider why this occurs. Store this run (Run 1).

4. Obtain the second aluminum disk from the accessory kit; determine its mass and diameter. Position this disk (cork pads down) over the sleeve of the screw holding the first disk to the pulley. Practice dropping the second disk onto the first so as to minimize any torque you might apply to the system (see Figure 2).

5. Begin the first disk rotating rapidly as before and begin collecting data. After a few seconds, drop the second disk onto the rotating disk and observe the change in both the θ-t and ω-t graphs. Store this run (Run 2).

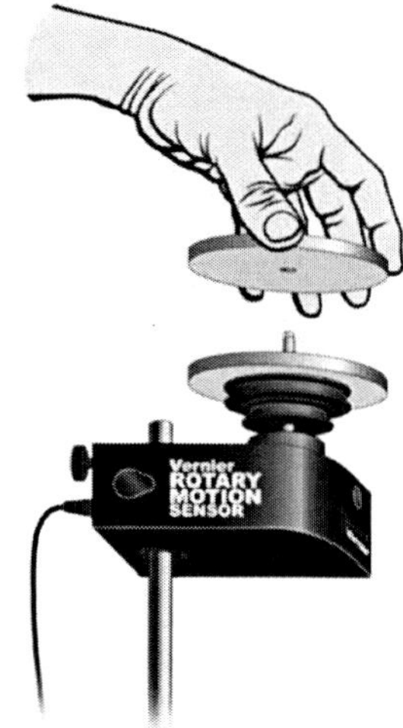

Figure 2

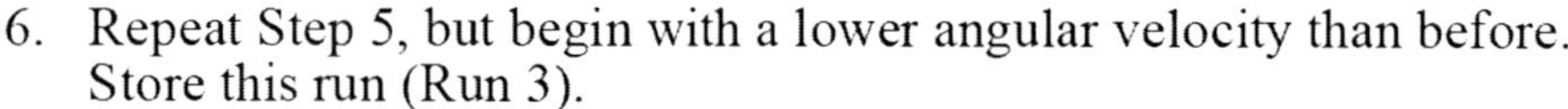

6. Repeat Step 5, but begin with a lower angular velocity than before. Store this run (Run 3).

7. Find the mass of the steel disk. Measure the diameter of both the central hole and the entire disk. Replace the first aluminum disk with the steel disk and hub and tighten the screw as before (see Figure 3).

8. Try to spin the steel disk about as rapidly as you did the aluminum disk in Step 3 and then begin collecting data. Store this run (Run 4).

9. Repeat Step 5, dropping the aluminum disk onto the steel disk after a few seconds. Store this run (Run 5) and save the experiment file in case you need to return to it.

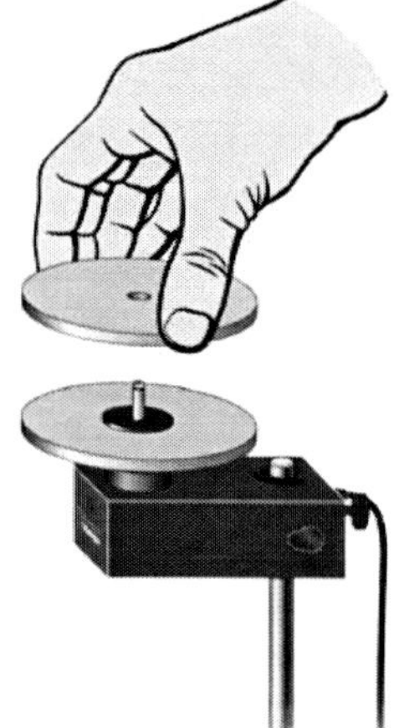

Figure 3

EVALUATION OF DATA

1. Use a text or web resource to find an expression for the moment of inertia for a disk; determine the values of I for your aluminum disks. With its large central hole, the steel disk should be treated as a cylindrical tube. Using the appropriate expression, determine the value of I for the steel disk.

2. Examine the ω-t graph for your runs with the single aluminum disk (Run1) and the steel disk (Run 4). Determine the rate of change of the angular velocity, ω, for each disk as it slowed. Account for this change in terms of any unbalanced forces that may be acting on the system. Explain the difference in the rates of change of ω (aluminum *vs.* steel) in terms of the values you calculated in Step 1.

3. Examine the ω-t graph for Run 2. Determine the rate of change of ω before you dropped the second disk onto the first. Record the angular velocity just before and just after you increased the mass of the system. Determine the time interval (Δt) between these two velocity readings.
   - In Logger *Pro*, drag-select the interval between these two readings. The Δx in the lower left corner gives the value of Δt.
   - In LabQuest App, drag and select the interval between these two readings and use the Delta function under Statistics to perform this task.
4. The angular momentum, L, of a system undergoing rotation is the product of its moment of inertia, I, and the angular velocity, ω.

   $$L = I\omega$$

   Determine the angular momentum of the system before and after you dropped the second aluminum disk onto the first. Calculate the percent difference between these values.
5. Use the initial rate of change in ω and the time interval between your two readings to determine $\Delta\omega$ due to friction alone. What portion of the difference in the angular momentum before and after you increased the mass can be accounted for by frictional losses?
6. Repeat the calculations in Steps 3–5 for your third and fifth runs.

EXTENSION

In this experiment, the moment of inertia of the rotating system was changed by adding mass. In what other way could one change the moment of inertia? Consider an example of how this is done outside the lab. Explain how this change in I produces a change in ω.

Experiment

14

INSTRUCTOR INFORMATION

Conservation of Angular Momentum

This lab is written with the assumption that the instructor will engage the students in discussions at critical junctures. These discussions can take place with the entire class or with individual lab groups. The icon at left indicates where these discussions should occur.

The Microsoft Word files for the student pages can be found on the CD that accompanies this book. See *Appendix A* for more information.

OBJECTIVES

In this experiment, the student objectives include

- Collect angle *vs.* time and angular velocity *vs.* time data for rotating systems.
- Analyze the θ-t and ω-t graphs both before and after changes in the moment of inertia.
- Determine the effect of changes in the moment of inertia on the angular momentum of the system.

During this experiment, you will help the students recognize that some portion of the difference in angular momentum of the system after the mass was increased can be accounted for by frictional losses.

REQUIRED SKILLS

In order for student success in this experiment, it is expected that students know how to

- Perform a linear fits to a graph. This is addressed in Activity 1.
- Determine the length of an interval in Logger *Pro* or LabQuest App.

EQUIPMENT TIPS

This experiment is designed to use the Vernier Rotary Motion Sensor and Rotary Motion Accessory Kit.

PRE-LAB DISCUSSION

This experiment should be performed only after students have a thorough grounding in rotational kinematics, so that they readily recognize that ω is the angular counterpart to linear velocity, v. They also need to recognize that moment of inertia, I, like mass, is a measure of the resistance to a change in the velocity. Students should also know that the linear momentum, $p = mv$, of a system remains constant unless the system is subject to an unbalanced external force.

LAB PERFORMANCE NOTES

When the Rotary Motion Sensor is connected to the data-collection interface and Logger *Pro* or LabQuest App is started, the default graph screen shows both angle *vs.* time (θ-t) and velocity *vs.* time (ω-t) graphs. It is instructive for the students to see both graphs during the collection of data. Later, during the evaluation, they can work with just the velocity-time graph.

SAMPLE RESULTS AND POST-LAB DISCUSSION

Step 1

Students should find that the moment of inertia for a disk is $\frac{1}{2}mr^2$, whereas that of a cylindrical tube is $\frac{1}{2}m(r_1^2 + r_2^2)$. For the aluminum and steel disks in the accessory kit, the moments are 0.0107 kg·m^2 and 0.0285 kg·m^2 respectively.

Step 2

Figure 1 shows the ω-t graphs for Runs 1 (aluminum) and 4 (steel). The rate of change of ω with respect to time is considerably smaller for the steel disk than for the aluminum one. Considering that the moment of inertia, I, for the steel disk much greater than that for the aluminum, it makes sense that frictional forces produce a smaller rate of change in ω. It is important that students recognize that this rate of change of ω also depends on the angular velocity. As ω decreases, so does its rate of change. However, during any short interval, $\Delta\omega/\Delta t$ is nearly constant.

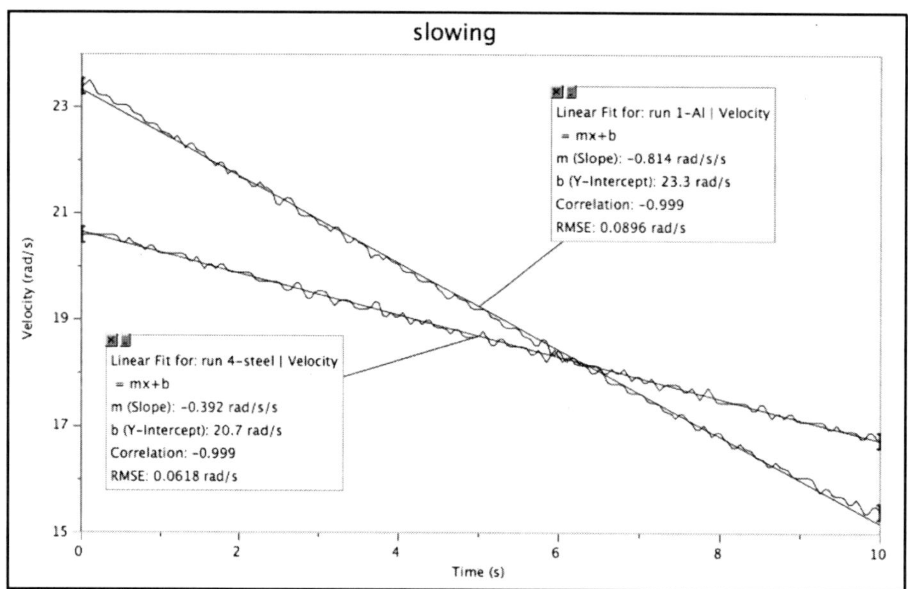

Figure 1 Slowing due to friction

Step 3

By now, students ought to know that when a graph appears linear, they can determine the rate of change by finding the slope in a linear fit. When students select a portion of the graph in Logger *Pro* and choose the Statistics tool, they can see both the maximum and minimum values of velocity. In the lower left corner of the graph they can also see Δx, which represents the time interval (see Figure 2).

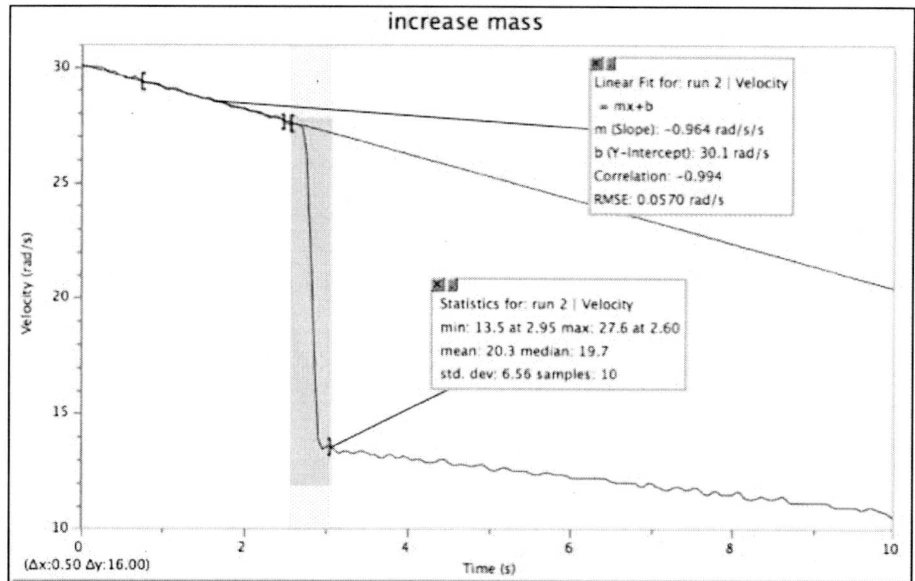

Figure 2 – ω-t graph for Run 2

For this graph, difference in time between the two velocity readings, Δt, is 0.50 s.

Step 4

The angular momentum of the original disk is $L = (0.0107\text{kg}\cdot\text{m}^2)(27.6\,\text{s}^{-1}) = 0.295\ \text{kg}\cdot\text{m}^2\cdot\text{s}^{-1}$. The value for the two-disk system is $L = 0.0214\text{kg}\cdot\text{m}^2\cdot 13.5\ \text{s}^{-1} = 0.289\,\text{kg}\cdot\text{m}^2\cdot\text{s}^{-1}$. The angular momentum after is 2.7% lower than the momentum of the system before the mass was increased.

Step 5

Friction provides a torque that reduces the angular velocity. One can estimate its impact by multiplying the rate of change in ω by the time interval between readings. In the example shown in Figure 2 for the aluminum disk:

$$\Delta\omega = \left(-0.964\ \text{rad}\cdot\text{s}^{-2}\right)\left(0.50\,\text{s}\right) = -0.482\,\text{rad}\cdot\text{s}^{-1}$$

The change in angular momentum due to friction alone is -0.00515 kg·m^2·s^{-1}, which accounts for most of the difference between the initial and final values. Students should expect minor discrepancies in this analysis due to the fact that velocity values displayed on the graph are calculated by the software.

Step 6

Similar calculations performed using data in Runs 3 and 5 yield the following results.

Run	ω_1 (rad/s)	ω_2 (rad/s)	L_1 (kg·m^2/s)	L_2 (kg·m^2/s)	% diff
3	16.4	8.1	0.175	0.173	–1.1%
5	19.8	14.3	0.564	0.561	–0.5%

EXTENSION

The moment of inertia depends both on the mass and its distribution from the axis of rotation. A change in the distance from this axis also produces a change in *I*. Figure skaters routinely do this during a spin. As they pull their arms closer to their body, their moment of inertia decreases; as a result, their angular rotation increases as the product of these quantities remains constant.

Simple Harmonic Motion

The Mathematical Model for Simple Harmonic Motion

INTRODUCTION

When you suspend an object from a spring, the spring will stretch. If you pull on the object and release it, it will begin to oscillate up and down. In this experiment, you will examine this kind of motion, perform a curve fit on the position-time graph, and relate the parameters of the equation with physical features of the system.

OBJECTIVES

In this experiment, you will

- Collect position *vs.* time data as a mass, hanging from a spring, is set in an oscillating motion.
- Determine the best fit equation for the position *vs.* time graph of an object undergoing simple harmonic motion (SHM).
- Define the terms amplitude, offset, phase shift, period and angular frequency in the context of SHM.
- Relate the parameters in the best-fit equation for a position *vs.* time graph to their physical counterparts in the system.
- Use deductive reasoning to predict the system mass required to produce a given value of angular frequency in the curve fit to SHM.

MATERIALS

Vernier data-collection interface
Logger *Pro* software
Vernier Motion Detector
ring stand and right angle clamp
spring
mass hanger and standard lab masses
wire basket

PRE-LAB INVESTIGATION

Attach a rod to a vertical support rod using a right-angle clamp. Hang a spring on the horizontal rod, as shown in Figure 1. Now hang a mass hanger from the spring as directed by your instructor. Assume that the bottom of the hanger is the zero position. Pull on the mass hanger slightly and release it. Observe the motion of the hanger. On the axes below, sketch a graph of the position of the hanger as a function of time.

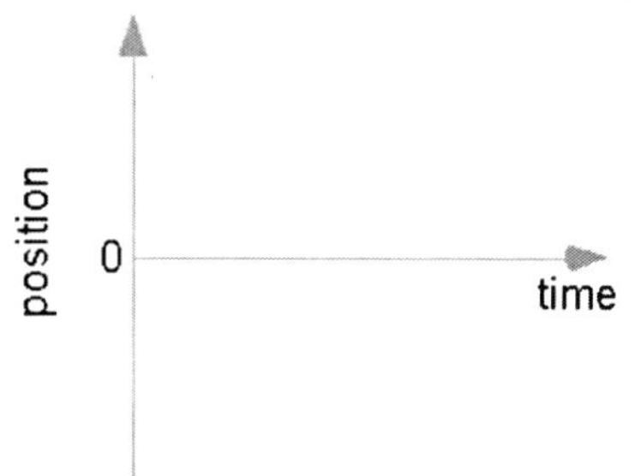

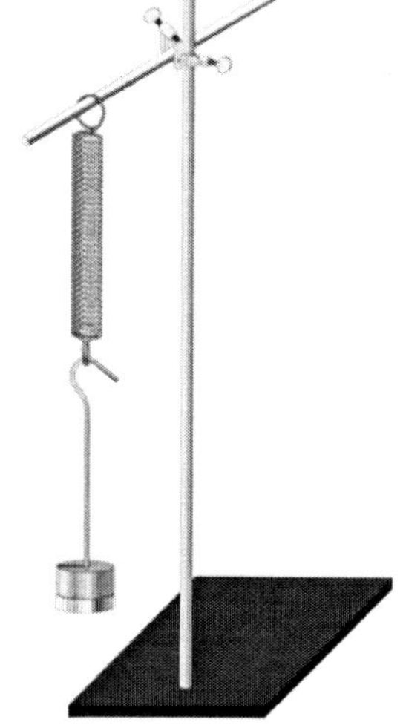

Figure 1

Compare your sketch to those of others in the class.

PART 1 – EXAMINATION OF SIMPLE HARMONIC MOTION

PROCEDURE

1. Connect the Motion Detector to the interface connected to a computer and start Logger *Pro*. Two graphs, position *vs.* time and velocity *vs.* time, will appear in the graph window. For this experiment, you will need only the position *vs.* time graph. Delete the velocity-time graph and choose Auto Arrange from the Page menu.
2. The default data-collection rate is appropriate.
3. If your motion detector has a switch, set it to Track.
4. Place the motion detector on the floor beneath the mass hanger. Place the wire basket over the motion detector to protect it.
5. Lift the hanger and weights slightly, then release. When the hanger is oscillating smoothly, begin collecting data.
6. If the position-time graph appears to be a smooth curve, begin the evaluation of data. If not, repeat until you obtain a smooth curve. If you are satisfied with the data, save this file.

EVALUATION OF DATA

1. Compare the position-time graph you obtained with the one you sketched in the Pre-Lab Investigation. In what ways are the graphs similar? In what ways do they differ? What function appears to describe the position-time behavior of an oscillating body?

2. Before you fit a curve to the position-time graph, turn off Connect Points and turn on Point Protectors. Since the hanger-mass system moves vertically, double-click on the **position** header in the data table and enter *y* as the short name for position.

3. Use the curve-fitting features of Logger *Pro* to fit a sine curve to your data. Write the equation of the representing the motion of the system. Be sure to record the values of the *A*, *B*, *C*, and *D* parameters in the curve fit. Later, you can compare these values to the ones you obtain in Part 2.

PART 2 – EXPLORING THE PARAMETERS OF THE SINE FIT

PROCEDURE

1. Create a new Logger *Pro* file. As before, you will need only the position-time graph. Increase the data-collection rate to 25 Hz. Now, select the Triggering tab and enable Triggering. In your class discussion, determine how to use this feature to control when Logger *Pro* starts data collection; make sure you understand why you have chosen this setting.

2. Hang the mass hanger and the additional masses assigned to you by your instructor from the spring. As before, place the motion detector on the floor beneath the mass hanger and place the wire basket over the motion detector to protect it.

3. Zero the motion detector. Set the masses on the end of the spring in motion, then start data collection. Logger *Pro* will begin collecting data at the appropriate point in the motion of the system. Autoscale the graph. If your position-time graph appears to be a smooth curve, move on to Step 4; otherwise repeat the run.

Exploring Parameter A

4. Turn off Connect Points and turn on Point Protectors, as before. Choose a sine curve fit as you did in Part 1. After you click Try Fit in the Curve Fit dialog box, note the value of *A* suggested by Logger *Pro* to fit the data. Within this dialog box, you can adjust the parameters of the sine fit. Adjustment is made by clicking the + and – buttons to the right of the parameter value. To the right of the + button, you will see a delta (Δ) button. Click this button to control the amount that you change the parameter with the + and – buttons. Change the delta value for the *A* parameter to 0.01. Now click the + and – buttons to see what effect this has on the test plot to fit the data. Cancel to return to the graph window. What aspect of the sine function does the *A* parameter appear to control?

5. Test your conclusion by producing a data set that has an *A* parameter of 0.10 m. Fit a sine curve to the graph and check the *A* parameter. When you have done so, choose Store Latest Run from the Experiment menu and save the experiment file.

Exploring Parameter D

6. Delete the curve fit to your data; perform a curve fit on Run 1. After clicking the Try Fit button, change the value of *D* in the sine fit by increments of ± 0.01 to see what attribute of the sine function is controlled by the *D* parameter.

7. Test your conclusion by producing a data set that has about the same value of A as before, but with a D parameter of –0.05 m. Choose a sine fit to check D. When you have done so, store this latest run, then choose Hide the Data Set for Run 2. Only your first run should appear on the graph.

Exploring Parameter C

8. Choose a sine function to fit the graph in your first run. As before, click the Try Fit button, then note whether Logger *Pro* used a value other than 0 for the C parameter. If so, substitute 0 for the value of C and compare this fit to the one originally suggested. If Logger *Pro* suggested 0 (or a value very close to 0), change C to 6.28 and note the effect on the curve fit.

9. Return the value of C to 0, then increase its value incrementally until you reach 1.6. By what fraction of a complete cycle has the test plot shifted from the original fit to the data? If you are unable to tell for sure, gradually increase C until you reach a value of 3.14 (π). Click Cancel to return to the graph window.

10. To control the value of C in an experiment, you must control the time at which you start collecting data. Choose Data Collection from the Experiment menu and click the Triggering tab. Select the Triggering check box and click On Keyboard to begin data collection.

11. Re-zero the motion detector with the mass at rest, then set the hanger-mass system in simple harmonic motion. Click the Collect button; no data will be collected until you press the space bar. When the hanger has reached the position you believe corresponds to a C value of 1.6, press the space bar. If the value of C in the sine fit does not match your expectations, repeat data collection until you have a run in which C is within $\pm$ 0.1 of 1.6. Store this latest run then hide the Data Set for Run 3.

Exploring Parameter B

You may have noticed that none of the changes you made to the system or the time you began collecting data had any effect on the value of B. Taking a closer look at B; its units must be radians/s in order for the sine function to operate on the argument ($Bt + C$). The B parameter is a measure of how frequently the hanger and mass oscillate.

12. Choose a sine function to fit the graph in your first run. As before, click the Try Fit button, then increase the value of B incrementally to see what effect this change has on the time required to complete one cycle. Now, change B to 6.28, and note the time to complete one cycle. Consider why this is the case.

13. Predict the effect on B of reducing the mass of the hanger and masses to half of its original value. To test your prediction, return the data-collection conditions to those you used when you explored parameters A and D. Re-zero the motion detector, set the lighter hanger-mass system in simple harmonic motion, then collect data.

14. Fit a sine function to your graph as before. Does the value of B agree with your prediction? Store this latest run and save the Logger *Pro* file; you will return to it later.

EVALUATION OF DATA

To help you evaluate your data, you can return to your Logger *Pro* file and, by selecting More on the vertical axis, you can choose to view the position for any of your saved runs.

Exploring Parameter A

The name given to the A parameter is *amplitude*. What physical aspect of the system did you need to control in order to produce a data set with an amplitude of 0.10 m? Given the function $y = A\sin(Bt + C) + D$, explain why A governs the maximum or minimum value of y in the graph.

Exploring Parameter D

Describe how you modified your experimental apparatus to produce a graph with the D parameter (known as the *offset*) equal to –0.05 m. After having made this change, how could you have produced a graph that was symmetrical about the time axis?

Exploring Parameter C

Considering how you set up Logger *Pro* to collect data, explain why the value of C was very nearly 0 or 2π in your first run. At what point in the cycle did you begin to collect data in order to make $C \sim \pi/2$? At what point in the cycle would you have had to begin collecting data in order to make the value of $C \sim \pi$? Explain.

Exploring Parameter B

The time required to complete a cycle is known as the *period, T*, of oscillation. Divide 2π by T and compare the value to B. The name given to the B parameter is *angular frequency*, ω. From what you know about trigonometry and the unit circle, explain why the period of the sine function was one second when you changed B to 6.28 rad/s.

From a consideration of the elastic and kinetic energy of the oscillating system, it can be shown that $\omega = \sqrt{k/m}$. This explains the fact that reducing the mass to half of its original value did not double the frequency. Calculate what hanging mass would oscillate at double the angular frequency you obtained in your first run. Return to your experimental set up and re-open (if necessary) your Logger *Pro* file. Hide all but your first run, then perform a new run to test your prediction. How close did your results come to your prediction?

Consider any other factors that may have an effect on the value you obtain for B. After your discussion, make the necessary adjustment to the hanging mass to test your prediction and perform another run. Save your Logger *Pro* file in case your instructor wishes to examine it.

EXTENSION – DAMPED HARMONIC MOTION

In this experiment, you stopped collecting data before frictional losses slowed the hanging mass appreciably. Set up another experiment in which you collect data for 10 seconds. Tape half of an index card to the bottom of the mass to provide greater air resistance. Collect data for a run and fit a sine function to the graph as you have done before. In what way does this function no longer match the position-time behavior of the oscillating body?

In a new data set use the Examine tool to record the time and position of the amplitude for each of the peaks in your graph. Fit an exponential function to the max-position *vs.* time data; record the *C* parameter of the exponential term "exp(–Ct)". Note that this *C* is different from the *C* in the sine fit. Now, return to your first data set, choose Curve Fit and select the sine function. After you choose Try Fit (so that Logger *Pro* suggests values for the parameters) select Define Function. Multiply the existing function by "exp(C*t)" using the value of *C* you recorded earlier; give the function a sensible name (such as sine-decay). Wait until Logger Pro has built this function, then click Try Fit. Wait again until you see a trace appear on the test graph. You may need to replace the values of *B* and *C* with the values from your sine fit to the graph. Discuss why this more complex function better describes the position-time behavior of an object undergoing damped harmonic motion.

Computer Experiment

15

INSTRUCTOR INFORMATION

Simple Harmonic Motion

The Mathematical Model for Simple Harmonic Motion

This lab is the first of two investigations of simple harmonic motion. These experiments have very different points of emphasis. In this experiment, students examine the connections between the physical system and the parameters used in a sine curve fit to the position-time data. In Experiment 16, students examine relationships between position-time, velocity-time, and acceleration-time graphs as well as the relationship between force and acceleration for a system undergoing SHM. This lab is written with the assumption that the instructor will engage the students in discussions at critical junctures. These discussions can take place with the entire class or with individual lab groups. The icon at left indicates where these discussions should occur.

This version of the experiment is designed for use when Logger *Pro* is used for data collection. The Microsoft Word files for the student pages can be found on the CD that accompanies this book. See *Appendix A* for more information.

OBJECTIVES

In this experiment, the student objectives include

- Collect position *vs.* time data as a mass, hanging from a spring, is set in an oscillating motion.
- Determine the best fit equation for the position *vs.* time graph of an object undergoing simple harmonic motion (SHM).
- Define the terms amplitude, offset, phase shift, period and angular frequency in the context of SHM.
- Relate the parameters in the best-fit equation for a position *vs.* time graph to their physical counterparts in the system.
- Use deductive reasoning to predict the system mass required to produce a given value of angular frequency in the curve fit to SHM.

During this experiment, you will help the students

- Distinguish between frequency (cycles/second) and angular frequency (rad/s).
- Understand how to use triggering in data collection in Logger *Pro*.
- Derive an expression relating the frequency of an object undergoing SHM to both the spring constant, k, and the mass.
- Understand that a portion of the mass of the spring must also be included in the calculation of the frequency of oscillation.

REQUIRED SKILLS

In order for student success in this experiment, it is expected that students know how to

- Zero a Motion Detector. This is addressed in Activity 2.
- Perform curve fits in Logger *Pro*. This is addressed in Activity 1.

EQUIPMENT TIPS

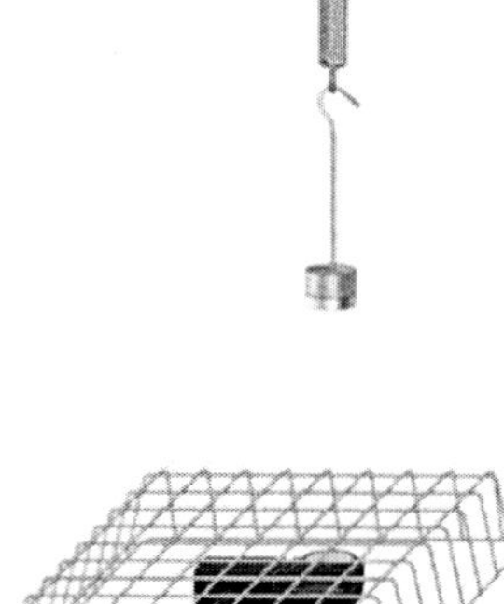

Figure 1

For this experiment, the spring can be attached directly to the rod, as shown in Figure 1. The mass that you use depends in part on the type of springs and hangers you have available for this experiment. Vernier sells a spring set appropriate for this experiment, order code SPRINGS. This accessory kit contains six springs that are 6 cm long and 0.8 cm in diameter. Three springs have a spring constant of 5 N/m and the other three have a spring constant of 15 N/m.

During the course of the experiment, students will need to reduce the hanging mass to one-fourth of its original value. As a result, it is important that you choose a starting mass that allows them to do so. For example, for the data included in this experiment, an additional 150 g of weights were added to a 50 g mass hanger. The total initial mass of 200 g was suspended from a 4.0 N/m spring. For springs with a larger spring constant, the starting mass will have to be greater; otherwise the period of oscillation will be too short. Use of larger masses will require that you make sure the support rod is securely attached to the table so that the system does not wobble while the mass is oscillating. A wire basket available at an office supply store can be used to protect the Motion Detector from falling masses (see Figure 1).

PRE-LAB DISCUSSION

It is likely that students will recognize that the motion they observe is periodic. However, they might draw a saw-tooth curve, focusing on the extremes of the motion, instead of sketching a sinusoidal curve.

LAB PERFORMANCE NOTES

When the Motion Detector is connected to the data-collection interface and Logger *Pro* is started, the default graph screen shows both position *vs.* time (y-t) and velocity *vs.* time (v-t) graphs. Students can delete the v-t graph and choose Auto Arrange from the Page menu to re-size the y-t graph.

If the motion detector has a range setting, set the switch to Track mode. Position the hanger and weights high enough so that when they drop to their lowest point in the oscillating motion they are not closer to the motion detector than the minimum distance[1].

Remind students to turn off the Connect Points feature and turn on Point Protectors for the graphs they produce in this experiment.

In the exploration of parameters A, D, and B (amplitude, offset, and frequency) in Part 2, you may need to show the students how to set triggering for the collection of data. Guide the students to understand that they want Logger *Pro* to begin collecting position-time data when the mass-hanger system is passing through the equilibrium position on its way up. At this point the motion detector reading is passing from negative values through 0 to positive values. This will ensure

[1] This is 15 cm for motion detectors with a switch and 45 cm for motion detectors without a switch.

that that their position-time data will look like a standard sine curve starting at $y = 0$ with positive slope.

In the exploration of parameter *D*, students will have to lower the rod holding the spring *after* they have zeroed the motion detector.

In the exploration of parameter *C*, students need to change the triggering setting to On Keyboard. Then, when they perceive that the mass hanger has reached the top of its oscillating motion, they should press the spacebar to begin collecting data. Getting this timing right so as to obtain a run with a *C* value ±0.1 rad of the suggested value of 1.6 rad (~ π/2) takes a bit of practice. Reassure students that there is nothing wrong with the equipment.

In the exploration of parameter B, students have to be able to reduce the system mass to one-fourth of its original value in an attempt to double the angular frequency. When they find that this reduction does not quite achieve this goal, students will need access to extra masses to add to the hanger to make B for their final run half of the value they obtained for their minimum mass.

SAMPLE RESULTS AND POST-LAB DISCUSSION – PART 1

This may be the first opportunity students have had to fit a sine curve to data they have collected. It is highly likely that sine fit will be shifted from the standard graph given by the equation $y = \sin\theta$ with which students are familiar (see Figure 2). When students obtain an equation of the form $y = A\sin(Bt + C) + D$, they are likely to have questions about the meaning of these parameters. Inform them that they will investigate these carefully in Part 2.

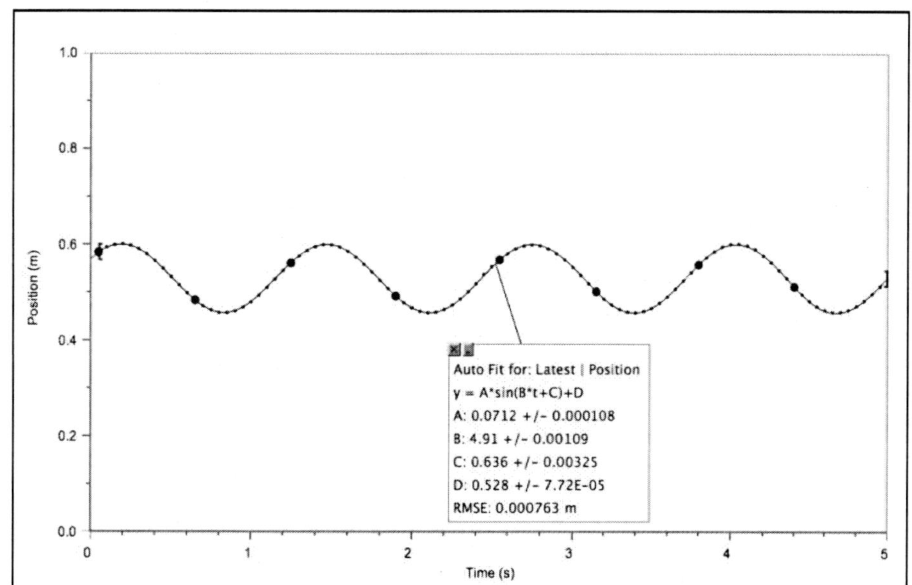

Figure 2 Sine curve for initial investigation of SHM

SAMPLE RESULTS AND POST-LAB DISCUSSION – PART 2

Exploring Parameter A

With the conditions specified for the collection of data in Part 2, students will generate a graph with which they should be more familiar. Figure 3 shows data after the motion detector is zeroed and correctly triggered. Next to each of the fields displaying the value of the parameters are up and down arrows and a button labeled with a delta (Δ). Clicking on this enables one to choose the amount by which the value is changed. Students should change this value to 0.01 when they examine A. When students change the value of A, they will see the amplitude of the test plot change. Students should realize that the output of the sine function has a range from –1 to +1. So, the A parameter, or *amplitude*, is a factor that scales the maximum or minimum value of y in the sine function.

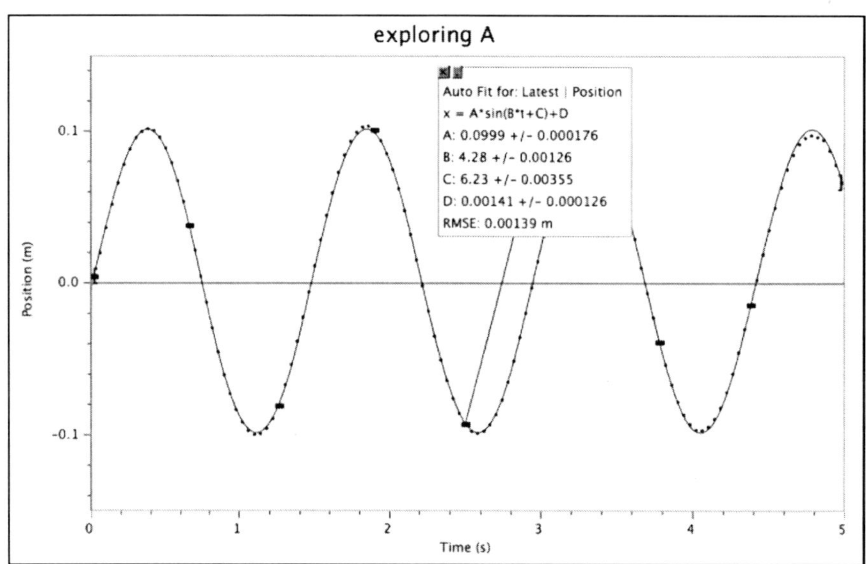

Figure 3 Exploring parameter A

Exploring Parameter D

As they did in the exploration of the A parameter, students should change the value of the increment to 0.01 when they examine D. Students should recognize that in order to obtain a value of –0.05 for D, they had to lower the rod supporting the force sensor, spring, and mass hanger by 5 cm. To return the sine curve to its initial position, they have to re-zero the motion detector before collecting data.

Exploring Parameter C

Depending on their trigonometry background, students may need some guidance in realizing that the argument of the sine function must be an angle, typically expressed in radians. The B parameter must have units of rad/s while C must be in radians. Parameter C governs the extent to which the sine curve is shifted left or right from its standard position. One complete cycle is equivalent to 2π rad, so any value of C that is an integer multiple of 2π rad will yield a function that starts at zero with positive slope. Data collection does not begin exactly when the mass hanger is in the equilibrium position. As a result, Logger *Pro* usually suggests a value of C that is close to 6.28.

A sine curve with C equal to 1.6 rad is shifted 1/4 of a cycle from its normal starting position; i.e., the mass hanger would have been at its maximum height when data collection began. Figure 4 shows both the original run and one in which $C = 1.67$ rad. To obtain a value of C equal to π rad, data collection would have to begin just as the hanger passes downward through the equilibrium position.

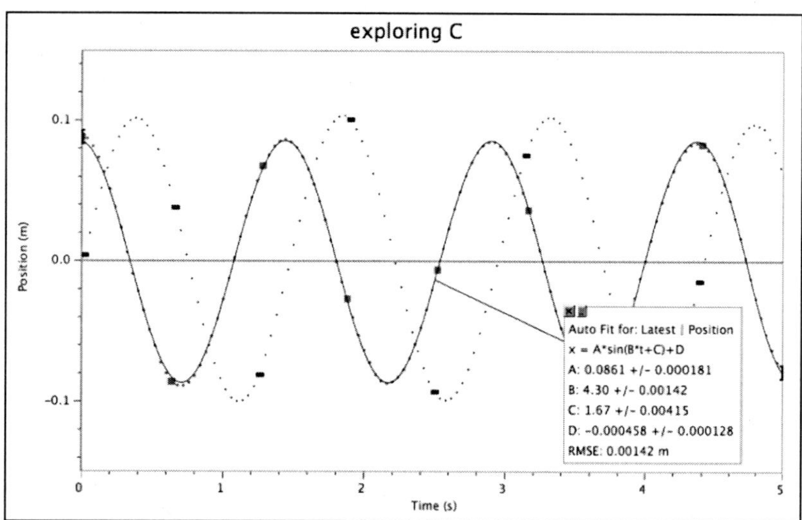

Figure 4 Exploring parameter C

Exploring Parameter B

The first step in the analysis of this parameter is to recognize that its units suggest that B is a measure of the rate of oscillation. An object completing one oscillation/second would have an angular frequency of 6.28 rad/s. The period, T is simply 2π rad/B. Because changes in the way the data were collected had no effect on the value of B, students are likely to suspect that B depends on physical features of the system. When students recognize that halving the mass does not double the frequency (see Figure 5), they are ready for a more in-depth analysis.

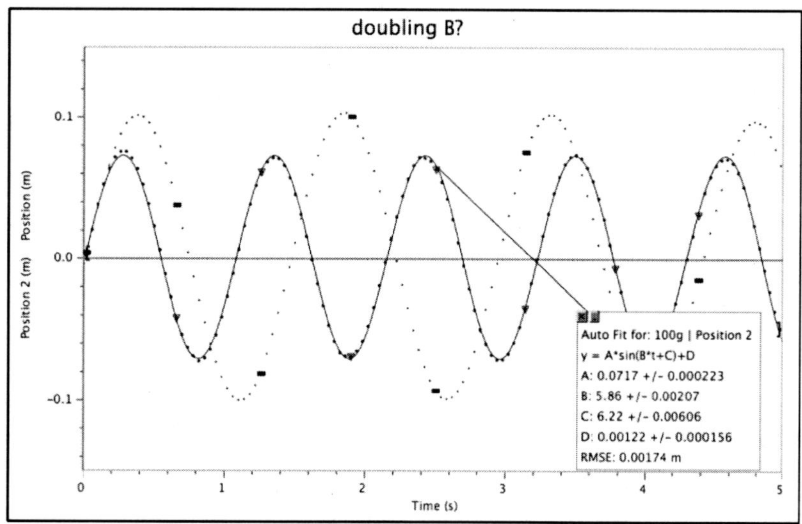

Figure 5 Exploring parameter B

The derivation of the dependence of angular frequency, ω, on the spring constant, k, and mass is left as an extension activity in Exp 16. For the purposes of this experiment, students need only know that the frequency is inversely proportional to the square root of the mass. They should predict that in order to double the value of B, one would have to reduce the hanging mass to one-fourth of its original value. Students are instructed to return to their apparatus to check their prediction. When they do so, they will find that they came closer to this goal than before, the new value of B is still less than twice the original value (see Figure 6).

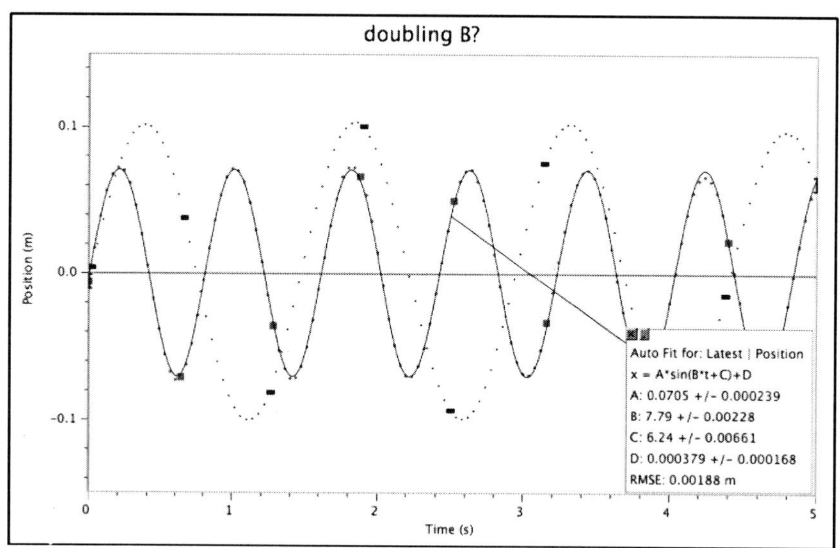

Figure 6 Mass reduced to 1/4 of original value

At this point, you will need to point out that some portion of the mass of the spring must also be taken into account when calculating the frequency. The general approximation is that one-third of the spring mass must be added to the hanging mass to determine the mass of the oscillating *system*.

Figure 7 shows graphs for masses of 64 g (hanger + 1/3 mass of spring) and 264 g (as close to 4× original mass as was practical). Note that B for the heavier mass is very nearly twice that of the lighter mass.

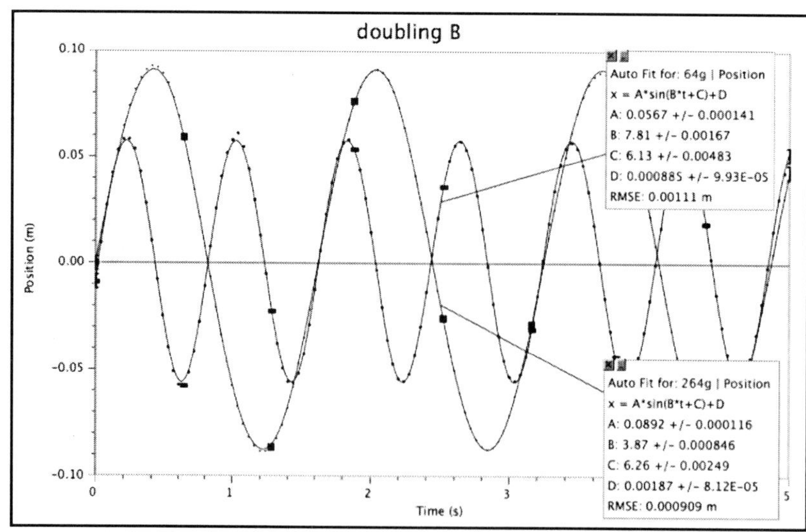

Figure 7 Including a portion of the spring mass

EXTENSION – DAMPED HARMONIC MOTION

Your decision to have students perform the extension will depend, in part, on their familiarity with the exponential function. When the air resistance is increased, the amplitude of the oscillating motion decreases noticeably with time. A simple sine fit to the data does not account for this decrease in amplitude. Examination of the peak amplitude for eight cycles yields the following data.

Time (s)	Position (m)
0.34	0.113
1.64	0.109
2.92	0.107
4.20	0.104
5.48	0.101
6.76	0.098
8.08	0.096
9.36	0.093

These data were entered into a new data set. When an exponential function is fit to the data, the equation $y = 0.0986\exp(-0.0248\,t) + 0.0149$ is obtained. The e^{-Ct} portion of the equation can be used as a factor to modify the amplitude of the original sine fit to the position-time data. Make sure that students recognize that the *C* in the exponential fit is not the same as the *C* in the original sine fit

Students must return to the original data set and fit a sine function to their data. Then, before they click OK, they need to click the Define Function button to bring up the User Defined Function window (see Figure 8).

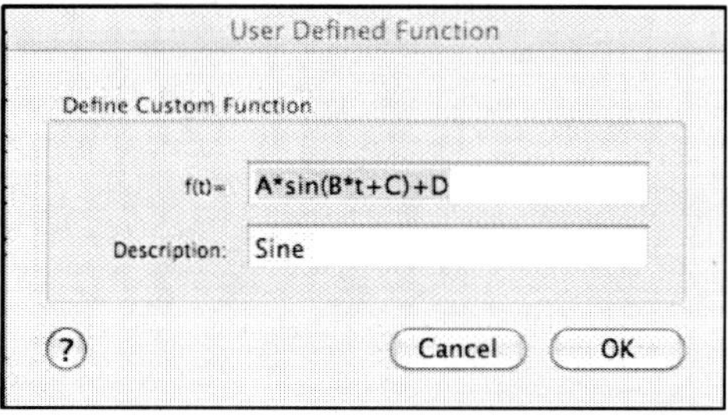

Figure 8

Because *D* was negligible, it was not included in the function shown in Figure 9.

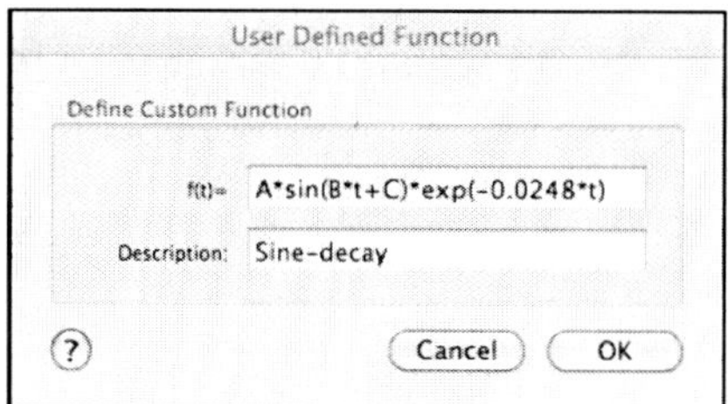

Figure 9

Students need to allow Logger *Pro* some time to first generate this function, then try to fit this function to the data. It's possible that the *B* and *C* values suggested will differ greatly from the original sine fit. If this is the case, replace the values suggested by Logger *Pro* with those provided earlier, then click OK. A sample curve fit is shown in Figure 10.

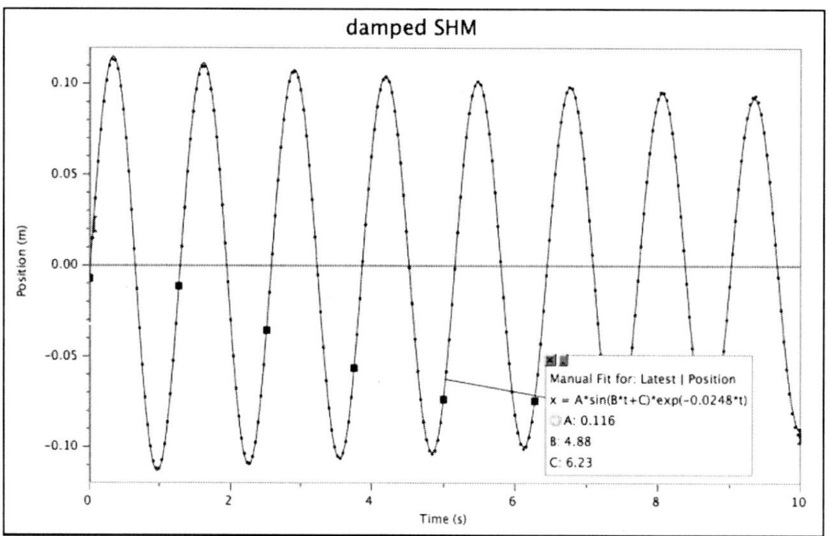

Figure 10 Damped SHM

The factor e^{-Ct} describes the exponential rate of decay of the sine fit to the position-time data caused by the air resistance provided by the card.

Simple Harmonic Motion

The Mathematical Model of Simple Harmonic Motion

INTRODUCTION

When you suspend an object from a spring, the spring will stretch. If you pull on the object and release it, it will begin to oscillate up and down. In this experiment, you will examine this kind of motion, perform a curve fit on the position-time graph, and relate the parameters of the equation to physical features of the system.

OBJECTIVES

In this experiment, you will

- Collect position *vs.* time data as a mass, hanging from a spring, is set in an oscillating motion.
- Determine the best fit equation for the position *vs.* time graph of an object undergoing simple harmonic motion (SHM).
- Define the terms amplitude, offset, phase shift, period and angular frequency in the context of SHM.
- Relate the parameters in the best-fit equation for a position *vs.* time graph to their physical counterparts in the system.
- Use deductive reasoning to predict the system mass required to produce a given value of angular frequency in the curve fit to SHM.

MATERIALS

Vernier LabQuest
LabQuest App
Vernier Motion Detector
ring stand and right angle clamp
spring
mass hanger and standard lab masses
wire basket

PRE-LAB INVESTIGATION

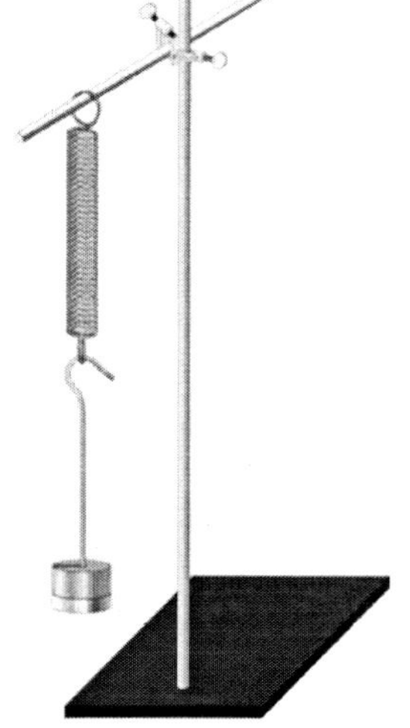

Figure 1

Attach a rod to a vertical support rod using a right-angle clamp. Hang a spring on the horizontal rod, as shown in Figure 1. Now hang a mass hanger from the spring as directed by your instructor. Assume that the bottom of the hanger is the zero position. Pull on the mass hanger slightly and release it. Observe the motion of the hanger. On the axes below, sketch a graph of the position of the hanger as a function of time.

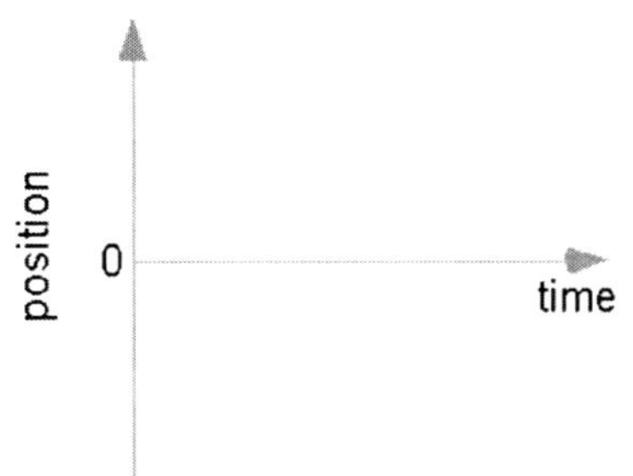

Compare your sketch to those of others in the class.

PROCEDURE

1. Connect the Motion Detector to LabQuest and start a new file in LabQuest App. On the Meter screen, tap Rate and increase the data-collection rate to 50 Hz. Enable Triggering. In your class discussion, determine how to use this feature to control when Logger *Pro* starts data collection; make sure you understand why you have chosen this setting.

2. Hang the mass hanger and the additional masses assigned to you by your instructor from the spring. Place the motion detector on the floor beneath the mass hanger and place the wire basket over the motion detector to protect it.

3. Zero the motion detector. Set the masses on the end of the spring in motion, then start data collection. LabQuest App will begin collecting data at the appropriate point in the motion of the system. If your position-time graph appears to be a smooth curve, store your run and move on to the Evaluation of Data; otherwise repeat the run.

EVALUATION OF DATA

Compare the position-time graph you obtained to the one you predicted in the Pre-Lab Investigation. In what ways are the graphs similar? In what ways do they differ? What function appears to describe the position-time behavior of an oscillating body?

Modeling a curve fit to the data

On the Graph tab choose to view only the position-time graph for this experiment. Since your position-time data appears sinusoidal, you will model a sine function $y = A\sin(Bt + C) + D$ to the data by choosing values of the parameters that will enable you to *manually* fit a curve to your position-time data. The following steps will help you determine the appropriate values of these parameters. You are likely to find that your first effort will produce a model equation that approximates your position-time data. Repeating this process will enable you to refine your equation to more closely match the data. Before you begin, tap on the position-time graph to determine the maximum and minimum values of position during your run.

Exploring Parameter D

1. Consider the range of values that the sine function returns when it operates on the argument ($Bt + C$). Under Graph options, choose 1.0 and –1.0 for the top and bottom values displayed on your graph to ensure that your data will appear in the graph window.

2. Choose Model from the Analyze menu, then select $y = A\sin(Bx + C) + D$ as the Equation; keep in mind that the x variable stands for time. The test plot displayed is a much larger version of your sinusoidal y-t data. Using the down arrow, reduce the value of D until the test plot is symmetrical about the time axis. *Do not* tap OK yet. What aspect of the sine fit does the D parameter appear to control? Consider that prior to data collection, you zeroed the motion detector.

Exploring Parameter A

3. While the test plot is now symmetrically placed on the time axis, it is much larger than the plot of your data. Try reducing the value of A until the maximum and minimum values of the test plot approximate those in the graph of your data. What aspect of the sine fit does the A parameter appear to control?

Exploring Parameter C

4. Note that at time $t = 0$, the y-value of your test plot is not 0. Apparently, the default value of 1.0 is not a good match to your data. Try reducing the value of C until the initial value of y equals 0; this also has the effect of making the position of the maximum value of the test plot more closely match one of the maxima in your data. Note this value of C. Now, try increasing C until the test plot is again increasing through 0 when $t = 0$.

5. Since the argument of a sine function must be an angle, the expression $Bt + C$ must have units of an angle measure. The values LabQuest App uses for the C parameter are given in radians. Considering what you know about radian measure and the unit circle, explain why a test plot with a value of $C = 0$ or 2π looks very much like a standard sine curve.

6. Considering how you set up LabQuest App to collect data, explain why it appears that the C parameter for your graph should be ~ 0.

Exploring Parameter B

7. Gradually increase the value of the B parameter. Note the effect this change has on the time required for the object to move through one complete cycle. As B approaches the value that best models your data, you may find that you need to make minor adjustments to C as well. When you think your test plot is a good approximation to the position-time data, tap OK. Record the value of the parameters of the sine fit to your data.

8. Choose Autoscale Once from the Graph menu. Use the Delta function from the Analyze menu and drag across the graph to determine the length of time required for the motion to go through one complete cycle. Do this for a couple of successive cycles and average the value of Δt; then turn off the Delta function. The time required to complete a cycle is known as the *period, T*, of oscillation.

9. Let's take a closer look at *B*; its units must be radians/second in order for the sine function to operate on the argument ($Bt + C$). The *B* parameter, known as *angular frequency, ω*, is a measure of how frequently the hanger and mass oscillate. Given that one cycle is 2π radians, divide 2π by *T* and compare this value to *B*.

10. On the zoomed-in view of the graph, model a sine function that more closely approximates your position-time data. Start with your original parameters, then make slight adjustments to *A, B* and *C*. Use what you learned in Step 9 to choose the value of B. Record the value of your improved parameters.

Revisiting the parameters

11. Return to your apparatus and perform a second run, this time with a smaller amplitude. Store this run. Model a sine fit to these data as you did before and compare the parameters for this run to those from your first run. Which one(s) were different? Explain.

12. Now, reduce the mass hanging from the spring to half of its original value. Perform a third run in which you attempt to keep the amplitude nearly the same as in your second run. When you obtain a run that meets this criterion, store it. Model a sine fit to these data as you did before and compare the parameters for this run to those from your second run. What effect did reducing the mass have on the value of *B*?

13. Save your file before you move on to the Extension.

EXTENSION – PHYSICAL FACTORS AFFECTING *B*

From a consideration of the elastic and kinetic energy of the oscillating system, it can be shown that $\omega = \sqrt{k/m}$. This explains the fact that reducing the mass to half of its original value did not double the frequency. Calculate what hanging mass would oscillate at double the angular frequency you obtained in your first run. Return to your experimental set up and re-open (if necessary) your LabQuest App file. Perform a new run to test your prediction. How close did your results come to your prediction?

Consider any other factors that may have an effect on the value you obtain for *B*. After your discussion, make the necessary adjustment to the hanging mass to test your prediction and perform another run.

INSTRUCTOR INFORMATION

Simple Harmonic Motion

The Mathematical Model of Simple Harmonic Motion

This lab is the first of two investigations of simple harmonic motion. These experiments have very different points of emphasis. In this experiment, students examine the connections between the physical system and the parameters used in a sine curve fit to the position-time data. In the next experiment, students examine relationships between position-time, velocity-time and acceleration-time graphs as well as the relationship between force and acceleration for a system undergoing SHM. This lab is written with the assumption that the instructor will engage the students in discussions at critical junctures. These discussions can take place with the entire class or with individual lab groups. The icon at left indicates where these discussions should occur.

This version of the lab is designed for use with the LabQuest as a standalone device. The Microsoft Word files for the student pages can be found on the CD that accompanies this book. See *Appendix A* for more information.

OBJECTIVES

In this experiment, the student objectives include

- Collect position *vs.* time data as a mass, hanging from a spring, is set in an oscillating motion.
- Determine the best fit equation for the position *vs.* time graph of an object undergoing simple harmonic motion (SHM).
- Define the terms amplitude, offset, phase shift, period and angular frequency in the context of SHM.
- Relate the parameters in the best-fit equation for a position *vs.* time graph to their physical counterparts in the system.

During this experiment, you will help the students

- Distinguish between frequency (cycles/second) and angular frequency (rad/s).
- Understand how to use triggering in data collection in LabQuest App.
- Derive an expression relating the frequency of an object undergoing SHM to both the spring constant, *k*, and the mass.
- Understand that a portion of the mass of the spring must also be included in the calculation of the frequency of oscillation.

REQUIRED SKILLS

In order for student success in this experiment, it is expected that students know how to

- Zero a Motion Detector. This is addressed in Activity 2.

EQUIPMENT TIPS

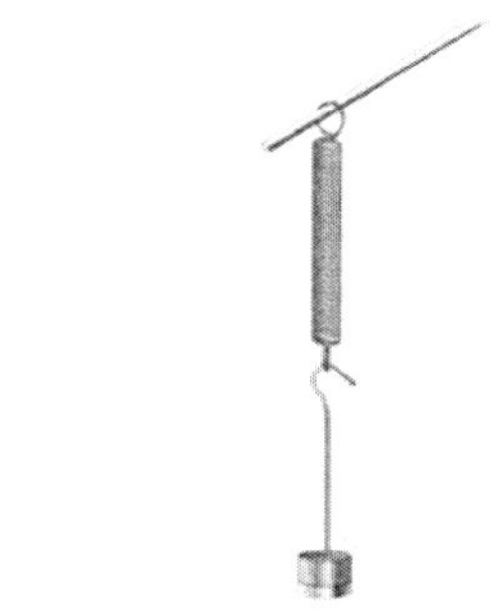

For this experiment, the spring can be attached directly to the rod, as shown in Figure 1. The mass that you use depends in part on the type of springs and hangers you have available for this experiment. Vernier sells a spring set appropriate for this experiment, order code SPRINGS. This accessory kit contains six springs that are 6 cm long and 0.8 cm in diameter. Three springs have a spring constant of 5 N/m and the other three have a spring constant of 15 N/m.

During the course of the experiment, students will need to reduce the hanging mass first to 1/2, then to 1/4 of its original value. So it is important that you choose a starting mass that allows them to do so. For example, for the data included in this experiment, an additional 150 g of weights were added to a 50 g mass hanger. The total initial mass of 200 g was suspended from a 4.0 N/m spring. For springs with a larger spring constant, the starting mass will have to be greater; otherwise the period of oscillation will be too short.

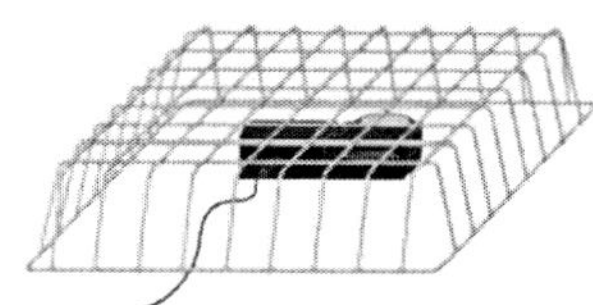

Figure 1

A wire basket can be used to protect the motion detector from falling masses (see Figure 1).

PRE-LAB DISCUSSION

It is likely that students will recognize that the motion they observe is periodic. However, they might draw a saw-tooth curve, focusing on the extremes of the motion, instead of sketching a sinusoidal curve.

LAB PERFORMANCE NOTES

When the Motion Detector is connected to LabQuest and LabQuest App is started, the default graph screen shows both position *vs.* time (y-*t*) and velocity *vs.* time (*v*-*t*) graphs. Students should chose to view only the *y*-*t* graph for this experiment.

Make sure that the motion detector is set to Track mode and that the initial position of the hanger and weights is high enough so that when they drop to their lowest point in the oscillating motion they are not closer to the motion detector than the minimum distance[1].

You may need to show the students how to set triggering for the collection of data. Guide the students to understand that they want LabQuest App to begin collecting position-time data when the mass-hanger system is passing through the equilibrium position on its way up. At this point the motion detector reading is passing from negative values through 0 to positive values. This will ensure that that their position-time data will look like a standard sine curve starting at $y=0$ with a positive slope.

In the exploration of parameter *B*, students have to be able to reduce the system mass to 1/2 of its original value in an attempt to double the angular frequency. In the extension activity, they will have to be able to further reduce the mass to 1/4 of its original value. When they find that this

[1] This is 15 cm for newer motion detectors with a switch and 45 cm for older ones without a switch.

reduction does not quite achieve this goal, students will need access to extra masses to add to the hanger to make *B* for their final run 1/2 of the value they obtained for their minimum mass.

SAMPLE RESULTS AND POST-LAB DISCUSSION

This may be the first opportunity students have had to model a sine curve to data they have collected. You may have to guide students in the use of the Model function under the Analyze menu.

Exploring Parameter D

When students rescale their graph and adjust the *D* parameter to 0, their screen should look something like Figure 2.

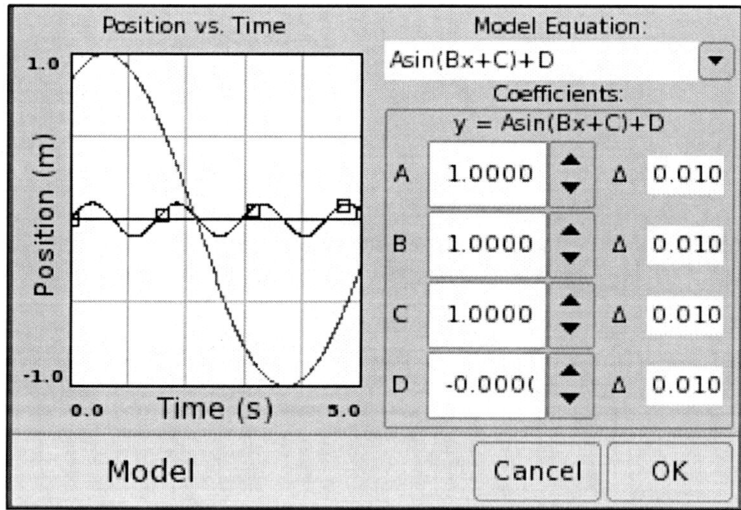

Figure 2 Adjusting parameter D

Exploring Parameter A

When students reduce the value of the *A* parameter, the amplitude of their test plot will approximate that of their data (see Figure 3). They should conclude that *A* controls the maximum and minimum values of the model function.

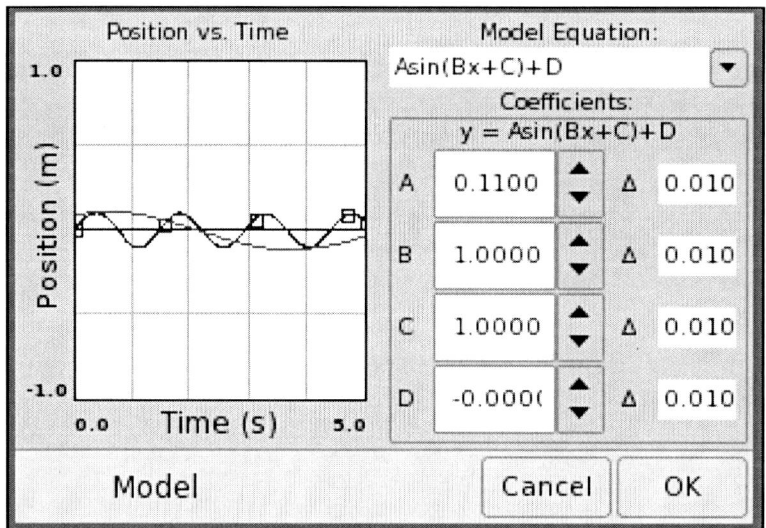

Figure 3 Adjusting parameter A

Exploring Parameter C

Depending on their trigonometry background, students may need some guidance in realizing that the argument of the sine function must be an angle, typically expressed in radians. The *B* parameter must have units of rad/s while *C* must be in radians. Parameter *C* governs the extent to which the sine curve is shifted from its standard position. One complete cycle is equivalent to 2π rads, so any value of *C* that is an integer multiple of 2π rads will result in a sine curve that is in standard position.

Since triggering was set so that data collection began as the mass hanger was ascending through 0, the value of *C* should be adjusted to zero (see Figure 4).

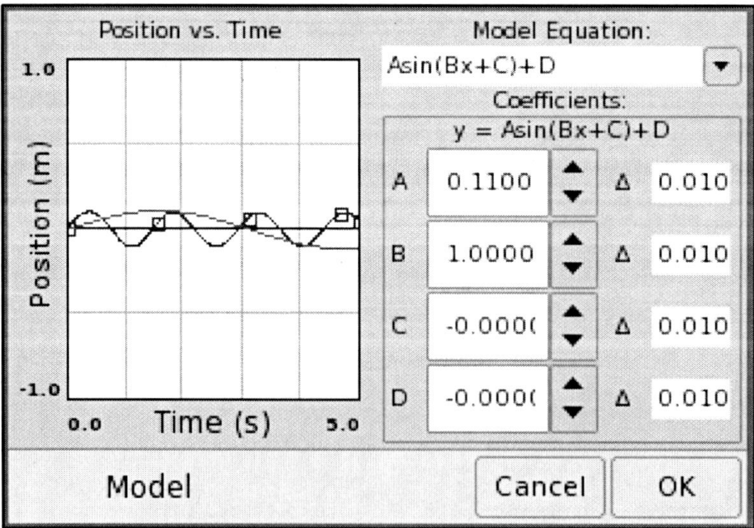

Figure 4 Adjusting parameter C

Exploring Parameter B

When the *B* parameter is increased, the period of the test plot decreases. Eventually, adjustments to *B* and *C* should result in a plot that nearly matches the position-time data (see Figure 5).

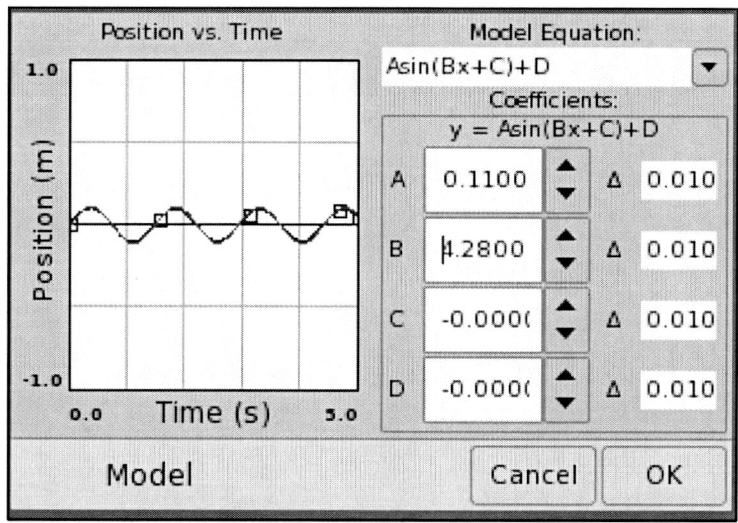

Figure 5 Adjusting parameter B

When students Autoscale the graph, they are likely to see that the parameters of their model are not quite right. To help them choose a better value of *B*, they should use the Delta function to obtain an estimate of the period, *T*, of the oscillation. Help students recognize that since there are 2π radians in one cycle, dividing 2π by *T* should yield the angular frequency in radians/s (see Figure 6).

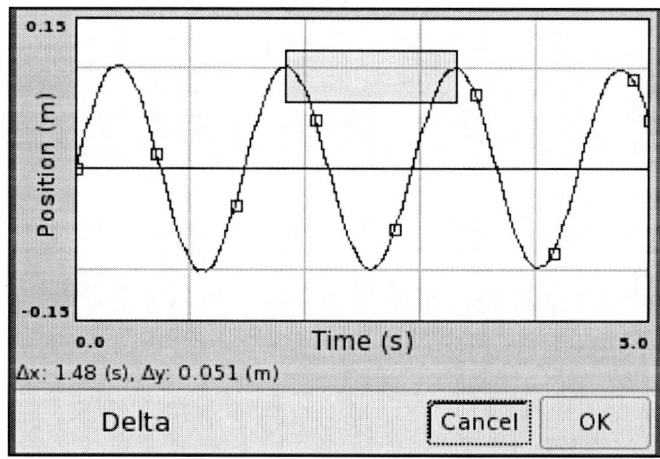

Figure 6 Use Delta to determine T

Using the value of *B* calculated from $\omega = 2\pi/T$, they should obtain a better fit to their data.

Revisiting the parameters

When students perform a second run, and model a sine fit to their data, they will find that all the parameters are the same except for *A*. This is due to the fact that they did not change the initial position, nor the time LabQuest App began collecting data. The spring and mass were also unchanged. Only the amplitude was different.

Students will find that a reduction in mass does results in an increase in the value of B; however, they may be surprised to find that reducing the mass to half of its original value does not double the angular frequency (see Figure 7).

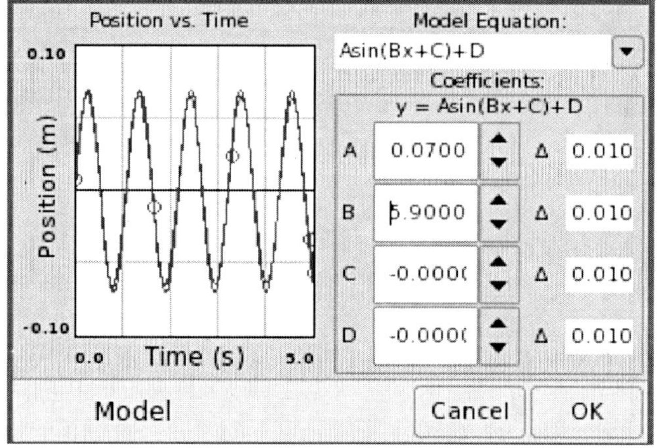

Figure 7 Halving the mass

This sets up the discussion that will take place in the Extension activity.

EXTENSION – PHYSICAL FACTORS AFFECTING *B*

The derivation of the dependence of angular frequency, ω, on the spring constant, k, and mass is left as an extension activity in Experiment 16. For the purposes of this experiment, students need only know that the frequency is inversely proportional to the square root of the mass. They should predict that in order to double the value of B, one would have to reduce the hanging mass to 1/4 of its original value. Students are instructed to return to their apparatus to check their prediction. When they do so, they will find that they came closer to this goal than before, the new value of B is somewhat less than twice the original value (see Figure 8).

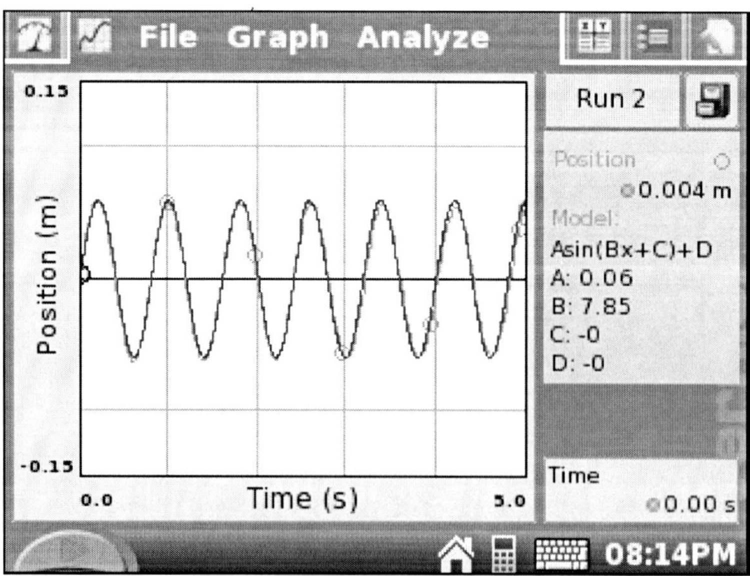

Figure 8 One-fourth of the mass

At this point, you will need to point out that some portion of the mass of the spring must also be taken into account when calculating the frequency. The general approximation is that 1/3 of the spring mass must be added to the hanging mass to determine the mass of the oscillating *system*.

Computer Experiment

16

Simple Harmonic Motion

Kinematics and Dynamics of Simple Harmonic Motion

INTRODUCTION

When you suspend an object from a spring, the spring will stretch. If you pull on the object, stretching the spring some more, and release it, the spring will provide a restoring force that will cause the object to oscillate in what is known as simple harmonic motion (SHM). In this experiment, you will examine this kind of motion from both kinematic and dynamic perspectives.

OBJECTIVES

In this experiment, you will

- Collect position *vs.* time data as a weight, hanging from a spring, is set in simple harmonic motion (SHM).
- Determine the best-fit equation for the position *vs.* time graph of an object undergoing SHM.
- Define the terms amplitude, offset, phase shift, period and angular frequency in the context of SHM.
- Predict characteristics of the corresponding velocity *vs.* time and acceleration *vs.* time graphs, produce these graphs and determine best-fit equations for them.
- Relate the net force and acceleration for a system undergoing SHM.

MATERIALS

Vernier data-collection interface
Logger *Pro*
Vernier Motion Detector
Vernier Dual-Range Force Sensor **or**
Wireless Dynamic Sensor System
ring stand and right angle clamp
spring
mass hanger and standard lab masses
wire basket

PRE-LAB INVESTIGATION

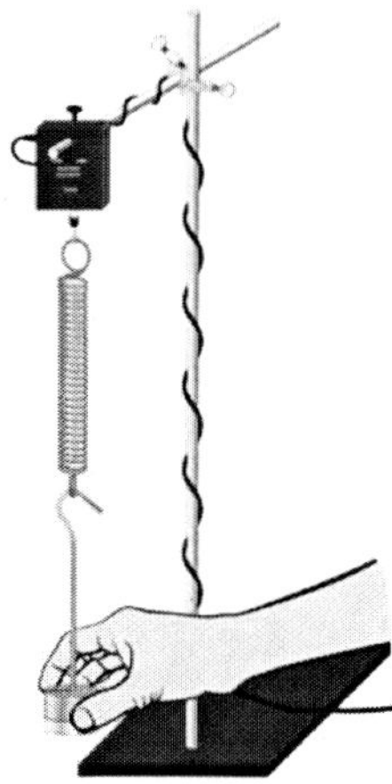

Figure 1

Attach a rod to a vertical support rod using a right angle clamp. Mount a Dual-Range Force Sensor (or WDSS) to the horizontal rod. Now hang a spring from the hook on the sensor and suspend a mass hanger and weights from the spring, as shown in Figure 1. Assume that the bottom of the hanger is the zero position. Pull on the mass hanger slightly and release it. Observe the motion of the hanger. On the axes below, sketch a graph of the position of the hanger as a function of time.

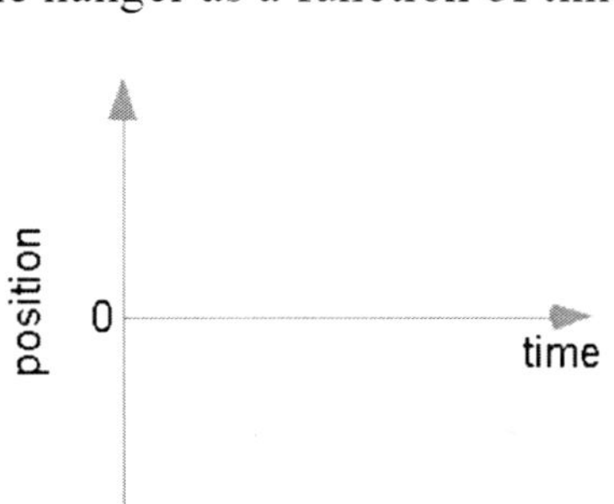

Compare your sketch to those of others in the class.

PROCEDURE

1. Connect the Motion Detector and the Dual-Range Force Sensor to the interface connected to a computer and start Logger *Pro*. Three graphs, force *vs.* time, position *vs.* time and velocity *vs.* time, will appear in the graph window. For now, delete all but the position-time graph. You will be able to insert the others later, when needed. Choose Auto Arrange from the Page menu to re-size the graph.

2. The default data-collection rate is appropriate; however, shorten the duration to 5 seconds.

3. If your motion detector has a switch, set it to Track. 

4. Hang the mass hanger and masses from the spring.
   Place the motion detector on the floor beneath the mass hanger. Place the wire basket over the motion detector to protect it.

5. Make sure the hanger is motionless, then zero both the force sensor and the motion detector.

6. Lift the hanger and weights a few centimeters, then release. When the mass hanger is oscillating smoothly, start data collection.

7. If the position-time graph appears to be a smooth curve, autoscale the graph and choose Store Latest Run from the Experiment menu. If not, repeat until you obtain a smooth curve.

8. Perform a second trial, except this time, make the initial displacement of the mass hanger different from what you did for your first trial. If the position-time graph does not appear to be a smooth curve, repeat until you obtain one.

EVALUATION OF DATA

Part 1 Exploration of SHM

1. Compare the position-time graphs you obtained with the one you sketched in the Pre-Lab Investigation. In what ways are the graphs similar? In what ways do they differ? What function appears to describe the position-time behavior of an oscillating body?

2. Before you fit curves to your position-time graphs, turn off Connect Points and turn on Point Protectors. Since the hanger-mass system moves vertically, double-click on the **position** header in the data table and use y as the short name for position.

3. Use the curve-fitting features of Logger *Pro* to fit a sine curve to the data for each of your runs. Write the equations that represent the motion of the system for each trial. Be sure to record the values of the A, B, C and D parameters in the curve fit.

4. Compare the values of the A parameter for each of your sine fits. What aspect of the y-t graph does A appear to describe? The name given to the A parameter is *amplitude.*

5. Unless you managed to begin collecting data at the instant the oscillating mass hanger was rising through the 0 position, your y-t graphs are likely to be shifted somewhat from the standard position of a graph of $y = \sin\theta$. Delete your curve fits for the y-t graphs and choose Curve Fit again for one of your runs. After you click Try Fit in the Curve Fit dialog box, switch to Manual as the Fit Type, then click OK. Doing so enables you to manipulate the parameters in the graph window. When you click on the value of C suggested by Logger *Pro* to fit the data, an arrow appears next to C as shown to the right. The up and down arrows on your keyboard allow you to change this value. Double-clicking on this window allows you to specify a value for the parameter or the amount of increment. Try increasing and decreasing the value of C by ± 0.1 to see what effect this has on the test plot used to fit your data. Return C to its original value.

   Manual Fit for: Run 2 | Position
   x = A*sin(B*t+C)+D
   A: 0.0639
   B: 4.90
   C: 1.12
   D: -0.00192

6. Since the argument of a sine function must be an angle, the expression $Bt + C$ must have units of an angle measure. The values Logger *Pro* uses for the C parameter are given in radians. What must be the units of B? Try changing the value of the suggested B parameter to see what effect this has on the number of cycles that appear in the test graph window. The B parameter is known as *angular frequency,* ω. Compare the values of B for your curve fits to the y-t data for both of your runs. Discuss the physical significance of this parameter before moving on to Part 2.

Part 2 Rates of change

In earlier experiments you investigated the relationship between position-time and velocity-time graphs for linear kinematics. In this part you will continue this investigation for the more complex motion of an oscillating body.

7. Hide the data set for Run 2. Choose Insert Graph and then do an Auto Arrange under the Page menu. Select More on the vertical axis of this graph, choose Velocity for the first run, then Autoscale this graph.

8. Select both graphs and choose Group Graphs (x-axes) to make sure that the time axes are aligned. Note the position of the mass hanger when its speed is at a maximum value and again when its speed is zero.

9. Click on the *y-t* graph to make it active and turn on the Tangent Tool. On the *v-t* graph turn on the Examine Tool. Move the cursor across the *y-t* graph; as you do so, compare the slope of the tangent to any point on the *y-t* graph to the value of the velocity on the *v-t* graph. Write a statement describing the relationship between these quantities. When you are finished, de-select these tools.

10. An object's velocity is the rate of change of its position with respect to time. Logger *Pro* does not *measure* the velocity of an object; rather it *calculates* it from the position-time data. Double-click on the column header for velocity to see the equation Logger *Pro* uses to determine velocity. Make sure you understand the function and its argument before you move on.

11. Since cosine is the derivative of the sine (that is, $\frac{d(\sin\theta)}{d\theta} = \cos\theta$), and velocity is the derivative of position, it seems reasonable to use the cosine to fit the velocity-time data. Select the *v-t* graph, choose Curve Fit, then select the sine function as before; this time, however, choose Define Function. In the User Defined Function window replace "sin" with "cos" and name the function "Cosine". After you choose Try Fit, check to see how closely the graph of this function matches that of your data. You may have to try a fit several times before you obtain a nice match. Before you click OK, replace the values of B and C suggested by Logger *Pro* with those used in the sine fit to your *y-t* graph; make sure the value of A is positive.

12. From what you know about the chain rule, determine the value of the coefficient of the cosine function when you take the derivative of your sine fit to the *y-t* graph. Compare this to the A parameter suggested by Logger *Pro* for the fit to the *v-t* graph.

13. Insert a new graph and use the Auto Arrange and Group Graphs features as you did in Steps 7 and 8. Choose acceleration for your first run as the vertical axis label and Autoscale the graph. With all three graphs selected, use the Examine tool to note how the position, velocity and acceleration of the hanger change at various times in a given cycle. For example, when the hanger is at its maximum height, what are the values of the velocity and acceleration? When you are done, turn off the Examine tool.

14. From what you know about velocity and acceleration, fit an appropriate function to the acceleration-time graph. In order to keep the keep the argument of the function the same as in the *y-t* and *v-t* graphs, what change do you have to make to parameter A? How does A compare to the value of the coefficient you obtain when you find the derivative of the function used to fit the *v-t* graph?

Part 3 The role of force

15. On your *v-t* graph replace velocity with force as your vertical axis label. Since you zeroed the force sensor before you began collecting data, this column in the data table ought to be labeled *net* force. Note how the net force acting on the mass hanger varies as its position changes from maximum to minimum. Explain why the net force responsible for SHM is called a *restoring force*.

16. Describe how the acceleration of the mass hanger varies as the net force varies through each cycle of SHM. Would you expect Newton's second law to apply to this type of motion?

17. To test whether the net force is proportional to the acceleration in this kind of motion, change the horizontal axis of the force-time graph to acceleration, then autoscale the graph. Perform a linear fit to the data. Relate the slope to any system parameter that was held constant.

18. Consider the components of the oscillating system when you try to explain any discrepancy between the value of the slope reported by Logger *Pro* and the constant of proportionality you may have expected.

EXTENSION

You should have noticed that the *B* parameter to the sine fit to your *y-t* data for each of your runs was the same. In this activity you will explore aspects of the physical system on which the angular frequency ω, depends.

As you saw in Part 3, Hooke's law describes the relationship between the restoring force and the position of the mass hanger, y. Substitution of this expression for force into Newton's second law yields $-k\,y = m\,a$. As you saw in Part 2, the acceleration is the second derivative of position with respect to time. The previous equation can be written as:

$$-k\,y = m\frac{d^2y}{dt^2}$$

This is a 2nd order differential equation. The solution to such an equation is a function. In this experiment, you have found a function for $y(t)$ that neatly describes the motion of the system. Substitution of this function in the equation above, rearranging and canceling like terms should enable you to derive an equation for ω in terms of k and m.

When you have done so, predict how you could change the angular frequency of the SHM by some simple factor (like doubling or halving). Go back to your experimental setup and test your prediction.

Consider any other factors that may have an effect on the value you obtain for B. After your discussion, make the necessary adjustment to the hanging mass to test your prediction and perform another run.

Computer Experiment
16

INSTRUCTOR INFORMATION

Simple Harmonic Motion

Kinematics and Dynamics of Simple Harmonic Motion

This is the second of two labs addressing simple harmonic motion (SHM). Students need to be sufficiently familiar with calculus that they can find the derivative of trig functions and can apply the chain rule when determining the derivative of a function. The lab is written with the assumption that the instructor will engage the students in discussions at critical junctures. These discussions can take place with the entire class or with individual lab groups. The icon at left indicates where these discussions should occur.

The Microsoft Word files for the student pages can be found on the CD that accompanies this book. See *Appendix A* for more information.

OBJECTIVES

In this experiment, the student objectives include

- Collect position *vs.* time data as a weight, hanging from a spring, is set in simple harmonic motion (SHM).
- Determine the best-fit equation for the position *vs.* time graph of an object undergoing SHM.
- Define the terms amplitude, offset, phase shift, period and angular frequency in the context of SHM.
- Predict characteristics of the corresponding velocity *vs.* time and acceleration *vs.* time graphs, produce these graphs and determine best-fit equations for them.
- Relate the net force and acceleration for a system undergoing SHM.

During this experiment, you will help the students

- Recognize that Logger *Pro* does not always suggest the simplest parameters in curve fits to periodic functions.
- Recognize the need to keep the argument of the trig function used to fit *y-t, v-t* and *a-t* graphs the same for the purposes of this experiment.
- Determine the derivative of the functions $y = A\sin(Bt + C) + D$ and $v = A\cos(Bt + C) + D$.
- Recognize that the net force acting on the hanger-mass system is the restoring force that produces SHM.
- Recognize that Newton's second law, $F_{net} = m\,a$, applies even when the acceleration is not constant.
- Understand why the constant of proportionality for the force *vs.* acceleration graph is greater than the mass of the hanger and weights.

REQUIRED SKILLS

In order for student success in this experiment, it is expected that students know how to

- Zero a Motion Detector. This is addressed in Activity 2.
- Group graphs and perform curve fits in Logger *Pro*. This is addressed in Activity 3.

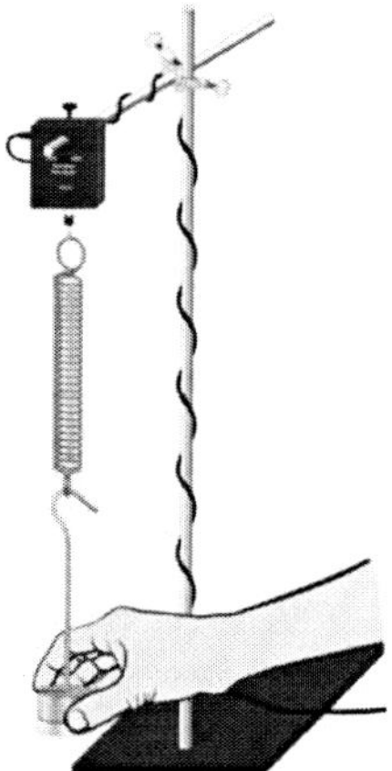

Figure 1

EQUIPMENT TIPS

The spring must be suspended from the hook on a force sensor (Dual-Range or WDSS) with the hanger and weights hanging freely from spring. Choose a system mass and spring combination that produces smooth oscillation that occurs slowly enough (a period of at least 0.5 s) that there will be enough data points to allow students to readily analyze the *y-t*, *v-t* and *a-t* graphs. Vernier sells a spring set appropriate for this experiment, order code SPRINGS. This accessory kit contains six springs that are 6 cm long and 0.8 cm in diameter. Three springs have a spring constant of 5 N/m and the other three have a spring constant of 15 N/m.

To protect the motion detector cover it with a wire basket like those found at office supply stores.

PRE-LAB DISCUSSION

It is likely that students will recognize that the motion they observe is periodic. However, they might draw a saw-tooth curve, focusing on the extremes of the motion, instead of sketching a sinusoidal curve.

LAB PERFORMANCE NOTES

When the Motion Detector is connected to the data-collection interface and Logger *Pro* is started, the default graph screen shows force-time (*F-t*), position *vs.* time (*y-t*) and velocity *vs.* time (*v-t*) graphs. Students are directed to delete the *F-t* and *v-t* graphs for now and choose Auto Arrange from the Page menu to re-size the *y-t* graph. They will insert the other graphs as needed.

If your motion detector has a switch, set it to Track mode. Position the hanger and weights high enough so that when they drop to their lowest point in the oscillating motion they are not closer to the motion detector than the minimum distance[1].

Remind students to turn off the Connect Points feature and turn on Point Protectors for the graphs they produce in this experiment.

Before they begin collecting data, it would be worthwhile for students to sketch a force diagram for the mass hanger. Point out that after they zero the force sensor, the reading will indicate the *net* force acting on the hanger-mass system.

[1] This is 15 cm for newer motion detectors with a switch and 45 cm for older ones without a switch.

SAMPLE RESULTS AND POST-LAB DISCUSSION – PART 1

If students have not performed Experiment 15, this may be the first opportunity they have had to fit a sine curve to data they have collected. It is highly likely that sine fit will be shifted from the standard graph given by the equation $y = \sin\theta$ with which students are familiar (see Figure 2). If they have performed 15 only a brief review of the meaning of the parameters to the sine fit to the *y-t* graph may be needed. Students can quickly move on to Parts 2 and 3. If not, when students obtain an equation of the form $y = A\sin(Bt + C) + D$, they are likely to have questions about the meaning of these parameters. This experiment affords the opportunity to examine the physical meaning of parameters *A, B,* and *C*.

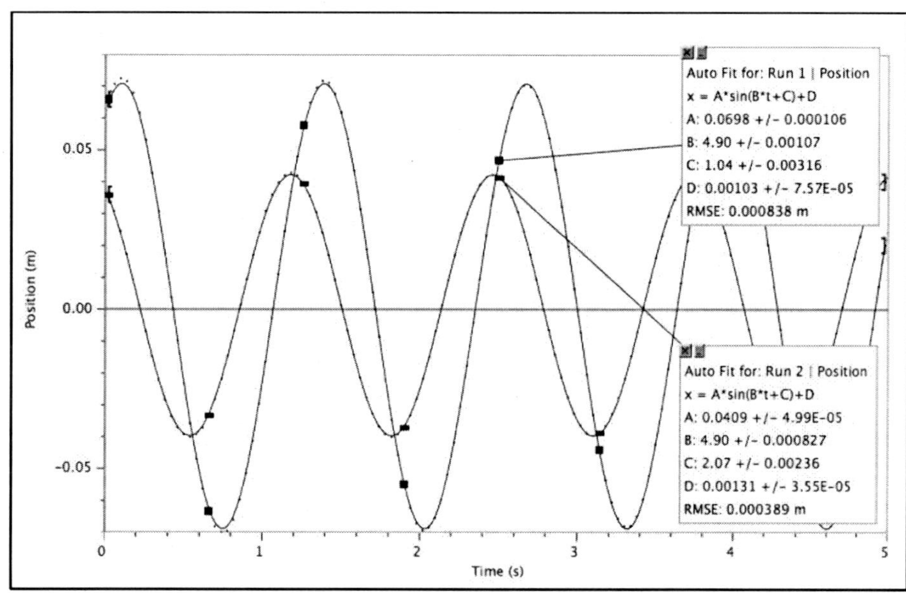

Figure 2 Sine fit to the y-t data

Step 4

The A parameter describes the extent to which the hanging mass oscillates above and below the zero position. It is known as the *amplitude.*

Step 5

Students can enter a specific value for a parameter or increment by clicking on the Manual Curve Fit dialog box to select it and then choosing Additional Object Options▶Curve Fit Options from the Options menu. An increment of 0.1 is reasonable for this exploration of *C*. Students should find that changing its value shifts the sine fit either left or right of the original fit to the *y-t* data. If students change the value of *C* by ± π, they will find that the test graph is shifted 180° out of phase with their original graph.

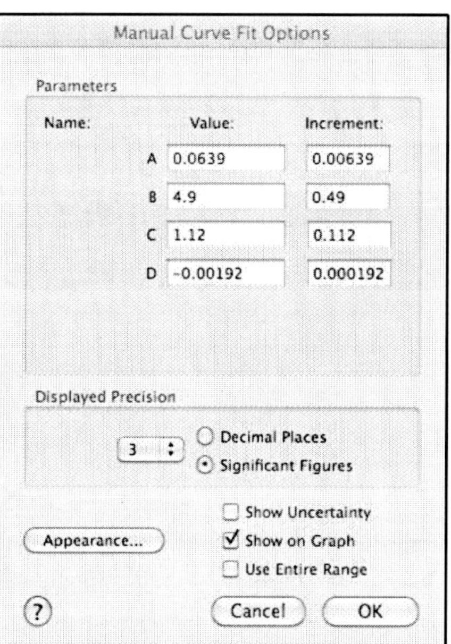

Step 6

The units of *B* must be rad/s in order for *Bt* to have units of radians. Students should note that the value of *B* is the same for both sine fits to their *y-t* graphs, despite that *A* and *C* are likely to be different. Explain to students that B depends on both the mass and the spring constant, neither of which changed when they did their second run.

SAMPLE RESULTS AND POST-LAB DISCUSSION – PART 2

Steps 7–8

With the *y-t* and *v-t* graphs grouped, students should note that the position of the mass hanger is nearly 0 when it is moving at its greatest speed. When the displacement is greatest, the velocity is nearly 0.

Steps 9–10

Students should note that the slope of the tangent to the *y-t* graph is the same as the velocity at the corresponding time on the *v-t* graph. If students understand the concept of instantaneous velocity, this should come as no surprise. They may not realize that Logger *Pro calculates* the velocity of the object from the position-time data. By choosing Settings from the File menu, students should find (unless these have been changed in an earlier experiment) that the default value for the number of points used in the derivative is 7. Reducing this value to 5 produces slightly better results; however, the default settings yield adequate results.

Step 11

Logger *Pro* uses a trial-and-error algorithm to determine parameters that fit a selected function to data. Sometimes, the trial graph suggested for a periodic function is more complex than required to match the data. If this occurs, students should choose Try Fit multiple times, if necessary, until they obtain a good match to their data. Even then, the B and C parameters may not match those of the original sine fit to the *y-t* graph. Manual override of suggested values with those in the original sine fit should produce a test graph that matches the *v-t* data nicely (see Figure 3).

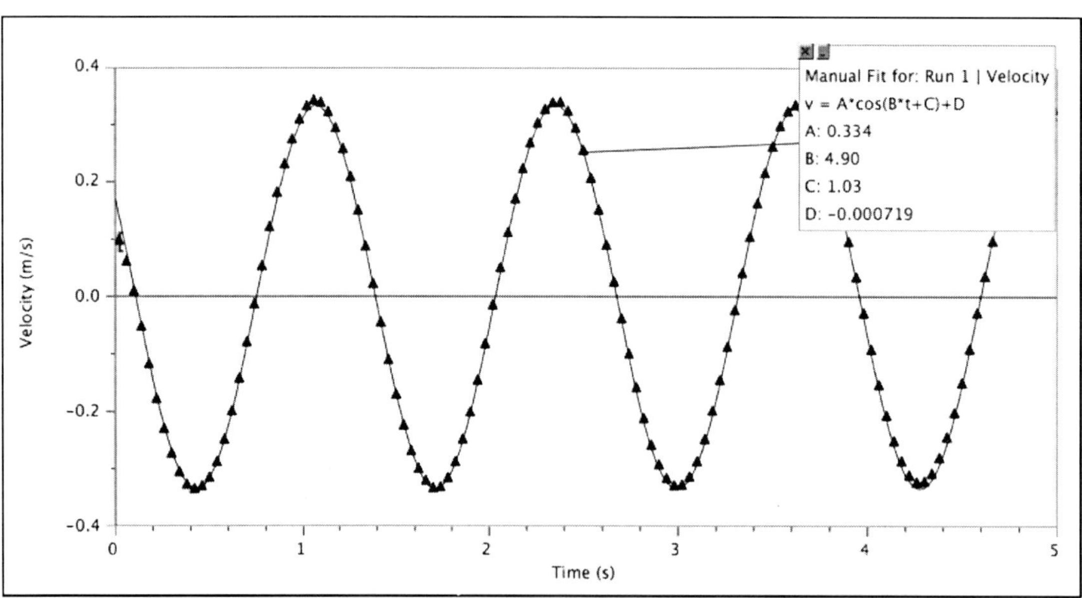

Figure 3 Cosine fit to the v-t data

Step 12

The derivative of $y = A\sin(Bt + C) + D$ is $v = AB\cos(Bt + C)$. Students should find that the product of A and B in their sine fit to the y-t data should match the suggested A parameter to the cosine fit to the *v-t* graph to within 5%. The agreement is poorer if higher numbers of points are used in derivative calculations.

Step 13

When students add the *a-t* graph to their graph window, and examine all three graphs, they should note that when the hanger is at its maximum height, the velocity of the hanger is 0, but that the acceleration has maximum magnitude in the opposite direction (see Figure 4).

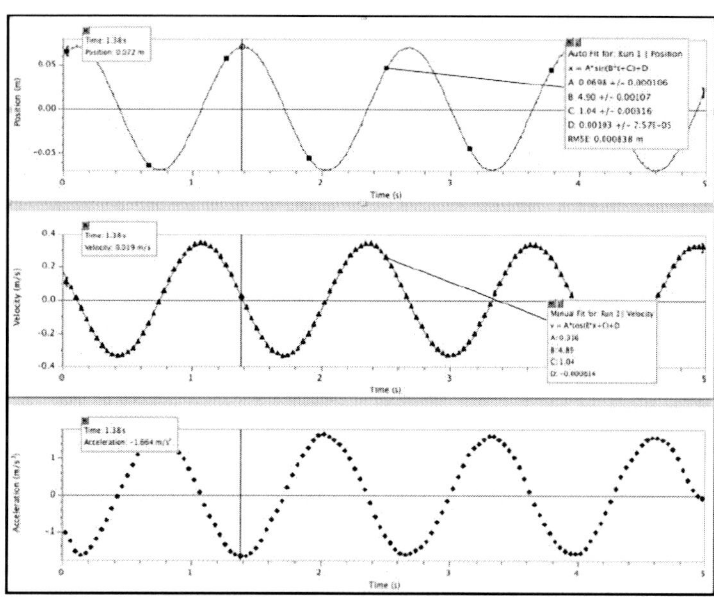

Figure 4 y-t, v-t, and a-t graphs

Step 14

Students should recognize that acceleration is the derivative of velocity. The derivative of a cosine function is the opposite of a sine function with the same argument. So, they should choose a sine fit to their acceleration-time data. As there are a number of parameters that would work, you may need to remind students that they should make the *A* parameter negative and adjust the *B* and *C* parameters until they match the original sine fit to the *y-t* graph. Since the derivative of $v = A\cos(Bt + C)$ is $a = -AB\sin(Bt + C)$, students should find a good match between the product of the *A* and *B* parameters of the cosine fit to the *v-t* data and the *A* parameter in the sine fit to the *a-t* data.

SAMPLE RESULTS AND POST-LAB DISCUSSION – PART 3

Step 15

When students examine how force varies with time and compare this behavior to the *y-t* graph, they should note that the net force opposes the displacement of the mass hanger; i.e., when the mass hanger has its maximum positive displacement, the net force acting on it has its greatest value in the opposite direction. In like manner, when the hanger is at its lowest position, the net force has its maximum value directed upwards. Hence the net force can be seen to be "restoring" the spring-hanger system to its equilibrium position.

Step 16

The *F-t* and *a-t* graphs appear to match nicely, leading one to suspect that the net force acting on the hanger is proportional to its acceleration.

Step 17

When a linear fit is performed to the graph of net force vs. acceleration, the units of the slope (N/m/s^2) suggest that the constant of proportionality is the mass of the oscillating system (see Figure 5).

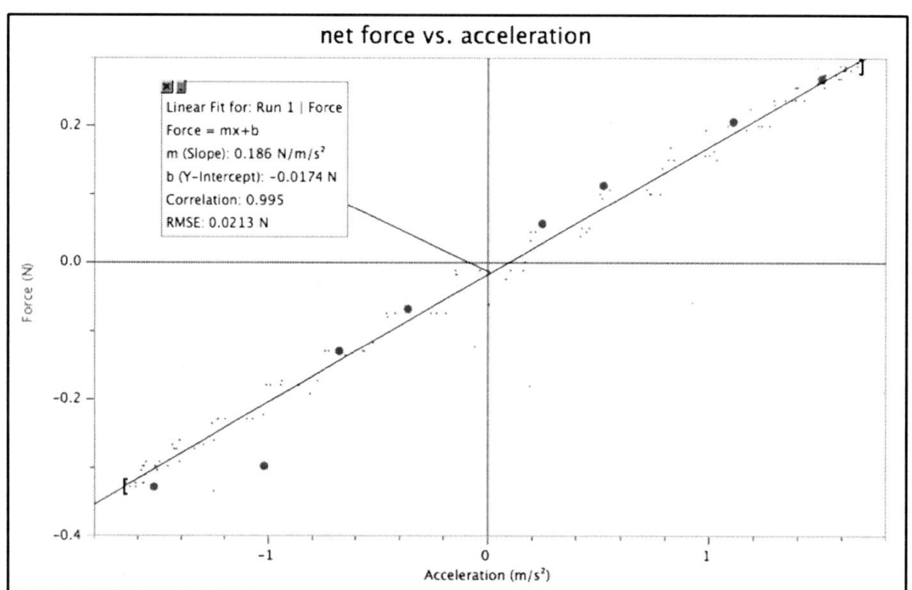

Figure 5 Net force vs. acceleration graph

Step 18

For the system in the sample data, the mass of the hanger and weights was 150 g; this is less than the value of the mass suggested by the slope of the F *vs.* a graph shown in Figure 5. Point out to students that the system accelerated by the net force included not only the hanger and weights, but also a portion of the mass of the spring. For a uniform spring, 1/3 of the spring mass must be added to the hanging mass to determine the mass of the oscillating *system*. You might have students look into this correction.

EXTENSION

The student version of the lab suggests that Newton's second law can be expressed as

$$-k\,y = m\frac{d^2y}{dt^2}$$

Students have found that $y = A\sin(Bt + C) + D$ describes the motion of the hanger-spring system in SHM. Before they attempt to solve the differential equation suggest that they use the standard form of the equation $y = A\sin(\omega\, t + \phi)$, where ω replaces *B*, ϕ replaces *C* and *D* is set to 0.

Substitution of this into the differential equation and simplification yields:

$$-k(A\sin(\omega t+\phi)=-mA\omega^2\sin(\omega t+\phi)$$

$$\frac{k}{m}=\omega^2$$

$$\omega=\sqrt{\frac{k}{m}}$$

From this students should conclude that changing the mass by a factor of four will either double or halve the angular frequency. When they test this prediction they will find that the experimental value of B will be close to, but not the same as their predicted value. Students may recall from Step 18 that they must take into account some portion of the mass of the spring when calculating the frequency.

LabQuest Experiment

16

Simple Harmonic Motion

Kinematics and Dynamics of Simple Harmonic Motion

INTRODUCTION

When you suspend an object from a spring, the spring will stretch. If you pull on the object, stretching the spring some more, and release it, the spring will provide a restoring force that will cause the object to oscillate in what is known as simple harmonic motion (SHM). In this experiment, you will examine this kind of motion from both kinematic and dynamic perspectives.

OBJECTIVES

In this experiment, you will

- Collect position *vs.* time data as a weight, hanging from a spring, is set in simple harmonic motion (SHM).
- Determine the best-fit equation for the position *vs.* time graph of an object undergoing SHM.
- Define the terms amplitude, offset, phase shift, period and angular frequency in the context of SHM.
- Predict characteristics of the corresponding velocity *vs.* time and acceleration *vs.* time graphs, produce these graphs and determine best-fit equations for them.
- Relate the net force and acceleration for a system undergoing SHM.

MATERIALS

Vernier LabQuest
LabQuest App
Vernier Motion Detector
Vernier Dual-Range Force Sensor
ring stand and right angle clamp
spring
mass hanger and standard lab masses
wire basket

PRE-LAB INVESTIGATION

Attach a rod to a vertical support rod using a right angle clamp. Mount a Dual-Range force sensor to the horizontal rod. Now hang a spring from the hook on the sensor and suspend a mass hanger and weights from the spring, as shown in Figure 1. Assume that the bottom of the hanger is the zero position. Pull on the mass hanger slightly and release it. Observe the motion of the hanger. On the axes below, sketch a graph of the position of the hanger as a function of time.

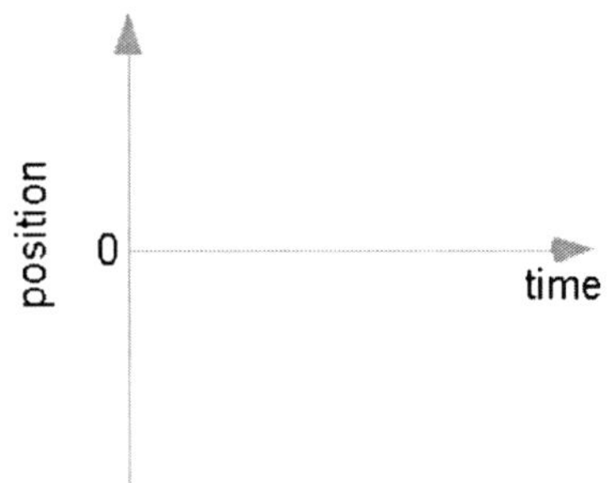

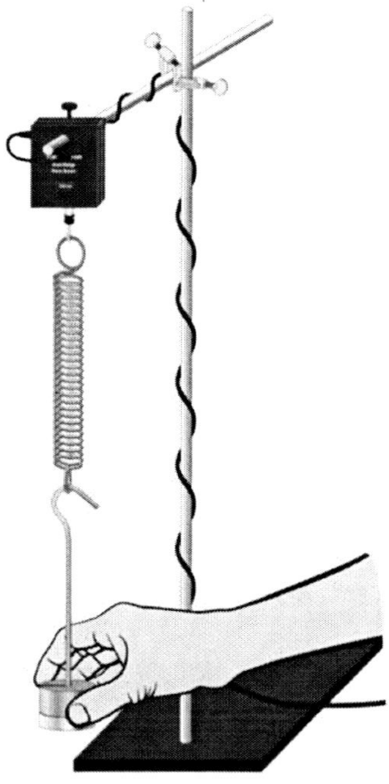

Figure 1

Compare your sketch to those of others in the class.

PROCEDURE

1. Connect the Motion Detector and the Dual Range Force Sensor to LabQuest and start a new file in LabQuest App. The default data-collection rate is appropriate; however, shorten the duration to 5 seconds.

2. If your motion detector has a switch, set it to Track. 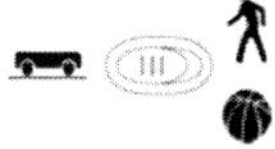

3. Place the motion detector on the floor beneath the mass hanger. Place the wire basket over the motion detector to protect it

4. Make sure the hanger is motionless, then zero both the force sensor and the motion detector.

5. Lift the hanger and weights about ten centimeters, then release. When the mass hanger is oscillating smoothly, begin collecting data.

6. If the position-time graph appears to be a smooth curve, store this run. If not, repeat until you obtain a smooth curve.

7. Perform a second trial, except this time, make the initial displacement of the mass hanger smaller than what you did for your first trial. If the position-time graph is not smooth, repeat until you obtain a smooth curve.

EVALUATION OF DATA

Part 1 Exploration of SHM

On the Graph tab choose to view only the position-time graph for your first run. Since your position-time data appears sinusoidal, you will model a sine function $y = A\sin(Bt + C) + D$ to the data by choosing values of the parameters that will enable you to *manually* fit a curve to your position-time data. The following steps will help you determine the appropriate values of these parameters. You are likely to find that your first effort will produce a model equation that approximates your position-time data. Repeating this process will enable you to refine your equation to more closely match the data. Before you begin, tap on the position-time graph to determine the maximum and minimum values of position during your run.

1. Compare the position-time graphs you obtained with the one you sketched in the Pre-Lab Investigation. In what ways are the graphs similar? In what ways do they differ?

2. Choose Model from the Analyze menu, then select $y = A\sin(Bx + C) + D$ from the drop-down Equation list; keep in mind that the x variable stands for time. The test plot displayed is a much larger version of your sinusoidal y-t data. In the following steps you will adjust the values of the parameters. *Do not* tap OK until directed to do so.

3. Because you zeroed the motion detector before you began collecting data, your position-time data should be centered about the time axis. Use the down arrow to reduce the value of D to 0. This should make the test plot symmetrical with respect to the time axis.

4. Next, reduce the A parameter to the value you observed for the maximum value of your position-time graph. What aspect of the y-t graph does A appear to describe? Tap Cancel, then test your conclusion by performing Steps 2–4 on your second run. Tap Cancel, return to your first run, and adjust D and A as before. *Do not* yet tap OK. The name given to the A parameter is *amplitude*.

5. Unless you managed to begin collecting data at the instant the oscillating mass hanger was rising through the 0 position, your *position-time* graph is likely to be shifted somewhat left or right from the standard position of a graph of $y = \sin\theta$. Try changing the value of the C parameter to see what effect this has on the position of the sine curve in the test graph window. Adjust C until the test plot appears to have the same initial y-value as your data.

6. Since the argument of a sine function must be an angle, the expression $Bt + C$ must have units of an angle measure. The values LabQuest App uses for the C parameter are given in radians. Now, take a closer look at B; its units must be radians/second in order for the sine function to operate on the argument ($Bt + C$). The B parameter, known as *angular frequency,* ω, is a measure of how frequently the hanger and mass oscillate.

7. Gradually increase the value of B and note what happens to the time required to complete one cycle– the *period* of oscillation. Continue increasing B until the test plot best matches your
y-t data. You may also need to fine tune your choice for C. Tap OK. Be sure to record the values of the parameters to the sine fit to your position-time data.

8. Use the Delta function from the Analyze menu and drag across the graph from one maximum to the next to determine the *period, T*, of oscillation. Given that one cycle is 2π radians, divide 2π by T and compare this value to B. Express angular frequency, ω, in terms of T. Make sure that you discuss the physical significance of this parameter before moving on to

Part 2 of the evaluation. At this point you may turn off the Model function for the position-time graph.

Part 2 Rates of change

In earlier experiments you investigated the relationship between position-time and velocity-time graphs for linear kinematics. In this part you will continue this investigation for the more complex motion of an oscillating body.

9. From the Graph menu, choose to show both graphs. Change the vertical axis of Graph 1 to velocity. Note the position of the mass hanger when its speed is at a maximum value and again when its speed is zero.

10. Turn on Tangent (found in the Analyze menu) and tap the position-time (*y-t*) graph at several places. Compare the slope of the tangent to any point on the *y-t* graph to the value of the velocity on the *v-t* graph. Write a statement describing the relationship between these quantities, then turn off Tangent.

11. The instantaneous rate of change of a function is known as its derivative. Since cosine is the derivative of the sine (that is, $\frac{d(\sin\theta)}{d\theta} = \cos\theta$), and velocity is the derivative of position, it seems reasonable to use the cosine to fit the velocity-time data. Choose to Model the *v-t* graph, then select the cosine function. Adjust the *D* parameter to 0 as before and reduce the *A* parameter to match the amplitude. Then, adjust the *B* and *C* parameters to the values used in the sine fit to the *y-t* data; this makes the arguments of the two functions the same.

12. Tap OK. From what you know about the chain rule, determine the value of the coefficient of the cosine function when you take the derivative of your sine fit to the *y-t* graph. Compare this to the *A* parameter you used to fit the cosine to the *v-t* graph.

13. Now change the vertical axis label of the position-time graph to acceleration and choose Autoscale Once to make sure that you can see the data points. Examine the *v-t* and *a-t* graphs to note how the velocity and acceleration of the hanger change at various times in a given cycle.

14. From what you know about velocity and acceleration, use the Model function to fit an appropriate function to the acceleration-time graph. Make sure that you keep the argument of the function the same as in the *y-t* and *v-t* graphs. What change do you have to make to parameter *A* in order to match the test plot to your *a-t* graph? How does *A* compare to the value of the coefficient you obtain when you find the derivative of the function used to fit the *v-t* graph? Tap OK to return to the Graph tab.

Part 3 The role of force

15. On your *v-t* graph replace velocity with force as your vertical axis label. Change the vertical axis label of the *a-t* graph to position. Since you zeroed the force sensor before you began collecting data, this column in the data table ought to be labeled *net* force. Note how the net force acting on the mass hanger varies as its position changes from maximum to minimum. Explain why the net force responsible for SHM is called a *restoring force*.

16. Change the vertical axis of the position-time graph back to acceleration (you may need to Autoscale to view the data points). Describe how the acceleration of the mass hanger varies

as the net force varies through each cycle of SHM. Would you expect Newton's second law to apply to this type of motion?

17. To test whether the net force is proportional to the acceleration in this kind of motion, choose to view just the force-time graph. Tap the y-axis label and select Acceleration. Turn off Connect Points in the Graph Options dialog box. From the Analyze menu, choose Curve Fit and perform a linear fit to the data. Relate the slope to any system parameter that was held constant.

18. Consider the components of the oscillating system when you try to explain any discrepancy between the value of the slope reported by LabQuest App and the constant of proportionality you may have expected.

EXTENSION

You should have noticed that the *B* parameter to the sine fit to your *y-t* data for each of your runs was the same. In this activity you will explore aspects of the physical system on which the angular frequency ω, depends.

As you saw in Part 3, Hooke's law describes the relationship between the restoring force and the position of the mass hanger, *y*. Substitution of this expression for force into Newton's second law yields $-ky = ma$. As you saw in Part 2, the acceleration is the second derivative of position with respect to time. The previous equation can be written as:

$$-ky = m\frac{d^2y}{dt^2}$$

This is a 2nd order differential equation. The solution to such an equation is a function. In this experiment, you have found a function for *y(t)* that neatly describes the motion of the system. Substitution of this function in the equation above, rearranging and canceling like terms should enable you to derive an equation for ω in terms of *k* and *m*.

When you have done so, predict how you could change the angular frequency of the SHM by some simple factor (like doubling or halving). Go back to your experimental setup and test your prediction.

Consider any other factors that may have an effect on the value you obtain for *B*. After your discussion, make the necessary adjustment to the hanging mass to test your prediction and perform another run.

LabQuest Experiment

16

INSTRUCTOR INFORMATION

Simple Harmonic Motion

Kinematics and Dynamics of Simple Harmonic Motion

This is the second of two labs addressing simple harmonic motion (SHM). Students need to be sufficiently familiar with calculus that they can find the derivative of trig functions and can apply the chain rule when determining the derivative of a function. The lab is written with the assumption that the instructor will engage the students in discussions at critical junctures. These discussions can take place with the entire class or with individual lab groups. The icon at left indicates where these discussions should occur.

This version of the lab is designed for use with the LabQuest as a standalone device. The Microsoft Word files for the student pages can be found on the CD that accompanies this book. See *Appendix A* for more information.

OBJECTIVES

In this experiment, the student objectives include

- Collect position *vs.* time data as a weight, hanging from a spring, is set in simple harmonic motion (SHM).
- Determine the best-fit equation for the position *vs.* time graph of an object undergoing SHM.
- Define the terms amplitude, offset, phase shift, period and angular frequency in the context of SHM.
- Predict characteristics of the corresponding velocity *vs.* time and acceleration *vs.* time graphs, produce these graphs and determine best-fit equations for them.
- Relate the net force and acceleration for a system undergoing SHM.

During this experiment, you will help the students

- Recognize the need to keep the argument of the trig function used to fit *y-t, v-t* and *a-t* graphs the same for the purposes of this experiment.
- Determine the derivative of the functions $y = A\sin(Bt + C) + D$ and $v = A\cos(Bt + C) + D$.
- Recognize that the net force acting on the hanger-mass system is the restoring force that produces SHM.
- Recognize that Newton's second law, $F_{net} = m\,a$, applies even when the acceleration is not constant.
- Understand why the constant of proportionality for the force *vs.* acceleration graph is greater than the mass of the hanger and weights.

REQUIRED SKILLS

In order for student success in this experiment, it is expected that students know how to z

- Zero a Motion Detector. This is addressed in Activity 2.
- Zero a Force Sensor.

EQUIPMENT TIPS

The spring must be suspended from the hook on a force sensor (Dual-Range or WDSS) with the hanger and weights hanging freely from spring. Choose a system mass and spring combination that produces smooth oscillation that occurs slowly enough (with a period of more than 0.5 s) that there will be enough data points to allow students to readily analyze the *y-t*, *v-t*, and *a-t* graphs. Vernier sells a spring set appropriate for this experiment, order code SPRINGS. This accessory kit contains six springs that are 6 cm long and 0.8 cm in diameter. Three springs have a spring constant of 5 N/m and the other three have a spring constant of 15 N/m.

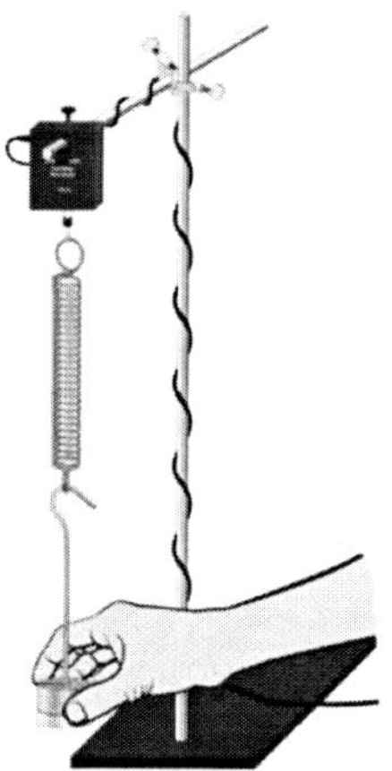

Figure 1

PRE-LAB DISCUSSION

It is likely that students will recognize that the motion they observe is periodic. However, they might draw a saw-tooth curve, focusing on the extremes of the motion, instead of sketching a sinusoidal curve.

LAB PERFORMANCE NOTES

When the motion detector is connected to the LabQuest and LabQuest App is started, the default graph screen shows force-time (*F-t*) and position *vs.* time graphs. Students will work with one graph in Part 1, then both graphs in Parts 2 and 3.

Make sure that the motion detector is set to Track mode and that the initial position of the hanger and weights is high enough so that when they drop to their lowest point in the oscillating motion they are not closer to the motion detector than the minimum distance[1].

Before they begin collecting data, it would be worthwhile for students to sketch a force diagram for the mass hanger. Point out that after they zero the force sensor, the reading will indicate the *net* force acting on the hanger-mass system.

SAMPLE RESULTS AND POST-LAB DISCUSSION – PART 1

If students have not performed Experiment 15 this may be the first opportunity they will have had to model a sine curve to data they have collected. You may have to guide students in the use of the Model function from the Analyze menu.

Step 2–4

When the value of the *D* parameter is reduced to 0 and the *A* parameter is reduced until the amplitude of the test plot approximates that of the data, students should obtain a screen like the

[1] This is 15 cm for newer motion detectors with a switch and 45 cm for older ones without a switch.

one in Figure 2. The A parameter describes the extent to which the hanging mass oscillates above and below the zero position. It is known as the *amplitude.*

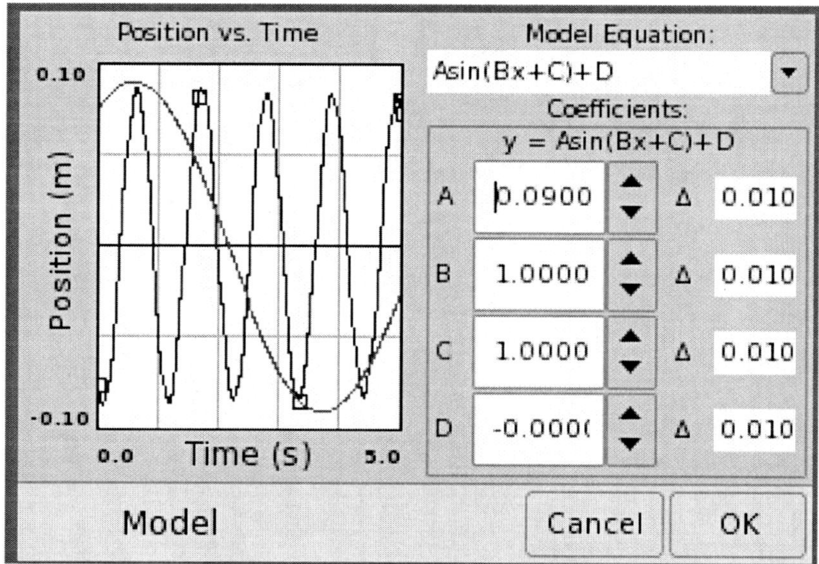

Figure 2 Adjust A and D parameters

Steps 5–7

Students should find that changing the value of C shifts the sine fit either left or right of the original fit to the y-t data. It is not easy to tell when the initial positions of the test plot and the data match. Students will find that as they increase B, the period, T, of the test plot decreases. When B approaches the optimum value, students will find that they have to "fine tune" the values of C and B to obtain the best match. It is clear, however, that the value of A should be adjusted to a value somewhat smaller than 0.09 (see Figure 3).

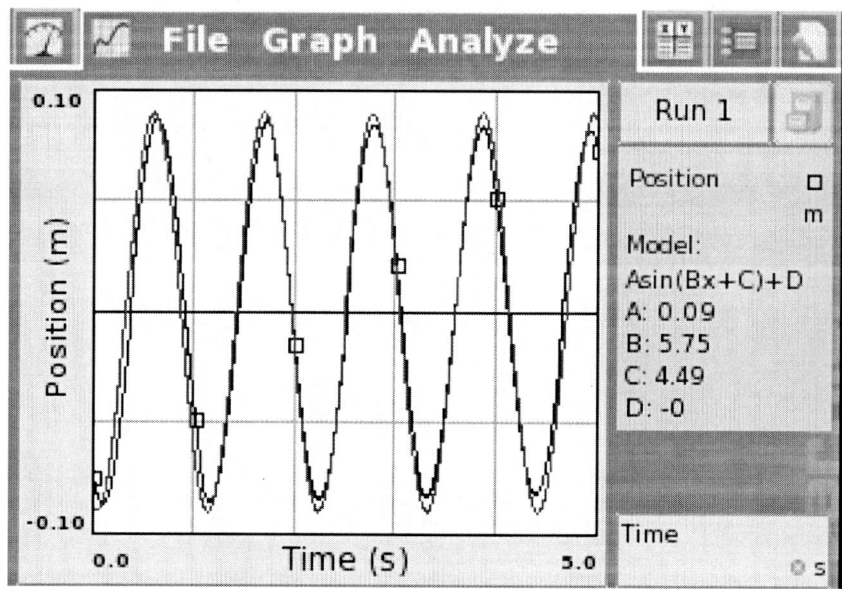

Figure 3 First pass at the model

Step 8

When students use the Delta function, they should be able to determine the value of the period of oscillation. Make sure that students recognize that there are 2π radians in one cycle. Dividing 2π by T yields the angular frequency, ω, in rad/s (see Figure 4).

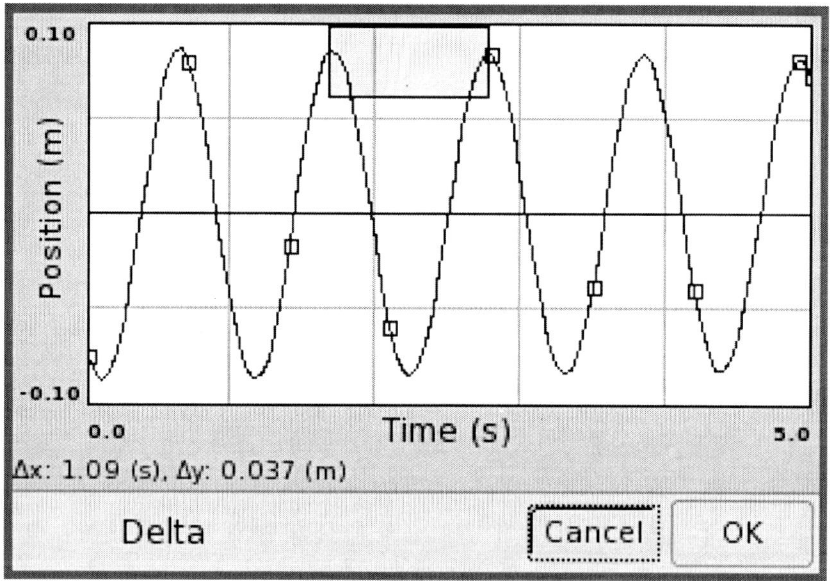

Figure 4 Using Delta to find T

For the graph shown above, $2\pi/T = 5.76$, which is very close to the value of B chosen in the model of the sine fit to the data.

SAMPLE RESULTS AND POST-LAB DISCUSSION – PART 2

Steps 9–10

When both the *y-t* and *v-t* graphs are displayed, students should note that the position of the mass hanger is nearly 0 when it is moving at its greatest speed. When the displacement is greatest, the velocity is nearly 0. Students should note that the slope of the tangent to the *y-t* graph is the same as the velocity at the corresponding time on the *v-t* graph. If students understand the concept of instantaneous velocity, this should come as no surprise.

Steps 11–12

When students model a cosine curve to their *v-t* data they should obtain a fit that looks like the one in Figure 5.

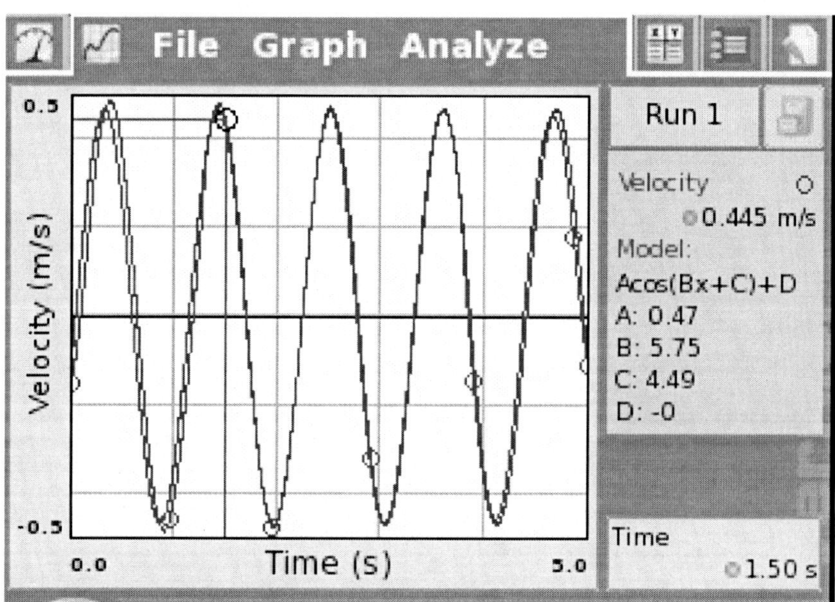

Figure 5 Cosine fit to v-t graph

The derivative of $y = A\sin(Bt + C) + D$ is $v = AB\cos(Bt + C)$. Depending on the quality of their fit to the *y-t* graph students should find that the product of *A* and *B* in their sine fit will match the suggested *A* parameter to the cosine fit to the *v-t* graph to within 5%. If the value of *A* in Figure 3 had been optimized to 0.086, the product of *A* and *B* would have been 0.49, comparing well with the A=0.47 in Figure 5.

Step 13

When students choose to view the *v-t* and *a-t* graphs, they should note that the acceleration of the hanger reaches its maximum value when the velocity is at its minimum. This occurs when the mass hanger is at either extreme of its motion.

Step 14

Students should recognize that acceleration is the derivative of velocity. The derivative of a cosine function is the opposite of a sine function with the same argument. So, they should choose a sine fit to their acceleration-time data. As there are a number of parameters that would work, you may need to remind students that they should make the *A* parameter negative and adjust the

B and *C* parameters until they match the original sine fit to the *y-t* graph. Since the derivative of $v = A\cos(Bt + C)$ is $a = -AB\sin(Bt + C)$, students should find a good match between the product of the *A* and *B* parameters of the cosine fit to the *v-t* data and the *A* parameter in the sine fit to the *a-t* data.

SAMPLE RESULTS AND POST-LAB DISCUSSION – PART 3

Step 15

When students examine how force varies with time and compare this behavior to the *y-t* graph, they should note that the net force opposes the displacement of the mass hanger; i.e., when the mass hanger has its maximum positive displacement, the net force acting on it has its greatest value in the opposite direction. In like manner, when the hanger is at its lowest position, the net force has its maximum value directed upwards. Hence the net force can be seen to be "restoring" the spring-hanger system to its equilibrium position.

Step 16

The *F-t* and *a-t* graphs appear to match nicely, leading one to suspect that the net force acting on the hanger is proportional to its acceleration.

Step 17

When a linear fit is performed to the graph of net force vs. acceleration, an examination of the units of the slope ($N/m/s^2$) suggest that the constant of proportionality is the mass of the oscillating system (see Figure 6).

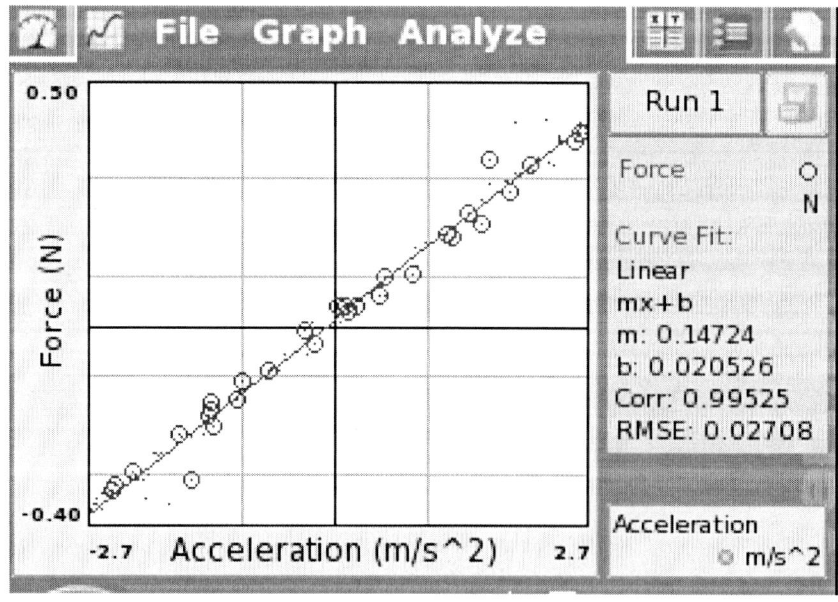

Figure 5 Net force vs. acceleration graph

Step 18

For the system in the sample data, the mass of the hanger and weights was 100 g; this is less than the value of the mass suggested by the slope of the *F vs. a* graph shown in Figure 6. Point out to students that the system accelerated by the net force included not only the hanger and weights, but also a portion of the mass of the spring.

EXTENSION

The student version of the lab suggests that Newton's second law can be expressed as

$$-k\,y = m\frac{d^2y}{dt^2}$$

Students have found that $y = A\sin(Bt + C) + D$ describes the motion of the hanger-spring system in SHM. Before they attempt to solve the differential equation suggest that they use the standard form of the equation $y = A\sin(\omega\, t + \phi)$, where ω replaces *B*, ϕ replaces *C*, and *D* is set to 0.

Substitution of this into the differential equation and simplification yields:

$$-k(A\sin(\omega t + \phi) = -mA\omega^2\sin(\omega t + \phi)$$

$$\frac{k}{m} = \omega^2$$

$$\omega = \sqrt{\frac{\mathrm{k}}{\mathrm{m}}}$$

From this students should conclude that changing the mass by a factor of four will either double or halve the angular frequency. When they test this prediction they will find that the experimental value of *B* will be close to, but not quite the same as their predicted value. Some students will recall from Step 18 that they must take into account some portion of the mass of the spring when calculating the frequency.

Experiment
17

Pendulum Periods

INTRODUCTION

The introductory treatment of the motion of a pendulum leaves one with the impression that the period of oscillation is independent of the mass and the amplitude, and depends only on the length of the pendulum. These relationships are generally true so long as two important conditions are met:

1. the amplitude is small (<< 1 radian), and
2. the mass of the system is concentrated at the end of the string.

In this experiment and the next you will examine the behavior of a pendulum in greater detail to see what occurs when these conditions are no longer true. You will examine the approximations made to simplify the analysis of the pendulum and determine when and why these approximations begin to break down. The first of these is the subject of this experiment; the second will be examined in Experiment 18.

OBJECTIVES

In this experiment, you will

- Collect angle *vs.* time data for a simple pendulum.
- Determine the best-fit equation for the angle *vs.* time graph.
- From an analysis of the forces acting on the pendulum bob, derive the equation describing the motion of the pendulum.
- Relate the parameters in the best-fit equation for the angle *vs.* time graph to their physical counterparts in the system.
- Determine the period of oscillation from an analysis of the angle *vs.* time graph.
- Account for the deviation from constant periods when the amplitude becomes large.

MATERIALS

Vernier data-collection interface
Logger *Pro* or LabQuest App
Vernier Rotary Motion Sensor
Vernier Rotary Motion Accessory Kit
vertical support rod and clamp
right-angle clamp
protractor
metric ruler or tape

PRE-LAB INVESTIGATION

Attach a rod to a vertical support rod using a right-angle clamp. Attach the aluminum rod from the Accessory Kit to the Rotary Motion Sensor, then attach the sensor to the horizontal rod, as shown in Figure 1. Now attach one of the heavy cylindrical weights from the accessory kit to the bottom of the rod as shown. Assume that when the rod is hanging vertically the weight is in the zero position. Pull on the weight slightly and release it. Observe the motion of the weight. On the axes below, sketch a graph of the angular position (in degrees or radians) of the weight as a function of time.

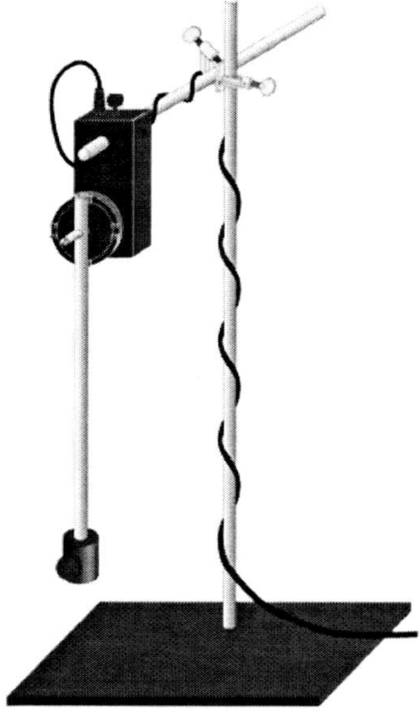

Figure 1

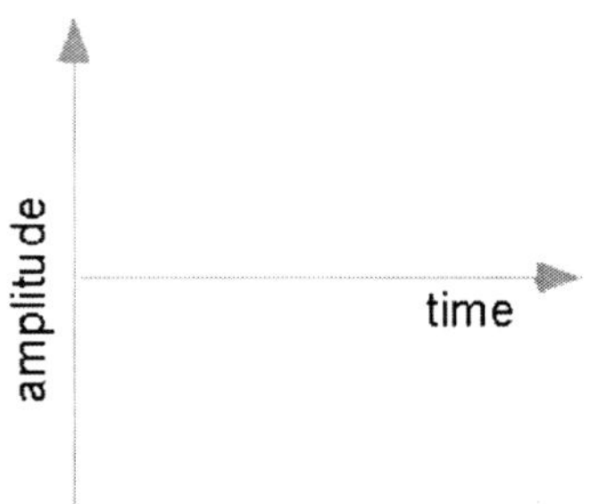

Compare your sketch to those of others in the class.

PART 1 – EXAMINATION OF THE MOTION OF A SIMPLE PENDULUM AT VARIOUS AMPLITUDES

PROCEDURE

1. Make sure that the vertical support rod for the Rotary Motion Sensor is securely attached to a bench or table. When the pendulum is set in motion the sensor should be stationary.

2. Measure the distance between the point of attachment of the rod to the 3-step pulley and the center of mass of the weight at the end of the rod. Record this value as the length, l, of the pendulum.

3. Connect the sensor to the interface and start the data-collection program. Two graphs: angle *vs.* time and velocity *vs.* time will appear in the graph window. For this experiment, you will need to view only the angle *vs.* time graph.

4. The default data-collection rate is appropriate. However, you should increase the resolution of the sensor by selecting the X4 mode.

   - In Logger *Pro*, choose Set Up Sensors from the Experiment menu. Once you select your interface, click on the icon for the RMV and then select X4 Mode.
   - In LabQuest App, tap on the meter window and then select the X4 Mode.

5. Because the default data-collection mode automatically resets the zero position when you start data collection, it is unnecessary to manually zero the sensor before collecting data. However, the bob must be motionless before you begin data collection.

6. Start data collection. Then, using a protractor to measure the angle, pull the rod through a 5° angle and release. Be sure that the pendulum swings freely for at least five seconds. Store this run.

7. Repeat Step 6 for amplitudes of 10°, 15° and 20°, storing each run. Save this file; you will return to it later in the experiment.

EVALUATION OF DATA

1. Compare the angle-time graph you obtained with the one you sketched in the Pre-Lab Investigation. In what ways are the graphs similar? In what ways do they differ? How does the angle-time graph compare to what you have seen for simple harmonic motion?

Determination of ω using Logger *Pro*

2. Before you fit a curve to the position-time graph, turn off Connect Points and turn on Point Protectors.

3. Drag-select that portion of the graph for your first run where the bob is swinging freely.Fit a sine curve to these data. Record the value of the *B* parameter to the sine fit. Repeat this process for your other runs.

4. The *B* parameter is the angular frequency, ω, for this oscillation. Does the value of ω appear to depend on the amplitude of the oscillation? Determine an average value of ω for your four runs.

Determination of ω using LabQuest App

2. Choose to view your first run. Choose the Delta function under the Analysis menu. Drag-select 4–5 cycles of the graph where the bob is swinging freely. The time (Δx) divided by the number of cycles gives the period of oscillation, *T*. Record this value. Repeat this process for your other runs.

3. The angular frequency, ω, for this oscillation is given by $\omega = \frac{2\pi}{T}$. Determine the value of ω for each of your runs. Record these values.

4. Does the value of ω appear to depend on the amplitude of the oscillation? Determine an average value of ω for your four runs.

Determination of the equation of motion for the pendulum.

5. Sketch a force diagram for the bob when it is displaced to one side. Write the expression for the restoring force, the component of the gravitational force that opposes the displacement through angle θ.

6. Write the Newton's second law equation describing the motion of the bob once it is released. Express the acceleration as the second derivative of the arc length, *s*, with respect to time.[1] Given that $s = l\theta$, express the acceleration in terms of θ, *l*, and *t*.

[1] Recall that $v = \frac{ds}{dt}$ and that $a = \frac{dv}{dt}$.

Keep in mind that the acceleration vector for the pendulum always acts in a direction opposing the displacement through angle θ.

7. Simplify your second equation and rearrange the terms so that you have set the equation equal to 0. You have now produced a 2nd order differential equation describing the motion of the pendulum bob. Check your equation with those of others in your class.

8. In your class discussion make sure that you determine and understand the simplification necessary to suggest a solution to the equation you have derived.

Relating ω to physical features of the system

9. The coefficient of t in your solution to the differential equation is the angular frequency, ω, of the motion. From the known values of g and l, calculate the value of ω you would expect for your pendulum. Compare this to the average value of ω you obtained from the curve fit to your data. Determine the percent difference between these values.

10. Slide the weight a few centimeters higher on the rod. Measure the new length, l, of the pendulum. Calculate the expected value of ω for this pendulum.

11. Collect angle-time data for another run using an amplitude between 10 and 20°. Determine ω as you did before. Compare this value to the one you calculated.

PART 2 – EFFECT OF AMPLITUDE ON PERIOD

PROCEDURE

1. Return the weight to its original position at the end of the aluminum rod. Re-open your experiment file from Part 1.

2. Collect angle *vs.* time data for the pendulum as before with an amplitude of 25°. Determine the period, T, of the oscillation.
   - In Logger *Pro*, perform a sine curve fit on the appropriate part of the angle-time graph. Record the value of ω for this run. Leaving the curve fit information window open speeds up this process for subsequent runs.
   - In LabQuest App, determine T as you did in Part 1.

3. Repeat step 2, increasing the amplitude by 5° each time until you reach 60°. It is unnecessary to store these runs. Beyond this angle, increase the amplitude by 10° until you reach 120°. Record the period for each run.

EVALUATION OF THE DATA

1. Disconnect the sensor from the interface and choose New from the File menu.
   - In Logger *Pro*, manually enter the frequency-amplitude data and use a calculated column to determine the period.
   - In LabQuest App, manually enter the period-amplitude data.

2. Examine the graph of period *vs.* amplitude. Be sure to scale the vertical axis from zero.

3. Examine the statistics on the first four data points for this graph. Perform a linear fit for this same portion of the graph. Note at what point the data show that the period is no longer independent of amplitude. Save this file.

4. Reflect on the simplifying assumption that you made in Part 1 that allowed you to solve the differential equation. Test the validity of your assumption by examining the relationship between θ and $\sin\theta$ for the range of values used in your experiment. Use radian measure for your angles. At what point does your approximation begin to break down?

EXTENSION

Because its period of oscillation is very nearly constant for small amplitudes, the simple pendulum has been used in timekeeping since the time of Huygens. You can test this reliability yourself by collecting angle-time data for a much longer time (say 1 or 2 minutes), then determine the period for short intervals within the duration of your experiment. Try this for angles for which your initial approximation is valid, then for a couple of angles outside the range of validity. What do you find?

Experiment

17

INSTRUCTOR INFORMATION

Pendulum Periods

This lab is the first of two investigations that more carefully examine the behavior of a pendulum. In the standard experiment one finds that the period of a pendulum is independent of the mass and the amplitude, and depends only on its length. These relationships are generally true because the approximations made in order to find a solution to the equation describing its motion depend on two important conditions:

1) the amplitude is small (<< 1 radian), and

2) the mass of the system is concentrated at the end of the string.

In this experiment and the next students will examine the behavior of a pendulum in greater detail to see what occurs when these approximations begin to break down. The first of these is examined in this experiment; the second will be examined in Experiment 18. The solution of the equation describing the motion of the pendulum involves some non-elementary calculus. With guidance, students should be able to follow the reasoning used to find the solution. A thorough treatment of the mathematics involved is provided in the Post-Lab Discussion.

This lab is written with the assumption that the instructor will engage the students in discussions at critical junctures. These discussions can take place with the entire class or with individual lab groups. The icon at left indicates where these discussions should occur.

The Microsoft Word files for the student pages can be found on the CD that accompanies this book. See *Appendix A* for more information.

OBJECTIVES

In this experiment, the student objectives include

- Collect angle *vs.* time data for a simple pendulum.
- Determine the best-fit equation for the angle *vs.* time graph.
- From an analysis of the forces acting on the pendulum bob, derive the equation describing the motion of the pendulum.
- Relate the parameters in the best-fit equation for the angle *vs.* time graph to their physical counterparts in the system.
- Determine the period of oscillation from an analysis of the angle *vs.* time graph.
- Account for the deviation from constant periods when the amplitude becomes large.

During this experiment, you will help the students

- Relate angular frequency, ω, and period, T.
- Recognize that a pendulum and a mass on a spring both undergo simple harmonic motion.
- Derive a 2nd order differential equation describing the motion of a pendulum, and recognize that one must make the approximation that $\theta \approx \sin\theta$ in order to find a solution.

- Recognize that $\theta(t) = \theta_0 \sin\left(\sqrt{\frac{g}{l}}t\right)$ is a reasonable solution to this equation, consistent with the graph of angle *vs.* time they have obtained.
- Recognize that the approximation $\theta \approx \sin\theta$ is valid only for small angles and why this limits the range of angles for which the period is independent of amplitude.

REQUIRED SKILLS

In order for student success in this experiment, it is expected that students know how to perform curve fits in Logger *Pro*. This is addressed in Activity 1.

EQUIPMENT TIPS

For this experiment, the Rotary Motion Sensor must be securely attached to support rods so that vibration of the sensor is minimized during data collection. The aluminum rod should be attached to the 3-step pulley, as shown in Figure 1.

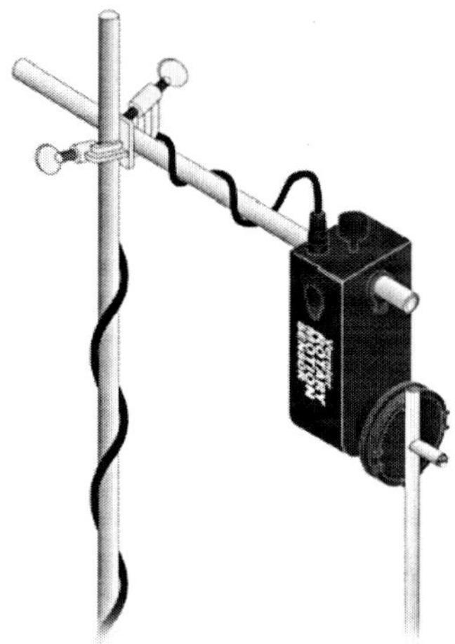

Figure 1

PRE-LAB DISCUSSION

It is likely that students will recognize that the motion they observe is periodic. If they have examined simple harmonic motion, they should sketch a sinusoidal curve.

LAB PERFORMANCE NOTES

When the Rotary Motion Sensor (RMV) is connected to the data-collection interface and Logger *Pro* or LabQuest App is started, the default graph screen shows both angle *vs.* time (θ-t) and velocity *vs.* time (v-t) graphs. Students can delete the v-t graph and use Auto Arrange from the Page menu to re-size the θ-t graph. In LabQuest App, they can elect to show Graph 1 only.

After multiple runs, the RMV tends to display a non-zero value for the angle in the live readout window. This should not be a concern because the software re-sets the zero when data collection begins. To obtain a graph that is symmetrical about the time axis, the pendulum should be motionless and hanging vertically when the Collect button is clicked; only then should it be set in motion. The default length of the experiment (10 s) provides sufficient time for students to measure the angle and release the pendulum after they begin data-collection. Students will obtain smoother curves if they use the X4 mode (1/4° resolution) to collect data.

- In Logger *Pro*, choose Set Up Sensors under the Experiment menu. Once they select their interface, they should click on the icon for the RMV and then select X4 Mode.
- In LabQuest App, tap on the meter window and then select the X4 Mode.

Remind students that they should save this experiment file for later use. Students using Logger *Pro* should turn off the Connect Points feature and turn on Point Protectors for the graphs they produce in this experiment. In LabQuest App, the default Connect Points feature should be used, since it aids in determining the period.

SAMPLE RESULTS AND POST-LAB DISCUSSION – PART 1

Step 1

If students have examined simple harmonic motion prior to performing this experiment, they should readily recognize that the motion of the pendulum is oscillatory in much the same way as the up and down motion of a mass suspended from a spring.

Determination of ω using Logger *Pro*

Students familiar with the analysis of simple harmonic motion should find fitting a sine curve to their data in Logger *Pro* to be straightforward. In Figure 2 below, the curve fit was performed on the portion of the angle-time graph where the pendulum was swinging freely. The B parameter in the equation $y = A\sin(Bt + C) + D$ yields the angular frequency, ω, which is the variable of interest of this experiment.

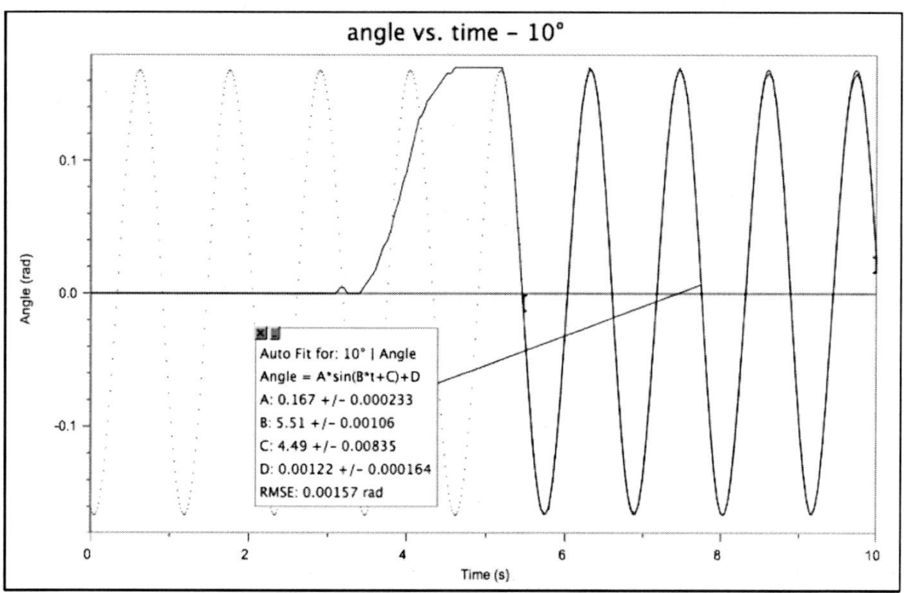

Figure 2 Sine curve fit to the angle-time data

Determination of ω using LabQuest App

In LabQuest App, students need to select several cycles of the angle-time graph and divide the time (Δx) by the number of cycle to obtain the period, T (see Figure 3). In this example, the period is 1.14 s.

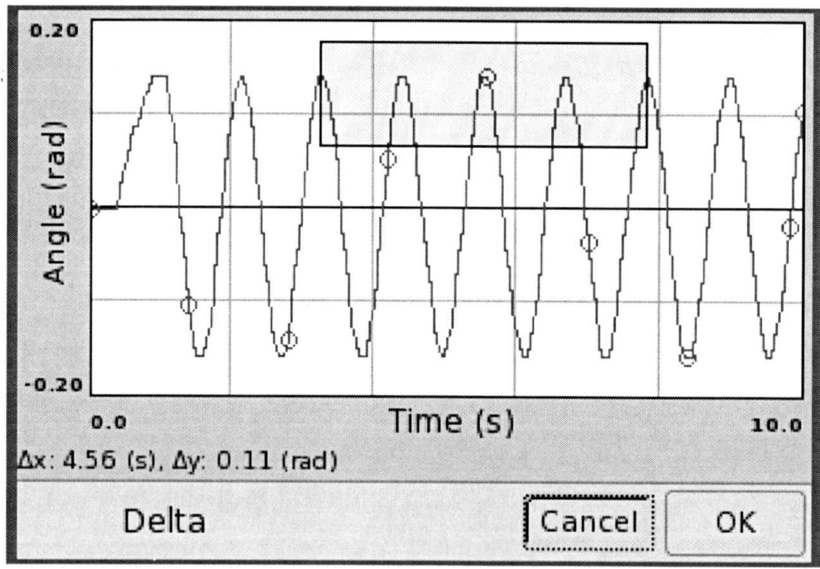

Figure 3 Using the Delta function to determine T

Students are asked to convert period, T, to angular frequency, ω, because that variable appears in the sine function that is a solution to the differential equation they will generate later.

Step 5

When students break up the gravitational force acting on the bob, mg, into components, they should find that the expression for the restoring force (so named because it acts in a direction opposing the displacement through angle θ) is given by $-mg\sin\theta$.

Steps 6–7

The Newton's second law equation describing the pendulum bob is $mg\sin\theta = ma$. The acceleration, a, is the rate of change of the velocity, which, in turn, is the rate of change in the bob's position as it moves along the arc length, s.

$$a = \frac{dv}{dt} = \frac{d^2s}{dt^2}$$

The arc length is $l\theta$. Substitution of $l\theta$ for s in the equation above yields $-mg\sin\theta = ml\frac{d^2\theta}{dt^2}$. Students ought to be able to simplify and rearrange terms to obtain the following equation

$$\frac{d^2\theta}{dt^2} + \frac{g}{l}\sin\theta = 0$$

Step 8

At this point, you will have to help them through the reasoning required to deal with this equation. It cannot be solved by elementary calculus techniques. However, if one makes the small-angle approximation $\sin\theta \approx \theta$, the equation becomes a straightforward 2nd order differential equation

$$\frac{d^2\theta}{dt^2} + \frac{g}{l}\theta = 0$$

It is unlikely that your students will have the experience to solve such an equation. Suggest that a time-honored approach to solving differential equations is "guess and check". It involves guessing a function $\theta(t)$ whose second derivative is the opposite of itself. This hint alone may not get your students closer to the solution. But, if you remind them of the equation they used to describe the position of a simple harmonic oscillator (a sine function) and that the second derivative of this equation, used to describe the acceleration is the opposite of the sine[1], they might be persuaded that $\theta(t) = A\sin(\omega t)$ is a function that could work. The second derivative of this function is

$$\frac{d^2\theta}{dt^2} = -A\omega^2 \sin(\omega t)$$

Substitution of these functions into the differential equation yields

$$-A\omega^2 \sin(\omega t) + \frac{g}{l}A\sin(\omega t) = 0$$

The "guessed" function is a solution if $\omega = \sqrt{\frac{g}{l}}$.

Step 9

Students should substitute known values for g and l into the equation above to determine the expected value of the angular frequency, ω. There is good agreement ($\approx$ 3%) between this calculated value and the value of ω obtained from the analysis of the angle-time data (5.51 rad/s).

$$\omega_{calc} = \sqrt{\frac{9.8\,\mathrm{m/s^2}}{0.34\,\mathrm{m}}} = 5.35\,\mathrm{rad/s}$$

[1] This analysis is done in Experiment 16.

SAMPLE RESULTS AND POST-LAB DISCUSSION – PART 2

Step 1

For the first four amplitudes (5° to 20°), the variation in the period is less than 0.4%. Beyond 25°, the period begins to become appreciably longer.

Steps 2–3

The graph in Figure 4 shows that after 25°, the period is no longer independent of the amplitude. At 60°, the deviation is nearly 7% and increases markedly with increasing amplitude.

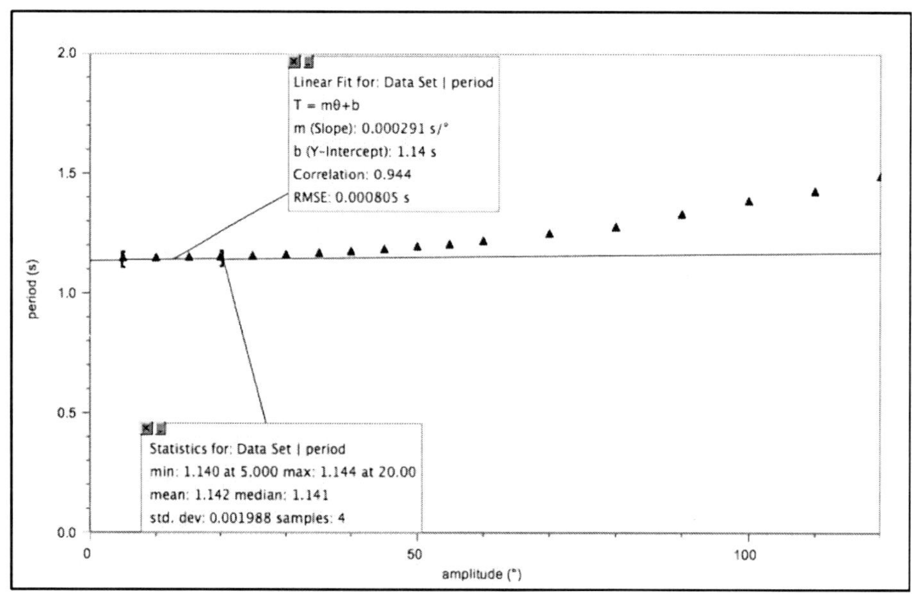

Figure 4 Period vs. *amplitude*

Step 4

The table below shows values for angles and the sine of these angles.

θ (°)	θ (rad)	sin θ
5	0.0872	0.0872
10	0.175	0.174
15	0.262	0.259
20	0.349	0.342
25	0.436	0.423
30	0.524	0.500
35	0.611	0.574
40	0.698	0.643
45	0.785	0.707
50	0.873	0.766
55	0.960	0.819
60	1.05	0.866
70	1.22	0.940
80	1.40	0.985
90	1.57	1.00

Clearly, the discrepancy between θ and *sin* θ (3% at 25°) makes the approximation used in the solution of the differential equation hard to justify as the amplitude increases beyond 25°.

EXTENSION

In the graph in Figure 5 below, a curve fit was performed for three 10-second intervals during the 60 s duration of the experiment (initial amplitude = 10°). Note that while the amplitude begins to decrease, the angular frequency (*B* parameter) remains constant.

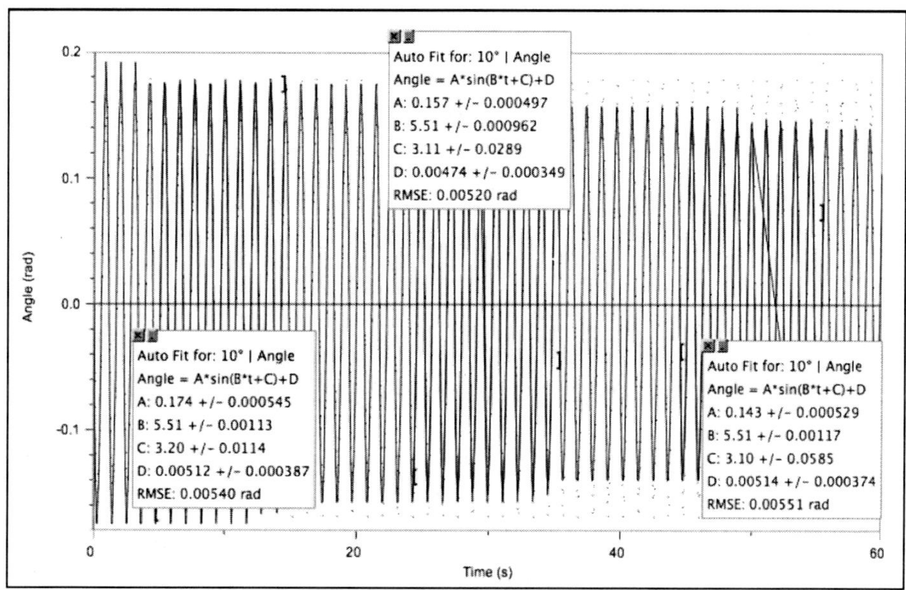

Figure 5

By contrast when this same analysis was performed on the angle-time graph for an initial amplitude of 50°, the increase in the period from beginning to end of the 60 s duration of the experiment was 1.5%. With an initial amplitude of 80°, the increase in the period was 5% over the 60 s interval. Clearly, for a pendulum to be useful in a timekeeping device, the amplitude must be kept relatively small. The motion is isochronous for small angles.

Experiment
18

Physical Pendulum

INTRODUCTION

The introductory treatment of the motion of a pendulum leaves one with the impression that the period of oscillation is independent of the mass and the amplitude, and depends only on the length of the pendulum. These relationships are generally true so long as two important conditions are met:

1) the amplitude is small (<< 1 radian), and
2) the mass of the system is concentrated at the end of the string.

Experiment 17 examined the non-ideal behavior of the pendulum when the amplitude was not kept small. In this experiment, you will investigate the effect on the behavior of a pendulum when the mass of the system can no longer be treated as a point mass at the end of a massless string.

OBJECTIVES

In this experiment, you will

- Collect angle *vs.* time data for a variety of physical pendulums.
- Determine the period of oscillation from an analysis of the angle *vs.* time graph.
- From an analysis of the torques acting on the system, derive the equation describing the motion of the physical pendulum.
- Compare this equation to the one that describes the motion of a simple pendulum.
- Relate the angular frequency, ω, of the system to its physical features.
- Compare the agreement between experimental and calculated values of ω determined by this treatment of the system with those obtained by treating the system as if it were a simple pendulum.

MATERIALS

Vernier data-collection interface
Logger *Pro* or LabQuest App
Vernier Rotary Motion Sensor
Vernier Rotary Motion Accessory Kit
vertical support rod and clamp
right-angle clamp
protractor
metric ruler or tape
balance

PRE-LAB INVESTIGATION

For a simple pendulum (a weight at the end of a light string) the analysis of forces acting on the weight yields an equation of motion in which the angular frequency is given by $\omega = \sqrt{g/l}$. **Note:** The mass of the bob does not appear in this expression.

Consider the physical pendulum system shown in Figure 1. The aluminum rod is certainly not a light string. Discuss why an analysis of forces alone is insufficient to determine the angular frequency of this system. What approach do you think would be more fruitful?

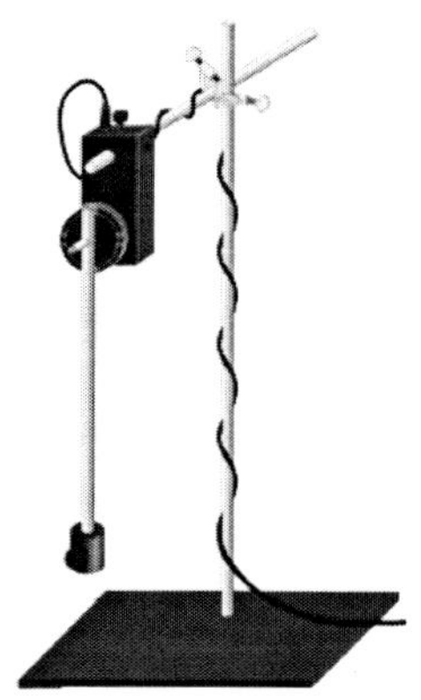

Figure 1

PART 1 – A CLOSER LOOK AT A SIMPLE PENDULUM

PROCEDURE

1. Find the mass of the aluminum rod and the cylindrical weight with thumbscrew. [1]
2. Set up the apparatus pictured in Figure 1. Make sure that the vertical support rod for the Rotary Motion Sensor is securely attached to a bench or table. When the pendulum is set in motion the sensor should be stationary.
3. Measure the distance between the point of attachment of the rod to the 3-step pulley and the center of mass of the weight at the end of the rod. Record this value as the radius, r, of the rotation of the weight.
4. Connect the sensor to the interface and start the data-collection program. Two graphs: angle *vs.* time and velocity *vs.* time will appear in the graph window. For this experiment, you will need to view only the angle *vs.* time graph.
5. The default data-collection rate is appropriate. However, you should increase the resolution of the sensor by selecting the X4 mode.
   - In Logger *Pro*, choose Set Up Sensors from the Experiment menu. Once you select your interface, click on the icon for the RMV and then select X4 Mode.
   - In LabQuest App, tap on the meter window and then select the X4 Mode.
6. Because the default data-collection mode automatically resets the zero position on Collect, it is unnecessary to manually zero the sensor before collecting data. However, the bob must be motionless before you begin data collection.

---

[1] If you have performed Experiment 17, open your saved data file from Part 1. Then once you have completed Step 1, you may proceed with the Evaluation of Data.

7. Start data collection. Then, using a protractor to measure the angle, pull the rod through a 5° angle and release. Be sure that the pendulum swings freely for at least five seconds. Store this run. Do not be too concerned if the graph is not a smooth curve; subsequent trials with larger amplitudes will produce smoother graphs.

8. Repeat Step 7 for amplitudes of 10°, 15° and 20°, storing each run. Save this file; you will return to it later in the experiment.

EVALUATION OF DATA[2]

Determination of ω using Logger *Pro*

1. Before you fit a curve to the position-time graph, turn off Connect Points and turn on Point Protectors.

2. Drag-select that portion of the graph for your first run where the bob is swinging freely. Use the curve-fitting features of Logger *Pro* to fit a sine curve to these data. Record the value of the B parameter to the sine fit. Repeat this process for your other runs.

3. The B parameter is the angular frequency, ω, for this oscillation. Does the value of ω appear to depend on the amplitude of the oscillation? Determine an average value of ω for your four runs.

Determination of ω using LabQuest App

1. Choose to view your first run. Choose the Delta function under the Analysis menu. Drag-select 4–5 cycles of the graph where the bob is swinging freely. The time (Δx) divided by the number of cycles gives the period of oscillation, T. Record this value. Repeat this process for your other runs.

2. The angular frequency, ω, for this oscillation is given by $\omega = \frac{2\pi}{T}$. Determine the value of ω for each of your runs. Record these values.

3. Does the value of ω appear to depend on the amplitude of the oscillation? Determine an average value of ω for your four runs.

Determination of the equation of motion for the pendulum.

4. The physical pendulum can be viewed as an extended body that rotates about the point at which the rod is connected to the pulley. For now, neglect the rod and treat the weight as a point mass. Sketch a force diagram for the weight at the end of the rod when it is displaced to one side. Write the expression for the torque that acts on the weight to restore it to its original position when released.

5. Write the Newton's second law equation describing the rotation of the weight about the pivot point once it is released. In other words, relate the torque from the gravitational force on the pendulum to its angular acceleration. Consider the distance to the center of mass and use that as the point of action for the gravitational force. Express the angular acceleration as the

[2] If you have performed Experiment 17, review your determination of ω for your runs with the amplitude ranging from 5° – 20°. Then move on to Step 4.

second derivative of the angle θ, with respect to time.[3] Keep in mind that the acceleration vector for the pendulum always acts in a direction opposing the displacement through angle θ.

6. Divide through by I in your second equation and rearrange the terms so that you have set the equation equal to 0. You have now produced a 2nd order differential equation describing the motion of the pendulum bob. Check your equation with those of others in your class.
7. In your class discussion make sure that you understand the simplification necessary to suggest a solution to the equation you have derived.
8. Using the expression for the moment of inertia, I, of a point mass, further simplify the equation above. Compare this to the equation of motion for a simple pendulum derived from a consideration of linear dynamics only.
9. Now, consider how the inclusion of the torque and moment of inertia of the rod would change the general equation of motion for the physical pendulum.

PART 2 – EFFECT OF POSITION OF THE WEIGHT

Before you move on to this part of the experiment, make sure that you understand how the angular frequency, ω, depends on both the total torque and the total moment of inertia of the *system*. You may need to review the expression for the moment of inertia of a rod.

Both the torque acting on the weight and its moment of inertia depend on its position on the rod. By contrast, both the torque and the moment of the inertia for the aluminum rod are constant. It is recommended that you set up a Logger *Pro* file to facilitate calculation of the torque and the moment of inertia of the system as well as the expected value of the angular frequency, ω. This file will enable you to compare the expected value to your experimentally determined value of ω for a variety of configurations of your physical pendulum. Your instructor may guide you in the design of this file. See the sample column headers below.

radius (m)	I-weight (kgm^2)	I-total (kgm^2)	τ-weight (m-N)	τ-total (m-N)	ω-torq (rad/s)	ω-meas (rad/s)	%-diff	ω-force (rad/s)	%-diff-2

The **%-diff** columns allow you to compare the agreement between the experimentally determined value of ω and those calculated by treating the system using either rotational or linear dynamics.

When you have set up the Logger *Pro* file, you are ready to test your predictions.

PROCEDURE

1. Re-open your experiment file from Part 1.

---

[3] Recall that $a = \dfrac{d^2\theta}{dt^2}$.

2. Move the weight up the rod. Measure the length between the pivot point and the center of mass of the weight and record this value.

3. Collect angle *vs.* time data for the pendulum as before using an amplitude of 10–15°. Determine the angular frequency as you did in Part 1.

4. Repeat Step 3, decreasing the length gradually until you have data for at least 8 different lengths.

EVALUATION OF THE DATA

1. Re-open the Logger *Pro* file you built earlier.

2. Enter your values for radius and ω-meas that you recorded.

3. Compare the %-differences between measured and calculated values of ω treating the system as a physical pendulum as opposed to a simple one.

PART 3 – EFFECT OF ADDING ANOTHER WEIGHT TO THE SYSTEM

PROCEDURE

1. Re-open your experiment file from Part 2.

2. Add the second cylindrical weight to the rod as shown in Figure 2. Determine the distance between the pivot point and the center of mass of this second weight; label this as radius-2.

3. Collect angle *vs.* time data for the pendulum as before using an amplitude of 10–15°. Determine the angular frequency as you did in Part 1.

4. Repeat Step 3 for several new positions of the two weights.

Figure 2

EVALUATION OF DATA

1. Re-open the Logger *Pro* file you created earlier. Label your first data set: One Mass on Rod.

2. Choose New Data Set from the Data menu. Label this set: Two Masses on Rod. Add a new manual column, radius-2, and two new calculated columns, I-weight2 and τ-weight2, with the appropriate equations. Modify the equations used to calculate I-total and τ-total.

3. Choose Table from the Insert menu. Choose Table Options from the Options menu. De-select your first data set, then select the second set to be displayed in the table.

4. Enter your measured values of radius-1 and radius-2.

5. How do your measured values of ω compare to the values you have calculated?

EXTENSION

In Part 3 of this experiment, one end of the rod was attached to the rotary motion sensor and the weights were placed at various positions near the other end. Consider how the system would behave if you were to attach the center of the rod to the sensor and place weights on either side, as shown in Figure 3. How would this configuration affect the moment of inertia of the system and the torque acting on it? Once you have made your predictions, you might try collecting data with the weights at different positions to see how these configurations affect the angular frequency of the pendulum.

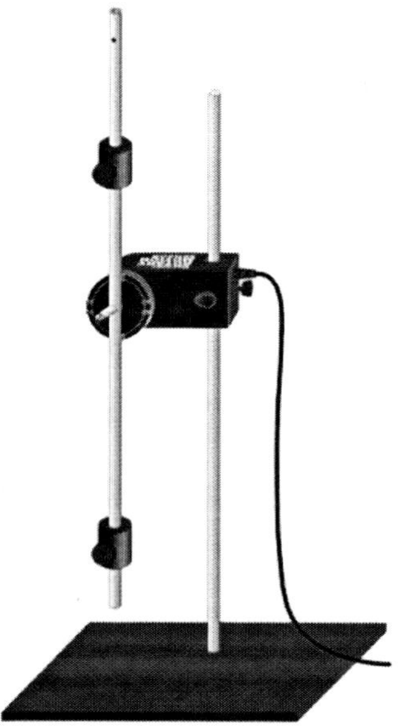

Figure 3

INSTRUCTOR INFORMATION

Experiment
18

Physical Pendulum

This lab is the second of two investigations that more carefully examine the behavior of a pendulum. In the standard experiment one finds that the period of a pendulum is independent of the mass and the amplitude, and depends only on its length. These relationships are generally true because the approximations made in order to find a solution to the equation describing its motion depend on two important conditions:

1) the amplitude is small (<< 1 radian), and
2) the mass of the system is concentrated at the end of the string.

In Experiment 17, students examine the non-ideal behavior of a pendulum when the amplitude is no longer kept small. In this experiment students use rotational dynamics to account for the behavior of a more complicated pendulum system – one which can no longer be modeled as a point mass at the end of a massless string. The solution of the equation describing the motion of the pendulum involves some non-elementary calculus. With guidance, students should be able to follow the reasoning used to find the solution. A thorough treatment of the mathematics involved is provided in the Post-Lab Discussion.

Calculations of the moments of inertia and torques acting on the pendulum are greatly facilitated by the use of Logger *Pro*. A sample Logger *Pro* file, 18 Phys Pend calculations.cmbl, is provided on the CD that accompanies this book for guidance. To save time, you may provide this file for students to use directly. If you exercise this option, make sure that students examine the equations for the calculated columns.

This lab is written with the assumption that the instructor will engage the students in discussions at critical junctures. These discussions can take place with the entire class or with individual lab groups. The icon at left indicates where these discussions should occur.

The Microsoft Word files for the student pages and a sample Logger *Pro* data-analysis file can be found on the CD that accompanies this book. See *Appendix A* for more information.

OBJECTIVES

In this experiment, the student objectives include

- Collect angle *vs.* time data for a variety of physical pendulums.
- Determine the period of oscillation from an analysis of the angle *vs.* time graph.
- From an analysis of the torques acting on the system, derive the equation describing the motion of the physical pendulum.
- Compare this equation to the one that describes the motion of a simple pendulum.
- Relate the angular frequency, ω, of the system to its physical features.
- Compare the agreement between experimental and calculated values of ω determined by this treatment of the system with those obtained by treating the system as if it were a simple pendulum.

During this experiment, you will help the students

- Relate angular frequency, ω, and period, T.
- Derive a 2nd order differential equation describing the motion of a pendulum, and recognize that one must make the approximation that $\theta \approx \sin\theta$ in order to find a solution.
- Recognize that $\theta(t) = \theta_0 \sin(\omega t)$, is a reasonable solution to this equation, consistent with the graph of angle *vs.* time they have obtained.
- Recognize that ω depends on both the sum of the torques and the moments of inertia acting on components of the system.
- Realize that the analysis of a physical pendulum using torques and moments of inertia better predicts ω than a treatment considering only force and mass.

REQUIRED SKILLS

In order for student success in this experiment, it is expected that students know how to

- Perform curve fits in Logger *Pro*. This is covered in Activity 1.

EQUIPMENT TIPS

For this experiment, the Rotary Motion Sensor must be securely attached to support rods so that vibration of the sensor is minimized during data collection. The aluminum rod should be attached to the 3-step pulley, as shown in Figure 1.

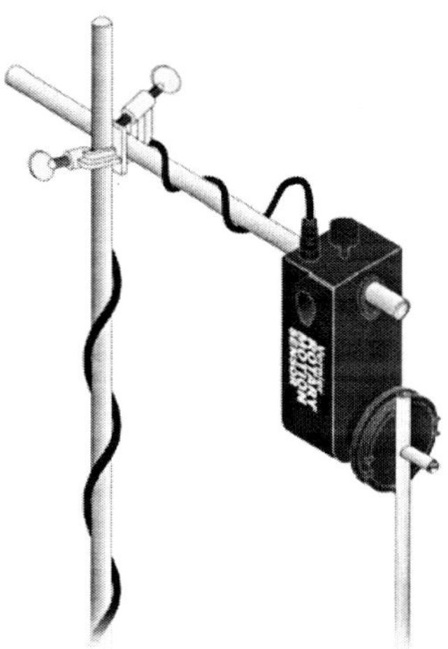

Figure 1

PRE-LAB DISCUSSION

Students should see that pretending that the mass of the connecting rod is negligible leads to problems in the analysis of the motion of the pendulum. By recognizing that the pendulum is an extended body, they should reason that the use of rotational dynamics would be a more fruitful approach to prediction the behavior of a physical pendulum.

LAB PERFORMANCE NOTES

When the Rotary Motion Sensor (RMV) is connected to the data-collection interface and Logger *Pro* or LabQuest App is started, the default graph screen shows both angle *vs.* time (θ-t) and velocity *vs.* time (v-t) graphs. Students can delete the v-t graph and use Auto Arrange from the Page menu to re-size the θ-t graph. In LabQuest App, they can elect to show Graph 1 only.

After multiple runs, the RMV tends to display a non-zero value for the angle in the live readout window. This should not be a concern because the software re-sets the zero when data-collection begins. To obtain a graph that is symmetrical about the time axis, the pendulum should be motionless and hanging vertically when the Collect button is clicked; only then should it be set in motion. The default length of the experiment (10 s) provides sufficient time for students to measure the angle and release the pendulum after they begin data-collection. Students will obtain smoother curves if they use the X4 mode (1/4° resolution) to collect data.

- In Logger *Pro*, choose Set Up Sensors from the Experiment menu. Once they select their interface, they should click on the icon for the RMV and then select X4 Mode.
- In LabQuest App, tap on the meter window and then select the X4 Mode.

Remind students using Logger *Pro* to turn off the Connect Points feature and turn on Point Protectors for the graphs they produce in this experiment. In LabQuest App, the default Connect Points feature should be used, since it aids in determining the period.

SAMPLE RESULTS AND POST-LAB DISCUSSION – PART 1

Determination of ω using Logger *Pro*

Students familiar with the analysis of simple harmonic motion should find fitting a sine curve to their data in Logger *Pro* to be straightforward. In Figure 2 below, the curve fit was performed on the portion of the angle-time graph where the pendulum was swinging freely. The B parameter in the equation $y = A\sin(Bt + C) + D$ yields the angular frequency, ω, which is the variable of interest of this experiment.

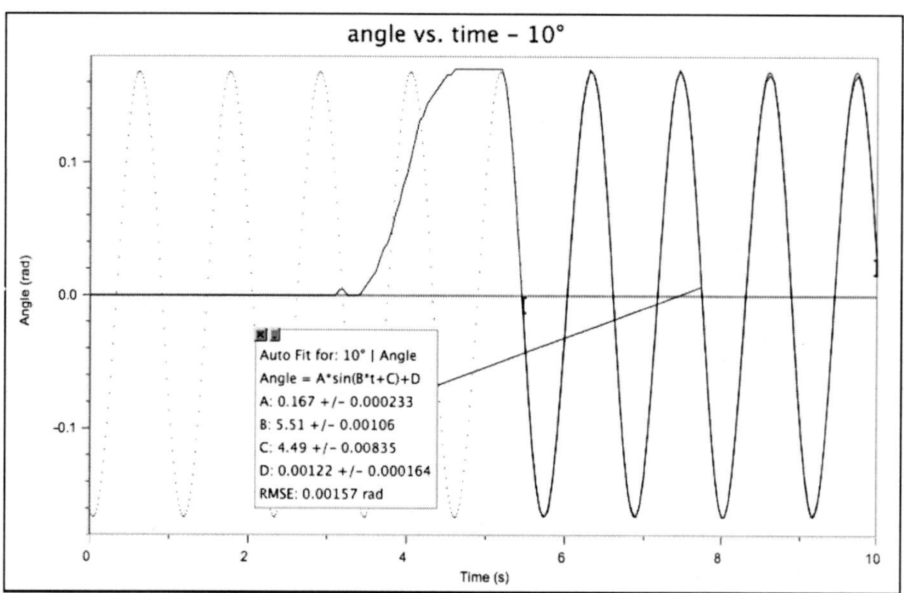

Figure 2 sine curve fit to the angle-time data

Determination of ω using LabQuest App

In LabQuest App, students need to select several cycles of the angle-time graph and divide the time (Δx) by the number of cycle to obtain the period, T (see Figure 3). In this example, the period is 1.14 s.

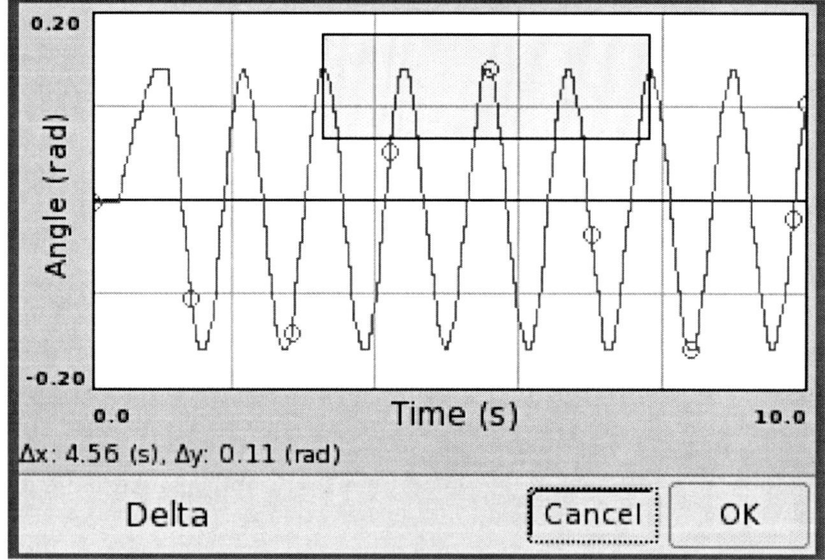

Figure 3 Using the Delta function to determine T

Students are asked to convert period, T, to angular frequency, ω, because that variable appears in the sine function that is a solution to the differential equation they will generate later.

Determination of the equation of motion for the pendulum.

If students have performed Experiment 17, they should find the analysis of the motion of the pendulum bob in terms of rotational dynamics to be entirely analogous to the treatment using linear dynamics.

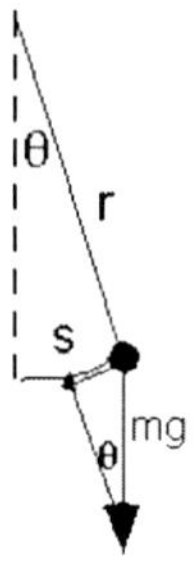

Step 4

When students break up the gravitational force acting on the bob, mg, into components, they should find that the expression for the restoring force (so named because it acts in a direction opposing the displacement through angle θ) is given by $-mg\sin\theta$. This force acts perpendicular to the connecting rod. The torque acting on the bob is $-rmg\sin\theta$.

Steps 5–6

The Newton's second law equation describing the rotation of the pendulum bob about the pivot point is $-rmg\sin\theta = I\alpha$. The angular acceleration, α, is the rate of change of the angular velocity, which, in turn, is the rate of change of the angle through which the bob rotates.

$$\alpha = \frac{d\omega}{dt} = \frac{d^2\theta}{dt^2}$$

Simplify and rearrange terms to obtain the following equation:

$$\frac{d^2\theta}{dt^2} + \frac{rmg}{I}\sin\theta = 0$$

Step 7

At this point, you will have to help them through the reasoning required to deal with this equation. It cannot be solved by elementary calculus techniques. However, if one makes the small-angle approximation $\sin\theta \approx \theta$, the equation becomes a straightforward 2nd order differential equation:

$$\frac{d^2\theta}{dt^2} + \frac{r\,mg}{I}\theta = 0$$

It is unlikely that your students will have the experience to solve such an equation. Suggest that a time-honored approach to solving differential equations is "guess and check." It involves guessing a function $\theta(t)$ whose second derivative is the opposite of itself. This hint alone may not get your students closer to the solution. But, if you remind them of the equation they used to describe the position of a simple harmonic oscillator (a sine function) and that the second derivative of this equation, used to describe the acceleration is the opposite of the sine[1], they might be persuaded that $\theta(t) = A\sin(\omega\, t)$ is a function that might work. The second derivative of this function is:

$$\frac{d^2\theta}{dt^2} = -A\omega^2\sin(\omega\, t)$$

Substitution of these functions into the differential equation yields

$$-A\omega^2\sin(\omega\, t) + \frac{r\,mg}{I}A\sin(\omega\, t) = 0$$

The "guessed" function is a solution provided that $\omega = \sqrt{\frac{r\,mg}{I}}$.

Step 8

The moment of inertia, I, for a point mass is mr^2; substitution of this expression into the equation above yields:

$$\omega = \sqrt{\frac{r\,mg}{I}} = \sqrt{\frac{r\,mg}{mr^2}} = \sqrt{\frac{g}{r}}$$

Note: In this system, distance between the pivot point and the center of mass, r, is the same as the length of a simple pendulum, l. Replacing r by l in the expression above yields the same equation that one obtains for the angular frequency of a simple pendulum using linear dynamics.

[1] This analysis is done in Experiment 16.

Step 9

For the physical pendulum, the expression for angular frequency must also include the torque acting on the rod $(\frac{1}{2}lm)$ as well as its moment of inertia $(\frac{1}{3}ml^2)$. The coefficient of θ in the simplified differential equation becomes:

$$\frac{(\frac{1}{2}lm + rM)g}{\frac{1}{3}ml^2 + Mr^2}$$

where m and M are the masses of the rod and weight, respectively, l is the effective length of the rod, and r is the distance from the center of mass of the weight to the pivot point.

SAMPLE RESULTS AND POST-LAB DISCUSSION – PART 2

Clearly, the use of Logger *Pro* or a spreadsheet greatly simplifies the calculations necessary to determine ω for a given configuration of the physical pendulum. The column headers for a Logger *Pro* file are given in the student version of this lab. You may choose to provide your students the Logger *Pro* file used to simplify the calculations; it can be found on the CD that accompanies this book.

The effective length of the rod, 0.320 m, takes into account the fact that its point of attachment to the pulley is 3.0 cm from one end. Its mass is 41.3 g. The moment of inertia of the rod is:

$$\tfrac{1}{3}ml^2 = \tfrac{1}{3}(0.0413\,\text{kg})(0.320\,\text{m})^2 = 0.00141\,\text{kgm}^2$$

If the mass of the rod is considered to be concentrated at its center; i.e., its center of mass, then the torque acting on the rod is:

$$\tfrac{1}{2}lmg = \tfrac{1}{2}(0.320\,\text{m})(0.0413\,\text{kg})(9.8\,\text{N/kg}) = 0.0648\,\text{mN}$$

The angular frequency, ω, is the square root of the expression in Step 9. Because these values do not change as the weights are moved to different positions on the rod, they can be simply added to the equations for the rMg and Mr^2 terms for the weight.

For the column ***ω-force***, the angular frequency is calculated by $\omega = \sqrt{\frac{g}{r}}$, where the system is treated as a simple pendulum (Step 8).

The percent difference columns allow students to compare the agreement between the values of ω calculated by the two different approaches and the value of ω determined experimentally.

In the table below are sample results for Part 2 in which students test their predictions when the weight is moved to different positions on the rod.

radius (m)	I-weight (kgm^2)	I-total (kgm^2)	τ-weight (m-N)	τ-total (m-N)	ω-torq (rad/s)	ω-meas (rad/s)	%-diff	ω-force (rad/s)	%-diff-2
0.340	0.00925	0.0107	0.267	0.331	5.58	5.51	1.18	5.37	–2.63
0.300	0.00720	0.00861	0.235	0.300	5.90	5.82	1.40	5.72	–1.83
0.270	0.00583	0.00724	0.212	0.276	6.18	6.03	2.40	6.03	–0.089
0.240	0.00461	0.00602	0.188	0.253	6.48	6.32	2.51	6.39	1.10
0.210	0.00353	0.00494	0.165	0.229	6.82	6.67	2.14	6.83	2.36
0.180	0.00259	0.00400	0.141	0.206	7.17	6.95	3.10	7.38	5.81
0.150	0.00180	0.00321	0.118	0.182	7.54	7.28	3.41	8.08	9.93
0.120	0.00115	0.00256	0.094	0.159	7.87	7.56	3.99	9.04	14.35
0.090	0.00065	0.00206	0.071	0.135	8.11	7.74	4.55	10.44	25.83

The data show that the use of the simple pendulum approximation works reasonably well (%-diff-2 < 3%) until the position of the weight is less than 0.20 m from the pivot point. After that point, ignoring the contribution from the rod introduces a very large error.

SAMPLE RESULTS AND POST-LAB DISCUSSION – PART 3

Once students know how to do the calculations to design and complete the table in Part 2, the modifications required to complete the calculations for Part 3 are minor. A new manual column, radius-2, and two new calculated columns, *I*-weight2 and *τ*-weight2, with the appropriate equations, are added. The equations used to calculate *I*-total and *τ*-total must also be modified. The column %-diff-2 is no longer needed because there is no simple way to use the simple pendulum approach when more than one weight is used.

The table with sample results for Part 3 is too large to display here. It can be found in the Logger *Pro* file on the CD that accompanies this book. Even better agreement (< 2%) between the calculated and experimental values of the angular frequency is found when two weights are used. Students should conclude that the use of rotational dynamics allows them to make reasonable predictions of the behavior of more complicated pendulum systems whereas linear dynamics is useful only so long as the pendulum is simple.

EXTENSION

Configuring a physical pendulum with a mass on either side of the pivot point creates a system that can be described with the same physics. However, the second mass increases the total moment of inertia of the system, while *reducing* the net torque. What will this do to the period of the system? Both the increased moment and the decreased torque contribute to increasing the period. You may choose to only have students consider this question qualitatively, rather than quantitatively. If students modify their calculations to account for the opposite sign of the torque from the second mass and compare to measurements, agreement will be good (< 5%) though not quite as good as in earlier parts. With reduced torque, the effects of friction are more pronounced.

Experiment

19

Center of Mass

INTRODUCTION

In the most of the previous experiments you have examined the motion of a single object as it underwent a variety of motions. You learned that an object subject to no external force moves at constant velocity. Suppose now that the system consists of two objects that undergo a collision. Clearly, the velocity of each object will change as a result of the collision. But if you were to consider both carts as a *system*, how would the center of mass of this *system* change, if at all, after a collision? In this experiment you will use video analysis techniques to examine the behavior of the center of mass of a *system* of moving objects.

OBJECTIVES

In this experiment, you will

- Use video analysis techniques to obtain position, velocity, and time data for two carts undergoing a variety of collisions.
- Analyze the position-time graphs for the individual carts and compare these to the position-time graph for the center of mass of the system.
- Compare the momentum of the system before and after collisions.

PRE-LAB QUESTION

Imagine two carts, connected by a 10 cm string, moving down a long dynamics track at constant velocity. Where would you imagine the center of mass of this system of carts to be? Now, suppose the leading cart is twice as massive as the trailing cart. Where would you expect the center of mass of this system to be? Sketch diagrams to support your answer to both questions.

PROCEDURE

1. Open the Logger *Pro* file, 19 Elastic equal mass.cmbl. In this file is a short video clip of an elastic collision between two carts of equal mass.

2. Enable Video Analysis by clicking the button in the lower right corner. This brings up a toolbar with a number of buttons (see Figure 1).

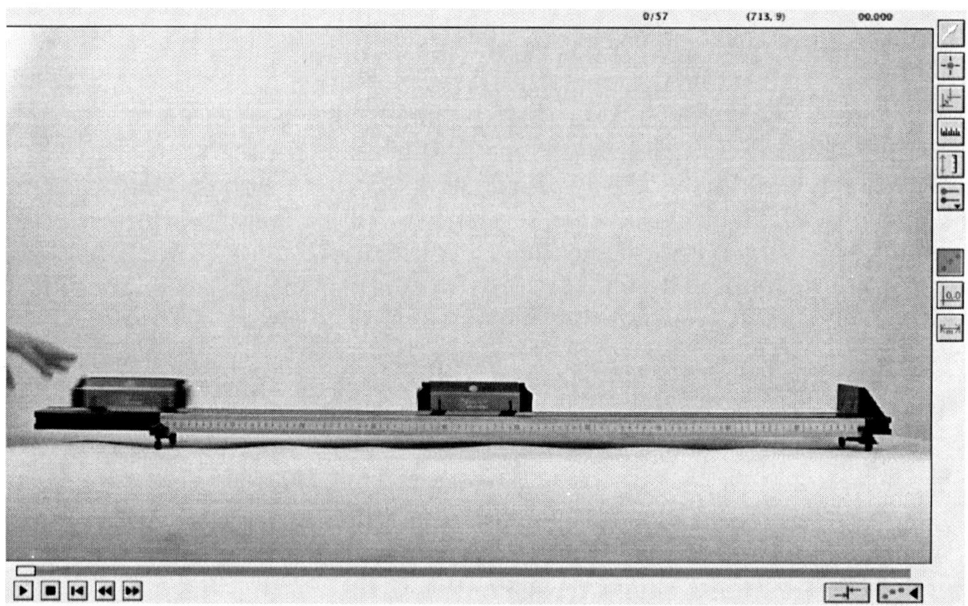

Figure 1 Video Analysis window

3. Click the **Set Origin** button (third from top), then click on the movie frame to set the location of the origin. If needed, this coordinate system can be rotated by dragging the yellow dot on the horizontal axis.

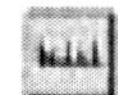

4. Click the **Set Scale** button (fourth from top), then drag across an object of known length in the movie. In this movie, the object of known length is a 1 m stick next to the track. When you release the mouse button, enter the length of the object; be sure the units are correct.

5. Select Movie Options from the Options menu. Check the two boxes under Video Analysis and choose to advance the movie 3 frames after adding a new point. Doing so will make the interval between frames 0.10 s.

6. Now click the **Add Point** button (second from the top). Click on the yellow dot on the left cart. Each time you mark the object's location, the movie advances three frames. Notice that position-time data are being plotted on the graph.

7. Continue this process until several frames after the carts collide. Should you wish to edit a point, click the **Select Point** button (top). This allows you to move or delete a mismarked point.

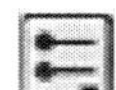

8. Now rewind the video to the first frame used for the left cart, click the Set Active Point button and choose Add Point Series so that you can mark the position of the right cart. **Note:** in this movie, the right cart is motionless for several frames, so these points will be superimposed upon one another. Marking the points will be easier if you turn off the Toggle Trails feature.

9. After you have completed the Evaluation of Data for this first movie, use the same procedure for the next two movies, 19 Inelastic unequal mass.cmbl and 19 Elastic unequal mass.cmbl.

EVALUATION OF DATA

Part 1 Elastic collision with *m1* = *m2*

1. Examine the graphs of *x*-position *vs.* time for both carts. Explain the shape of the graphs for each cart. Use the linear fit tool to find the velocity of each cart during that portion of the movie in which it was moving. How do these velocities compare?

2. You can simplify your data table by choosing Table Options from the Options menu and selecting to display only the columns for time and x-position for the two carts. Choose Auto Arrange from the Page menu to facilitate working with the data table.

3. Use the Draw Prediction tool from the Analyze menu to sketch a position-time graph of the Center of Mass of the *system*. If you are unsure about how the location of the CoM changes with time, you can create a calculated column to determine the CoM for the two cart system.

4. Return to the Video Analysis window. Click the Set Active Point button and choose Add Center of Mass Series. Logger *Pro* opens a dialog box where you could enter the masses of the two carts. Since they are the same in this part, it doesn't matter if you enter values here. Later, when the masses are different, you can enter the actual masses. Click OK to add the Center of Mass Series. Logger *Pro* marks the position of the Center of Mass for the frames where you have marked the positions of the two carts.

5. On the vertical axis of the position-time graph choose More and add the Center of Mass plot to your graph. How does the plot of *x*(CoM) *vs.* time compare with your predicted sketch?

6. What relationship appears to exist between the velocity of the CoM and the average value of the velocities of the carts when they were moving? Explain.

Part 2 Inelastic collision with *m1* > *m2*

1. As you did in Part 1, examine the graphs of *x*-position *vs.* time for both carts. Use the linear fit tool to find the velocity of cart 1 *before* the collision. Use what you have learned from conservation of momentum to predict the velocity of the coupled carts *after* the collision. Now graphically determine the velocity of the carts after the collision. Account for any discrepancy between the predicted and experimental values.

2. Choose the Draw Prediction tool from the Analyze menu to sketch a position-time graph of the Center of Mass of the *system*.

3. Return to the Video Analysis window and, as you did in Part 1, and add the Center of Mass Series so that you can view this plot on your graph.

4. Examine the position-time graph. How does the plot of *x*(CoM) *vs.* time compare with your predicted sketch? In what ways is this graph similar to the one you obtained in Part 1? In what ways does it differ? Account for your findings.

Part 3 Elastic collision with *m1* < *m2*

1. As you did in Part 1, examine the graphs of x-position *vs.* time for both carts. Use the linear fit tool to find the velocity of cart 1 before and after the collision. Use conservation of momentum to predict the velocity cart 2 *after* the collision. Now graphically determine the velocity of cart 2 after the collision. Account for any discrepancy between the predicted and experimental values.

2. Choose the Draw Prediction tool from the Analyze menu to sketch a position *vs.* time graph of the Center of Mass of the *system.*

3. Return to the Video Analysis window and, as you did in Part 1, and add the Center of Mass Series so that you can view this plot on your graph

4. Examine the position-time graph. How does the plot of x(CoM) *vs.* time compare with your predicted sketch? In what ways is this graph similar to the one you obtained in Part 2? In what ways does it differ?

Part 4 Consolidation of your findings

1. What do the CoM motions have in common? How do you account for this similarity?

2. How is the velocity of the CoM related to the total momentum of the system?

3. Based on your analysis of the motion of the carts in the video clips, write an expression for the center of mass (CoM) of a system of two bodies.

EXTENSION

Using existing data, determine the total kinetic energy of the system before and after the collisions. Is kinetic energy conserved in each type of collision? If not, explain any change.

Experiment

19

INSTRUCTOR INFORMATION

Center of Mass

This lab is written with the assumption that the instructor will engage the students in discussions at critical junctures. These discussions can take place with the entire class or with individual lab groups. The icon at left indicates where these discussions should occur.

The Microsoft Word files for the student pages can be found on the CD that accompanies this book. See *Appendix A* for more information. The analysis called for in this lab require Logger *Pro* 3.8.4 or later.

OBJECTIVES

In this experiment, the student objectives include

- Use video analysis techniques to obtain position, velocity, and time data for two carts undergoing a variety of collisions.
- Analyze the position-time graphs for the individual carts and compare these to the position-time graph for the center of mass of the system.
- Compare the momentum of the system before and after collisions.

During this experiment, you will help the students

- Apply principles of conservation of momentum to the analysis of the motion of the center of mass (CoM) of a system.
- Determine an expression for the CoM of a system of two bodies.

REQUIRED SKILLS

In order for student success in this experiment, it is expected that students know how to

- Determine the velocity of an object during an interval by performing a linear fit on the relevant portion of a position-time graph.

EQUIPMENT TIPS

There are several movies provided on the CD accompanying this book that students can analyze. Each cart is marked with a yellow dot.

Figure 1

For a challenge, students can produce their own video clips. To do this, students will need either a digital video camera or a digital still camera that shoots in movie mode (most do), a tripod and some object they can use for scaling. Tips for producing movie clips for video analysis can be found in the Teacher Information of Experiment 6.

If students produce their own videos, you will need to make sure that the magnets and Velcro patches have been installed on the carts. Refer to the booklet that comes with the Vernier Dynamics System for instructions on those procedures.

PRE-LAB DISCUSSION

This experiment should be performed only after students have a thorough grasp of impulse and momentum. For this discussion you will need two standard carts with the neodymium magnets placed in the cart end caps to allow elastic collisions without the carts physically touching one another. Start the discussion by setting a cart in motion on a level track, pointing out that, so long as the sum of the forces acting on the cart is zero, its velocity should remain constant. Next, suggest that they imagine a pair of equal mass carts, tied together with a string, moving on the track. Ask them to consider where the center of mass of the *system* is as the coupled carts are moving. This will seem trivial to most students. Then pose the question, "Suppose that the lead cart is twice as massive as the other; how would that change the center of motion of the system?" After the students have considered these questions and proposed solutions, place a cart in the center of the track, then set another cart in motion so that it will collide elastically with the first. Ask the students if they can visualize the center of mass of the system of carts before and after the collision. This task is much more difficult. Inform the students that in this experiment they will use video analysis techniques to examine the behavior of the center of mass of a pair of carts undergoing a variety of collisions.

LAB PERFORMANCE NOTES

Detailed instructions about performing the video analysis are provided in the student version of this experiment. If students are still uncertain how to do this, you may refer them to Tutorial 12 in the Experiments folder in Logger *Pro*.

It takes some practice to launch a cart so that a smooth collision occurs. Students need to exercise greater care when they launch both carts at one another. Encourage students to try to add the 500 g mass to either cart and to practice both elastic and inelastic collisions.

Because of frictional forces, some slowing of the carts is inevitable. To minimize the effect of friction on their evaluation of data, students can shorten the interval used to determine the slope of the linear fit to obtain a better value for the velocity of the carts.

SAMPLE RESULTS AND POST-LAB DISCUSSION – PART 1

Step 1

Students should obtain a graph like the one in Figure 2 below. These data are representative of typical video analysis experiments.

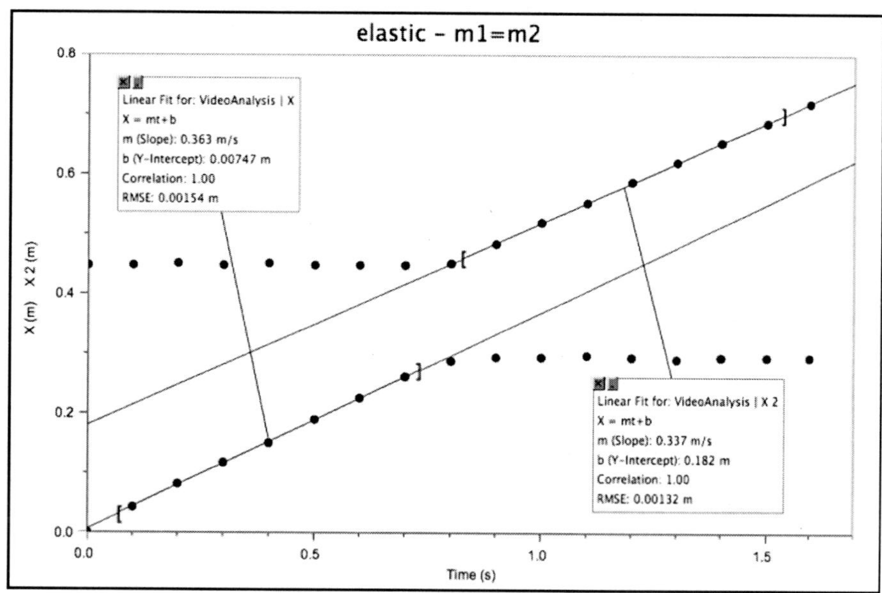

Figure 2

The velocity of cart 2 after the collision is about the same as that of cart 1 before. The decrease in velocity after the collision is due to frictional losses and the fact that the collision was not completely elastic.

Step 3

If students find it hard to obtain the desired precision with the Draw Prediction tool, they can print the position-time graph and then sketch their prediction of the position-time behavior of the center of mass on this. The symmetry of the graph in Figure 2 suggests that the plot of x_{CoM} *vs.* time should be a straight line midway between the two *x-t* graphs. Ask the students to justify the shape and placement of their predicted plot without appealing to a formula for CoM.

Steps 4–6

Students should obtain a graph similar to the one in Figure 3.

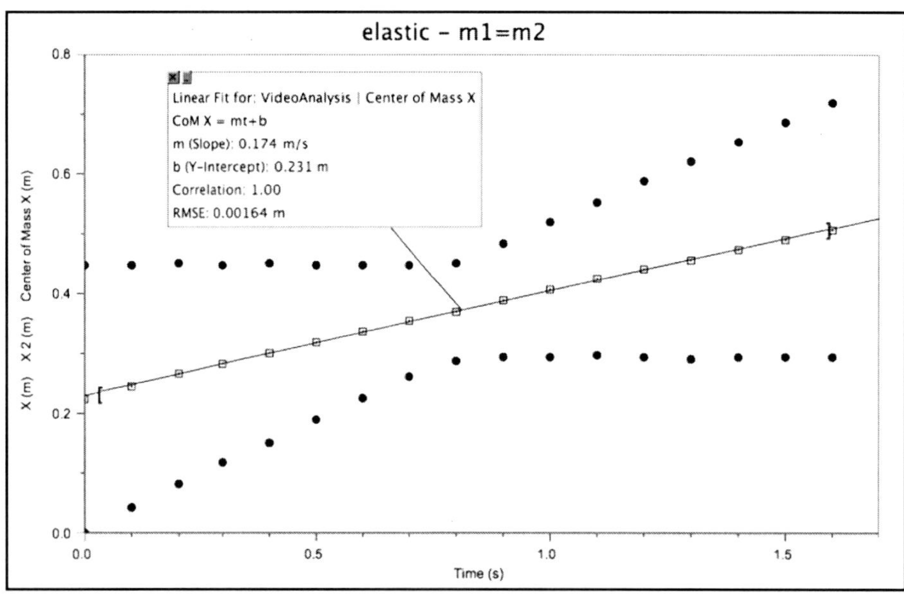

Figure 3

The plot of x_{CoM} *vs.* time is linear and positioned midway between the other position-time graphs. Students should note that the slope of this plot is ½ of the value of the average velocity of the individual carts. Get students to do more than simply appealing to an equation when they explain this relationship. In this case the mass of the system is twice as great as that of each cart, so it is reasonable that the system would be moving with half of the speed of an individual cart.

SAMPLE RESULTS AND POST-LAB DISCUSSION – PART 2

Step 1

Students should obtain a graph similar to Figure 4.

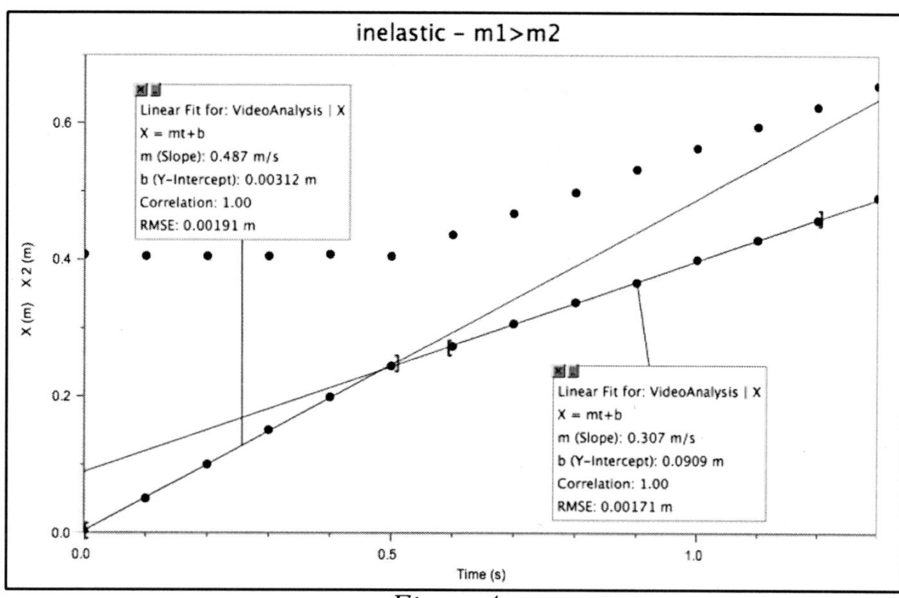

Figure 4

From conservation of momentum, students should predict that the velocity of the coupled carts would be found by

$$m_1v_1 + m_2v_2 = (m_1 + m_2)v'$$
$$1.010\,\text{kg}\cdot 0.487\,\text{m/s} + 0 = 1.52\,\text{kg}\cdot v'$$
$$0.324\,\text{m/s} = v'$$

The final velocity obtained from the slope of the graph is about 5% lower than expected, due mostly to frictional losses.

Steps 3–4

When students complete marking the CoM of the system, they should obtain a graph similar to the one shown in Figure 5.

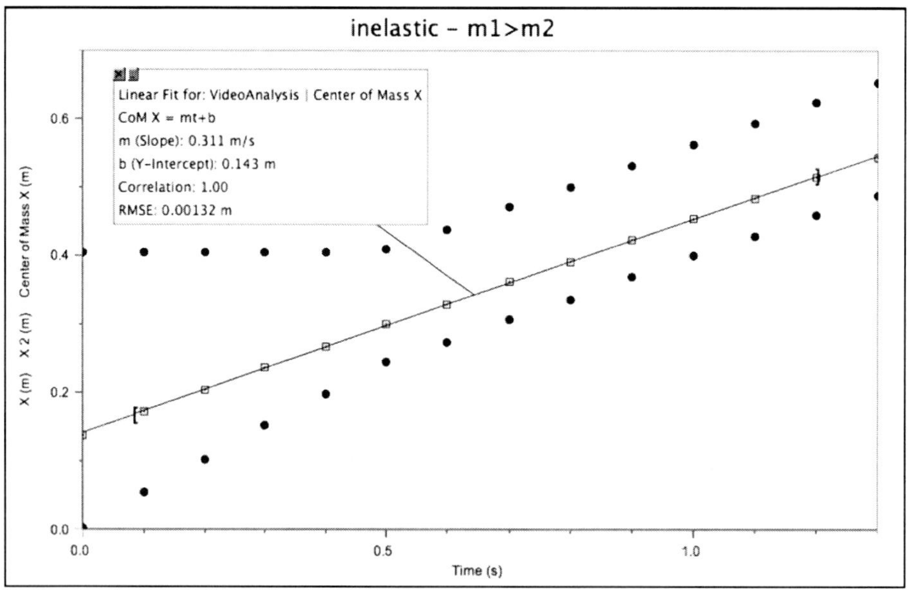

Figure 5

The plot of x_{CoM} *vs*. time is linear and positioned closer to the *x-t* plot of the more massive cart. The velocity of the CoM is only slightly less than 2/3 of the initial velocity of the more massive cart. Students that didn't consider the different masses might have predicted a position like that in the previous case. More careful analysis reveals two important points:

- the initial velocity of the CoM is not the simple average of the initial velocities of the carts; rather it is a weighted average with the more massive cart making a greater contribution to the average value, and
- the velocity of the CoM does not change as a result of the collision. Postpone discussion of why this is the case until students examine the third collision.

SAMPLE RESULTS AND POST-LAB DISCUSSION – PART 3

Step 1

Students should obtain a graph like that shown in Figure 6.

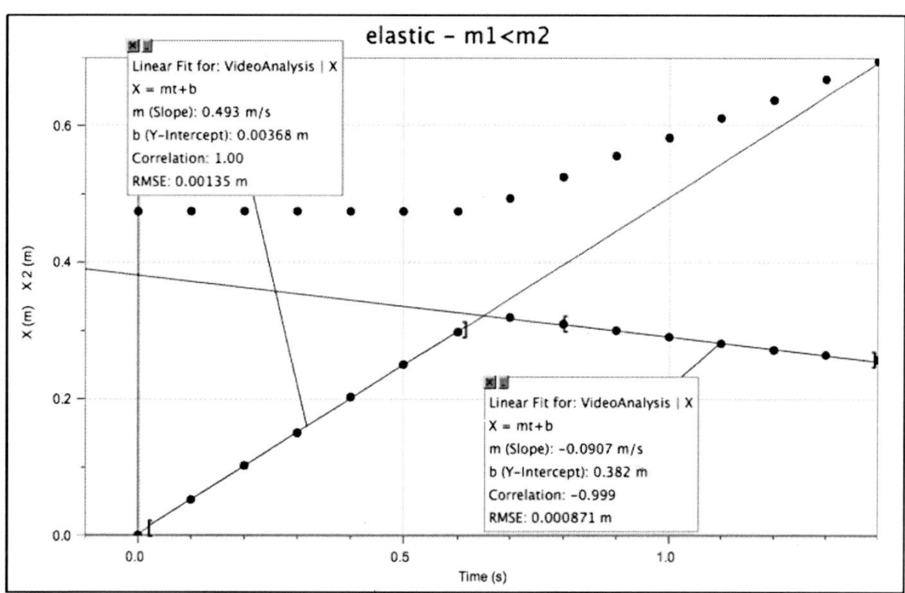

Figure 6

From conservation of momentum, students should be able to calculate the velocity of cart 2 after the collision.

$$0.51\,\text{kg}\cdot 0.493\frac{\text{m}}{\text{s}}+0=0.51\,\text{kg}\left(-0.0907\frac{\text{m}}{\text{s}}\right)+1.01\,\text{kg}\cdot v_{2f}$$

$$\frac{(0.251+0.046)\dfrac{\text{kg m}}{\text{s}}}{1.01\,\text{kg}}=v_{2f}$$

$$0.294\frac{\text{m}}{\text{s}}=v_{2f}$$

The slope of the position-time graph cart 2 after the collision reveals that its final velocity is 0.286 m/s–within 3% the expected value.

Step 2

Students might be uncertain about the shape of the plot of x_{CoM} *vs*. time, especially after the collision. However, if they consider their analysis of the first two collisions, students should be able to predict that the plot would be linear and that its position would be closer to the *x-t* graph of the more massive cart 2.

Steps 3–4

When students complete marking the position of the CoM for the system, they should obtain a graph similar to the one in Figure 7.

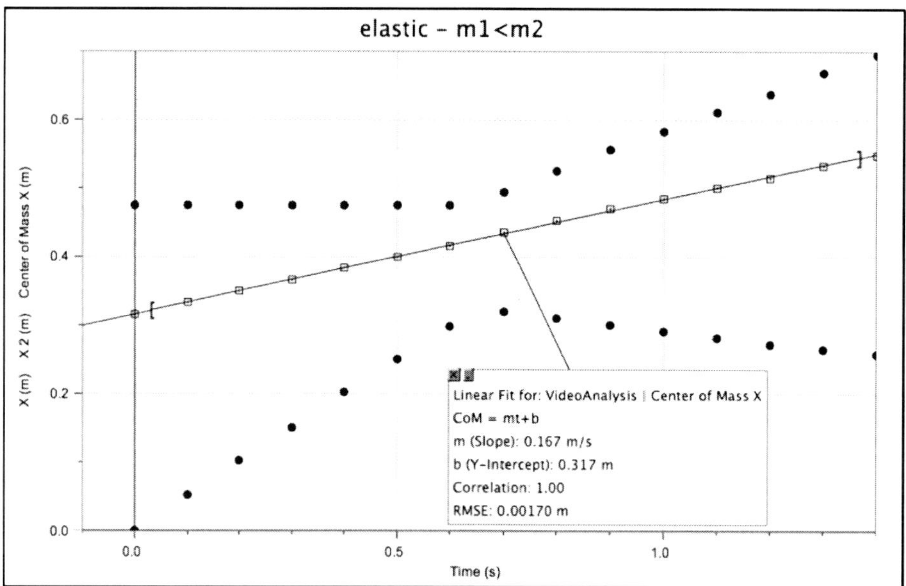

Figure 7

The plot of x_{CoM} *vs*. time is linear and appears closer to the *x-t* graph for cart 2, but that is due to the fact that, in this collision, cart 2 was more massive than cart 1. As students found in Part 2, the velocity of the CoM after the collision is the weighted average of the velocities of the two carts. Students should note that, once again, the velocity of the CoM is not changed by the collision.

SAMPLE RESULTS AND POST-LAB DISCUSSION – PART 4

1. In each of the collisions, the CoM of the system moved with constant velocity (neglecting minor frictional losses). Because there are no external forces acting on the system, its acceleration should be zero.

2. From what students have learned about impulse and change in momentum, they should conclude that the momentum of the *system* in each of the collisions is constant. When the system momentum is divided by the sum of the masses of the carts, the velocity of the CoM is obtained.

3. The momentum of the system at a given instant is given by $p = m_1v_1 + m_2v_2$. The velocity of the CoM appears to be the weighted average (by mass) of the velocities of the carts. Since velocity is the time rate of change of position, it is reasonable to expect that the *position* of the CoM is also the weighted average (by mass) of the positions of the cart at a given instant. Therefore the expression for the CoM of a two-body system should be:

$$CoM = \frac{m_1x_1 + m_2x_2}{m_1 + m_2}$$

EXTENSION

To answer this question, one could re-open the Logger *Pro* files for this experiment and add an additional calculated column: **system kinetic energy.** It should have the equation:

$$E_k = (m_1 v_1^2 + m_2 v_2^2)/2$$

Students will find that for the elastic collision of the two equal-mass carts, the system kinetic energy remains relatively constant. Some variation does occur – due to the difficulty in precisely determining the position of each cart –but this is minor. For an inelastic collision, however, the kinetic energy of the system is as much as 1/3 lower after the collision than it was before. This is consistent with results obtained in experiment 11 – Conservation of Momentum.

Appendix

A

Using the CD

The student preliminary activity pages can be found on the CD that accompanies this book.

The CD located inside the back cover of this book contains the following folders:

- **Advanced Physics – Mechanics Student Word**
- **Supplemental Materials**

Using the *Advanced Physics with Vernier - Mechanics* Word-Processing Files

Start Microsoft Word, and then open the file of your choice from the Advanced Physics – Mechanics Student Word folder. Files can be opened directly from the CD or copied onto your hard drive first.

This provides a way for you to edit the student experiments to match your lab situation, your equipment, or your style of teaching. The files contain all figures, text, and tables in the same format as printed in *Advanced Physics with Vernier - Mechanics.*

Appendix

B

Using Logger *Pro* to Transfer Data to a Computer

You may elect to use the Vernier Logger *Pro* program to transfer data from LabQuest to a computer. Logger *Pro* has many graphing features, such as labels and units for axes, autoscaling, and modification of axes. Printed graphs will have a better resolution and appearance than printed screens of the LabQuest display. Data tables can be displayed and printed with side-by-side columns and headings. Logger *Pro* also provides advanced data-analysis features, such as curve fitting, statistical analysis, and calculated spreadsheet columns.

The directions below are for the latest version of Logger *Pro* (version 3.7 or newer).

Transferring Data from LabQuest

Logger *Pro* can open files saved in the LabQuest App.

1. Connect LabQuest to your computer with a USB cable.
2. Start Logger *Pro* on your computer.
3. Choose LabQuest Browser ▶ Open from the File menu.
4. You will see a standard file selection dialog showing the files available on your LabQuest. Select the file name you want, and click Open. Logger *Pro* will open the LabQuest file, displaying any data, graphs, and notes.

Appendix

C

Vernier Products for Advanced Physics

The Vernier Software & Technology products required to perform the *Advanced Physics with Vernier - Mechanics* experiments are described in this appendix.

LabQuest

Vernier LabQuest provides a portable and versatile data-collection device for any class studying. It can be used as a computer interface, as a standalone device, or in the field. It has built-in graphing and analysis software and a vivid color touch screen. It is compatible with existing Vernier sensors. It has a rechargeable, high-capacity internal battery. It also has a built-in temperature sensor and microphone.

LabQuest Mini

The Vernier LabQuest Mini is a low-cost data-collection interface that connects to the USB port of a computer and had five sensor ports.

LabPro

Vernier LabPro offers another option for data collection. A wide variety of Vernier probes and sensors can be connected to each of the four analog channels and two sonic/digital channels. LabPro is connected to a computer using a serial or USB port.

Data-Collection Software for *Advanced Physics with Vernier - Mechanics*

Computer

Logger *Pro* is the data-collection software for collecting data on a computer. Logger *Pro* software now comes with a site license for both Windows and Macintosh, so you only need to order one copy of Logger *Pro* for your school or college department.

LabQuest

LabQuest App is the data-collection application used to collect data when using LabQuest as a standalone device.

Vernier Products for *Advanced Physics - Mechanics*

Item	Order Code
Vernier LabQuest interface	LABQ
Vernier LabQuest Mini interface	LQ-MINI
Vernier LabPro interface	LABPRO
Dual-Range Force Sensor	DFS-BTA
Motion Detector	MD-BTD
Photogate	VPG-BTD
Rotary Motion Sensor	RMV-BTD
Wireless Dynamics Sensor System (WDSS)	WDSS
Advanced Physics with Vernier – Mechanics lab manual	PHYS-AM

Vernier Accessories

Vernier Dynamics System	VDS
Bumper and Launcher Kit	BLK
Cart Friction Pad	PAD-VDS
Picket Fence	PF
Cart Picket Fence	PF-CART
Ultra Pulley Attachment	SPA
Pulley Bracket	B-SPA
Centripetal Force Apparatus	CFA
Rotary Motion Accessory Kit	AK-RMV
Springs	SPRINGS

Vernier Sensors for *Advanced Physics - Mechanics*

Dual-Range Force Sensor

This low-cost force sensor has two ranges: –10 to +10 N and –50 to +50 N. It can be easily mounted on a ring stand or dynamics cart, or used as a replacement for a spring scale. Use it to study friction, simple harmonic motion, impact in collisions, or centripetal force.

Motion Detector

The Motion Detector is a sonar device that emits ultrasonic pulses at a rate adjustable between 10 and 50 times per second. The time it takes for the reflected pulses to return is used to calculate distance, velocity, and acceleration. The range is 0.15 to 6 meters.

Photogate

Photogates can be used to study free fall, rolling objects, air track collisions, pendula, etc. This photogate can be easily mounted on a ring stand.

Rotary Motion Sensor

The Rotary Motion Sensor lets you monitor angular motion precisely and easily, and it is direction sensitive. Use it to collect angular displacement, angular velocity, and angular acceleration data. It can also be used to measure linear position to a fraction of a millimeter.

Wireless Dynamic Sensor System (WDSS)

The Wireless Dynamics Sensor System combines a 3-axis accelerometer, altimeter, and force sensor into one unit that communicates wirelessly with your computer using Bluetooth® wireless technology. You can also use it as a standalone data logger.

Appendix

D

Equipment and Supplies

A list of equipment and supplies for all the experiments is given below. The amounts listed are for a class of up to 30 students working in groups of two, three, or four students in a classroom. The materials have been divided into **nonconsumables** and **consumables**. Most consumables will need to be replaced each year. Most nonconsumable materials may be used many years without replacement. Some substitutions can be made.

Nonconsumables

Item	Amount	Experiment
balance	1	13, 14, 18
ball	8	6
basket, wire	8	15 (both), 16 (both),
books	various	1
bracket for Motion Detector	8	1, 10A
bracket for Motion Detector	16	11A
bumper and launcher kit	8	3, 7, 8, 9, 10A
camera, digital (either video camera or still camera with movie mode) with cable to transfer file to computer	8	6, 19
Cart Friction Pad	8	3
clamp, right-angle	8	12B, 15 (both), 16 (both), 17, 18
dynamics cart	8	1, 3, 4, 8, 9, 10A, 10B, 11A, 11B, 12B
dynamics cart	16	5, 19
dynamics track	8	1, 4, 5, 7, 9, 10A, 10B, 11A, 11B, 19
magnets, neodymium (included with Vernier Dynamics System)	16	11A, 11B, 19
mass, 500 g	8	11A, 11B, 19
mass hanger	8	4, 12B, 13, 15 (both), 16 (both),
masses, hooked or slotted	8 sets	4, 9, 12B, 13, 15 (both), 16 (both),
measuring tape, metric	8	12B
meter stick	8	6, 19
pad, foam	8	2
plunger cart	8	11A, 11B
protractor	8	17, 18
ring stand	8	2, 12B, 13, 14, 15 (both), 16 (both),
ruler, metric	8	14, 17, 18

spring	8	15 (both), 16 (both),
tripod	8	6, 19
Velcro patches (included with Vernier Dynamics System)	16	11A, 11B, 19

Consumables

Item	Amount	Experiment
string	15 meters	5, 10A, 10B, 12B, 13
rubber band	8	5

Sensor Information and Sensor Check

Dual-Range Force Sensor (Vernier order code: DFS-BTA)

Sensor Information

1. The calibration loaded with an auto-ID Dual-Range Force Sensor will work for the experiments in this book. If you want to improve the calibration, it is easy to recalibrate using a two point calibration. The first point is usually with no force applied. Select the calibration option in the program you are using and remove all force from the sensor. Enter **0** as the first known force. Now apply a known force to the senor. The easiest way to do this is to hang a labeled mass from the hook on the end of the sensor. Enter the weight of the mass (**Note:** 1 kg applies a force of 9.8 newtons).We recommend using a 1 kg mass (9.8 N) for this second calibration point. Be careful not to exceed the selected range setting.

2. The sensor is sensitive enough to measure the weight of the sensor hook. To minimize this effect, simply place the sensor in the orientation in which it will be used (horizontal or vertical) and choose zero in the software.

Sensor Check

Confirm that the setting on the sensor box is appropriate for the experiment. In general, you should use the +/-10 N range if you can. If the forces exceed ten newtons, use the +/-50 N range.

Motion Detector (Vernier order code: MD-BTD)

Sensor Information

1. This Motion Detector emits short bursts of ultrasonic sound waves from the gold foil of the transducer. These waves fill a cone-shaped area about 15 to 20° off the axis of the centerline of the beam. The Motion Detector then "listens" for the echo of these ultrasonic waves returning to it.

2. The Motion Detector has a Sensitivity Switch, which is located under the pivoting Motion Detector head. Slide the Sensitivity Switch to the right to set the switch to the "normal" setting. This setting is best used for experiments such as studying the motion of a person walking back and forth in front of the Motion Detector, a ball being tossed in the air, pendulum motion, and any other motion involving relatively large distances or with objects that are poor reflectors, e.g., coffee filters. The other sensitivity setting, which we call "Track", works well when studying motion of carts on tracks like the Vernier Dynamics System, or motions in which you want to eliminate stray reflections from objects near to the sensor beam.

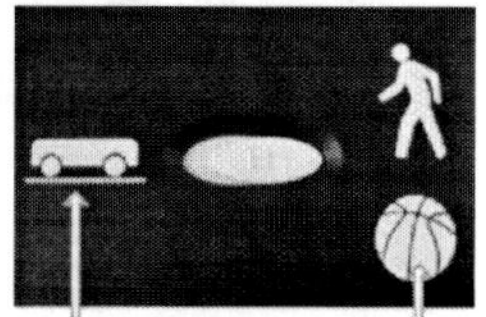

Sensor Check

The most frequently reported problems with a Motion Detector are (1) that the Motion Detector does not work beyond a certain distance or (2) that the graph is very noisy. There are a number of ways to troubleshoot these situations.

1. See if the Sensitivity Switch makes a difference. Simply set the Sensitivity Switch to the other position and retry the experiment. This change may solve the problem.

2. The Motion Detector does not work beyond a certain distance, e.g., it does not detect anything beyond 1.2 m. Here are some things to check if you have this problem:
   a. Check for movable objects (textbooks, ring stands, etc.) in the cone of the ultrasound near the maximum reading you obtain. If possible, move these objects out of the measurement cone. It may not take a very large object to cause problems.
   b. Check for a stationary object (chair, table, etc.) in the cone of the ultrasound. This object may be detected when you are trying to study an object further away. It may not take a very large object to cause problems. If you have trouble with a stationary object causing unwanted echoes, try setting the equipment up so that the objects are not in the cone or placing a cloth over the object. This minimizes the ultrasound reflection.
   c. The cone of ultrasound extends downward from the center line. This can cause problems if you are using the Motion Detector on a hard, horizontal surface. In these cases, try pivoting the head of the Motion Detector to aim it slightly upward.

Photogate (Vernier order code: VPG-BTD)

Sensor Information

1. The Vernier Photogate can be used as a traditional photogate for objects traveling between the arms of the gate, and also as a laser gate for objects passing outside of the arms of the gate. A mechanical shutter is used to block the internal gate, switching the device to laser gate mode. The laser gate mode requires a visible pen laser (not supplied).

2. **Laser Safety Note**: Do not align the external laser gate by sighting by eye. Follow all safety precautions indicated by the laser manufacturer.

Sensor Check

To tell if your Photogate is working correctly, try the following.

1. Is the shutter positioned properly? A mechanical shutter is used to block the internal gate, switching the device to laser gate mode.

2. Connect the Photogate to an interface and run the data-collection program. On the main screen, confirm that blocking and unblocking the gate is working OK by using your hand to block the gate.

3. If the Photogate is not working with your picket fence, make sure that the protective plastic covering the picket fence has been removed.

Rotary Motion Sensor (Vernier order code: RMV-BTD)

Sensor Information

The Rotary Motion Sensor uses a quadrature optical (incremental type) encoder to measure the amount and direction of rotation. The encoder, which is attached to the rotating sensor shaft, consists of a coded pattern of opaque and transparent sectors. The quadrature encoder produces two pulse output patterns 90° apart in phase. The position of the shaft is determined by counting the pulses. The phase relationship between the output signals determines the direction of rotation.

Sensor Check

1. Attach the Rotary Motion Sensor to the Vernier data-collection interface and start the data-collection software.
2. If you are using LabPro, open the Rotary Motion Angular.cmbl file from the Probes and Sensors folder in Logger *Pro*.
3. Start data collection.
4. Rotate theaxle one revolution counter clockwise.
5. The reading should advance from zero to approximately 6.28 radians.

Wireless Dynamics Sensor System (Vernier order code: WDSS)

Sensor Information

1. The Wireless Dynamics Sensor System (WDSS) allows you to take data from a three-axis accelerometer, force sensor, and altimeter, using a Bluetooth® wireless connection to your computer.
2. The WDSS is powered by its internal high-capacity Lithium-Ion battery pack. The battery pack is designed to provide power for 20 hours of use. The battery pack should be ready for an hour of use after only about 10 to 15 minutes of charging.
3. The WDSS will report any of force, x,y, and z acceleration, and altitude, as selected.
4. The sensors of the WDSS generally do not need to be calibrated. If you ever want to calibrate them, you can.

Sensor Check

Below you will find troubleshooting trips.

No LEDs come on when I turn on the WDSS.

- The WDSS rechargeable batteries may be dead.
- The rechargeable battery pack may have been removed from the WDSS. Check this.

When I try to charge the WDSS, the charging LED does not go on.

- The WDSS battery is already fully charged.
- Power is not getting to the power supply. Test the outlet.
- Battery is defective or of the wrong type or missing.

My Wireless Dynamics System does not show when I choose Experiment▶Connect Interface▶Wireless▶Scan for Wireless Devices

- Repeat the process of trying to connect several times.
- Make sure you are using Logger *Pro* 3.4.5 or newer. Older versions will not work with the WDSS. You can check the version by choosing About Logger *Pro* from the Help menu.
- Make sure your WDSS is fully charged.
- Make sure the WDSS is in range for the Bluetooth radio communication. There should not be a metallic barrier between the computer and the WDSS. There is extensive information on installing Bluetooth radios and establishing radio communications with WDSS at http://www.vernier.com/bthelp.
- If you computer has built-in Bluetooth, make sure it is turned on.

- If your computer has a removable Bluetooth transceiver, make sure it is plugged in.
- If your computer has a removable Bluetooth transceiver, remove it and plug it in again.
- Turn the WDSS power switch off and back on and then reconnect.
- Exit Logger *Pro*, restart it, and then reconnect.
- Quit Logger *Pro* and reboot your computer, then restart Logger *Pro*.

My WDSS does not show when I use my Windows computer.

- Use Windows XP, Service Pack 2 or a newer operating system. If you have an older version of Windows XP, you must upgrade to SP2.
- Use the Control Panels to try to learn more about your computer's Bluetooth radio. If you have WidCOMM drivers, remove them and install Microsoft Bluetooth drivers.

My WDSS does not show when I use my Macintosh computer.

- Use Mac OS 10.3.9 or newer.

The Connection LED turns red when I try to start live (connected with the computer) data collection:

- WDSS failed self test. Turn it off and on. If that fails, contact Vernier Software & Technology.

My acceleration (or force) data seem strange when I am studying a high frequency signal.

- If the frequency of the changes in the force or acceleration signals is higher than your sampling rate, you may well be seeing the results of *aliasing*. The best example of aliasing is seen in movies or television where a car wheel appears to be turning backwards. This happens because the camera only looks at the wheel at certain times and it happens to catch the wheel a bit earlier in its rotation each time around. Another example would be a pendulum where you "sample" just slightly faster than one period. You would keep seeing it a bit earlier in its cycle making it look like it was moving very slowly in the opposite direction than it is really moving. If you happened to sample it at an interval exactly the same as its period, you could even be tricked into thinking the pendulum bob was standing still. Similar affects can show up when you sample with the WDSS at a rate which is slower than or only a little faster than the frequency of the force or acceleration signal, you may get a confusing result. Increase the sampling rate.

There is some strange noise on my acceleration signals.

- Be sure to check to see if the noise is real, perhaps caused by a motor vibration or the wobble of a wheel on a rolling cart. Toy cars with hard wheels on a rough surface can generate great amounts of acceleration noise. Sometimes the noise from surface irregularities is larger than the acceleration being studied. Large, soft wheels are better than small hard wheels and heavier vehicles will tend run more smoothly as well. Fans can be another source of acceleration noise (vibration). Sometimes these noise signals can be aliased to show up in unexpected ways.

Index

(by Experiment Number)